Land Degradation: Creation and Destruction

D1534368

The Natural Environment
Series Editors: Andrew Goudie and Heather Viles

This new Blackwell series aims to provide readers with accessible, up-to-date accounts of the physical and natural environment of the Earth in the past and in the present, and of the processes that operate upon it.

Published
Oceanic Islands
Patrick D. Nunn

Land Degradation: Creation and Destruction
Douglas L. Johnson and Laurence A. Lewis

Forthcoming
Rock Slopes
Robert Allison

Caves
David Gillieson

The Changing Earth:
Rates of Geomorphological Processes
Andrew Goudie

Drainage Basin Form, Process and Management
K. J. Gregory and D. E. Walling

Deep Sea Geomorphology
Peter Lonsdale

Holocene River Environments
Mark Macklin

Wetland Ecosystems
Edward Maltby

Arctic and Alpine Geomorphology
Lewis A. Owen, David J. Evans and Jim Hansom

Humid Tropical Environments
Alison Reading, Andrew Millington and Russell Thompson

Weathering
W. B. Whalley, B. J. Smith and J. P. McGreevy

Land Degradation: Creation and Destruction

Douglas L. Johnson and Laurence A. Lewis

BLACKWELL
Oxford UK & Cambridge USA

First published 1995

Blackwell Publishers
238 Main Street
Cambridge, Massachusetts 02142
USA

108 Cowley Road
Oxford OX4 1JF
UK

Library of Congress Cataloging-in-Publication Data

Johnson, Douglas L.
    Land degradation / Douglas L. Johnson and Laurence A. Lewis.
        p.   cm. – (The natural environment)
    Includes index.
    ISBN 0–631–17997–6 (acid-free paper). – ISBN 0–631–19244–1
(pbk.: acid-free paper)
    1. Land degradation–Environmental aspects.   2. Man–Influence on nature.   I. Lewis, Laurence A.   II. Title.   III. Series.
GE140.J64   1995
333.73'137–dc20
                                                                94–27799
                                                                     CIP

British Library Cataloguing in Publication Data

A CIP catalogue record for this book is available from the British Library.

Typeset in 10 on 11½ pt Sabon/Helvetica by TecSet Ltd, Wallington, Surrey.
Printed in the Great Britain by T. J. Press Ltd, Padstow, Cornwall.

This book is printed on acid-free paper

This book is dedicated to the memory of
Vincent Nyamulinda
our former friend, colleague and advisee,
who provided us with many critical insights
into agriculture and land degradation in Rwanda
and the wider world

Contents

List of Figures

List of Plates

List of Tables

Series Editors' Preface

In this series we attempt to avoid the narrow systematic specialization of so much geographical and environmental work. We also attempt to incorporate humans into our consideration of environments, for they increasingly modify environments and create landscapes, while at the same time being influenced by them. In this series we shall be asking authors to take a range of major environment types and to give them a broad integrated treatment that also incorporates their relationships to human affairs.

A.S.G.
H.A.V.

Preface

Land degradation is an ancient problem. It likely began hundreds of thousands of years ago when humans first controlled the use of fire in order to hunt game, create artificial warmth, and prepare food. Through the millennia, control of fire gave humankind access to an increasingly wide range of habitats, initiated major environmental modifications, and encouraged more intensive use of local resources. The greater control over and pressure upon local habitats that these changes implied was accelerated by the agricultural revolution that began approximately 10,000 years ago. Success in agriculture often required significant changes in the vegetation properties of the utilized area. In particular, the substitution of annual crops for perennial vegetation acted as a catalyst for a multitude of environmental modifications that frequently resulted in degraded land resources. As more energy flowed from the environment into those segments of the ecosystem controlled by humankind and its domesticated plants and animals, both the intensity and spatial scale of human land use, and degradation, increased. The industrial revolution of the past three centuries, with its increasingly powerful technology has, in many cases, simply amplified trends long present in the human use of the Earth. In other instances, technology has initiated a new array of environmental problems often far removed spatially from the initiators of the degradational processes. Much has been written about this process of increasingly unwise exploitation of the Earth's resources, from Plato through George Perkins Marsh to Paul Ehrlich and an ever increasing host of contemporary commentators.

Given the plethora of publications in recent years addressing land degradation problems and advocating sustainable development, we risk redundancy in adding another volume to the discussion. However, our personal experiences of living in a post-industrial society, as well as many of our professional research encounters while engaged in fieldwork in Africa, Latin America, and the Middle East, convinced us that there are still fresh perspectives on the land degradation issue that merit consideration. Not least of these insights is the realization that there are numerous instances in which people purposely, occasionally with great success, degrade a portion of the environment in order to produce a positive advantage by which to meet specific societal needs. We define as *creative destruction* those intentional alterations of the existing habitat that impoverish one facet of the environment while erecting new, stable systems that are of greater benefit to humankind than the natural ecosystems that they replace. While land degradation is usually not an intended outcome of human activities, there are enough exceptions to this generalization to make exploration of their characteristics worthwhile. Of equal interest is the appearance of such "positive degradation" in a wide array of socioeconomic formations across both time and space. Indeed, the variety of these positive cases of land degradation is as impressive as the more common and more frequently recounted instances of unintentional negative land degradation.

A second inspiration for this volume arises from the conviction that land degradation is rooted in an interaction between physical and human systems. Much of the recent scholarship on land degradation stresses the political economy and structural features of environmental decay. This critique provides a welcome and important contribution to the environmental debate by emphasizing long neglected structural aspects of the problem. In this corrective development, the role of environmental processes in interaction with human systems has sometimes been lost. Many of the processes that contribute to land degradation are as much physical as they are human in nature. This is particularly true of accelerated soil erosion, where the strategies for curtailing and reversing soil loss already exist but often are socially and politically difficult to implement. In other instances, use of many lands proceeds without managerial recognition that physical processes link regions together, and that changes implemented in one area may have an adverse impact on distant sites. Exploration of the many critical spatial and process interactions among physical and human systems that contribute to land degradation in both an historical and a contemporary context seems useful. By so doing, this book attempts to provide insight into both the causes of land degradation and the reasons why it has been so difficult to control.

Douglas L. Johnson and Laurence A. Lewis

Acknowledgments

Many people have contributed to this volume. Recognizing all of them by name is impossible in a reasonable space, but the contributions of several warrant special mention. Students in many of our classes, and in particular in the 1991 Land Degradation seminar, contributed in important ways to the development of the ideas and case studies featured in this book. Tom Whitmore and Pat Benjamin commented provocatively and helpfully on early version of several of the chapters, without being responsible for the final results. The reseach librarians at Clark University's Goddard and CENTED Libraries, particularly Mary Hartman, Ed McDermott, Jeanne Kasperson, Cindy Noe, and Irene Walch, were unfailingly helpful and patient in support of this research. No acknowledgment of debts incurred in the course of research would be complete without mention of the encouragement and support provided by our families, whose contributions to the completion of this book are greater than even they realize. To all who have helped, we offer heartfelt thanks.

We would also like to thank the following individuals and organizations who have kindly given permission for the reproduction of copyright material:

Figure 1.1 reproduced from A. N. Strahler, *A Geologist's View of Cape Cod*, 1966, by permission of Doubleday, a division of Doubleday, Dell Publishing Group, Inc.; figure 1.2 reproduced from D. Henning and H. Flohn. "World map of desertification", in *UN Conference on Desertification*, Publication A/CONF. 74/31, 1977, by permission of the UN Environmental Program, Nairobi, Kenya; figure 1.3 reproduced from Donald Worster, *Dust Bowl: The Southern Plains in the 1930s* (p. 30, Map of area subject to severe erosion 1935–40), copyright © 1979 Oxford University Press, by permission of Oxford University Press, Inc.

Figure 2.1 reproduced from E. S. Deevey Jr, "The human population", *Scientific American*, 203(3), 194–204, copyright © 1960 by Scientific American, Inc. – all rights reserved; figure 2.4 reproduced from Y. Orev, *Desertification Control Bulletin*, No. 16, 1988, by permission of the Editor and the UN Environmental Program, Nairobi, Kenya; figures 2.5, 2.6, 5.8, and 5.9 reproduced from Billie Lee Turner et al. (eds), *The Earth as Transformed by*

Human Action, copyright © 1990 Cambridge University press, by permission of Cambridge University Press: figure 2.7 reproduced from W. H. al-Khashab, *The Water Budget of the Tigris and Euphrates Basin*, Department of Geography Research Paper No. 54, by permission of The University of Chicago Press, copyright © 1958 W. H. al-Khashab.

Figure 3.1 reproduced from D. K. Northington and J. R. Goodin, *The Botanical World*, 1984, by permission of Mosby-Year Book, Inc.; figure 3.8 reproduced from D. Henning and H. Flohn, "Climate aridity index map", in *UN Conference on Desertification*, 1977, by permission of the UN Environmental Program, Nairobi, Kenya; figures 3.9 and 3.10 reproduced by permission of Guy Foster; figures 3.12 and 7.6 reproduced from L. A. Lewis and L. Berry, *African Environments and Resources*, 1988, by permission of Routledge and Unwin Hyman; figures 3.13 and 7.7 reproduced from L. A. Lewis, *Land Degradation and Rehabilitation*, 3(4), 241–6, copyright © 1992 John Wiley & Sons Ltd, by permission of John Wiley & Sons, Ltd.

Figure 4.2 reproduced from B. J. Skinner and S. C. Porter, *Physical Geology*, copyright © 1987 B. J. Skinner and S. C. Porter, by permission of John Wiley & Sons, Inc.; figure 4.3 reproduced from D. J. Pratt and M. D. Gwynne, *Rangeland Management and Ecology in East Africa*, copyright © 1977 D. J. Pratt and M. D. Gwynne, by permission of Hodder & Stoughton Ltd/New English Library Ltd.

Figure 5.1 derived from D. Munro (ed.), *Cambridge World Gazetteer*, copyright © 1988 W & R Chambers Ltd and Cambridge University Press, by permission of Cambridge University Press; figure 5.6 reproduced from James Walls, *Land, Man and Sand: Desertification and its Solution*, copyright © 1980 Macmillan Publishing Company, by permission of Macmillan Publishing Company; figure 5.7 reproduced from N. Fenneman, *Physiography of the Eastern United States*, copyright © 1938 Neville Fenneman, by permission of McGraw-Hill, Inc.

Figures 6.1 and 6.3 reproduced from Richard W. Wilkie and Jack Tager, eds, *Historical Atlas of Massachusetts* (Amherst: The University of Massachusetts Press, 1991), copyright © 1991 The University of Massachusetts Press; figure 6.5 reproduced from G. S. Kust, *Desertification Control Bulletin*, 1992, by permission of the Editor and the UN Environmental Program, Nairobi, Kenya; figure 6.6 reproduced by permission of José Molinelli.

Figure 7.1 redrawn with the permission of the Canadian Museum of Civilization, from "!Kung Bushman subsistence: an input–output analysis" by Richard B. Lee, in David Damas, ed., *Ecological Essays, Proceedings of the Conference on Cultural Ecology, Ottawa, August 3–6, 1966*, National Museums of Canada, Bulletin No. 230 Ottawa, 1969: figure 7.3 reproduced from H. C. Conklin, *Transactions of the New York Academy of Sciences*, 2nd series, 17, 133–42, 1954, by permission of the Executive Editor, New York Academy of Sciences; figure 7.4 reproduced from

S. Hecht et al., *Human Organization*, 47(1), 25–35, by permission of the Editor; figure 7.8 first published in *Applied Geography*, 3(1), January 1983, p. 47, and reproduced here with the permission of K. Ruddle and Butterworth-Heinemann, Oxford, UK; figure 7.11 reproduced by permission of the Editor, *Geographical Magazine*; figures 7.12–14 reproduced from H. Smits, *Land Reclamation in the Former Zuyder Zee in the Netherlands*, 1988, by permission of the Ministry of Public Works and Water Management, Directorate Flevoland; figure 7.15 reproduced from M. Smiles and A. H. L. Huiskes, *Ambio*, *10(4)*, 158–65, by permission of the Editor-in-Chief.

ONE

Land Degradation: Human and Physical Interactions

Introduction

Ever since the formation of the Earth, change has been the rule rather than the exception on the planet's continental and other land areas. Land areas are continually being altered due to both tectonic forces (energy derived from the internal heat of the Earth) and exogenic processes (those that result from the action of wind, water, and ice, and from human activities). It is due to tectonic processes that mountains increase their magnitude, such as is occurring in the Andes, the Alps, and the Zagros Mountains along the Persian Sea. Lands also are uplifted from beneath oceanic areas, as occurred in recent geologic time in the southeastern United States, a process that resulted in the emergence of Florida. Similarly, volcanic eruptions create new lands, such as the island of Surtsey (Iceland), and destroy areas, for example as occurred during the Tamboro (Indonesia) eruption in 1815 when the mountain lost approximately 1,300 m (4,265 ft) of its upper portion. Conversely, terrestrial zones are also always exposed to exogenic forces that erode and ultimately lower surface areas. The lowering of lands takes place once tectonic processes that are uplifting land areas decrease in magnitude compared to the exogenic processes that are eroding these lands. The Appalachians and Piedmont areas in the eastern United States, the Urals in Russia, and the shield area of northeastern Brazil and Guyana represent tectonically stable areas that are gradually being lowered due to erosional processes. It is the net balance between the tectonic and exogenic processes that determines whether specific areas are growing or being destroyed.

Examination of the Earth, as a natural system (devoid of the role of human activities on geologic processes) clearly documents that the dynamic interaction between tectonic and exogenic forces culminates in a complex set of results. At one end of the spectrum, lands slowly evolve, allowing a complex array of Earth science relations to change in a largely stable manner. The evolution of a lacustrine area to a wetland over hundreds or thousands of years, or the reduction of a highland slowly into a lowland through fluvial erosion and slow mass movements, represent examples of the natural geomorphic evolution of landscapes. In these cases, change largely becomes evident only through a historical examination of the area's attributes, such as its lithology, vegetation, and soil properties. In opposition to these slow, chronic changes are rapid, acute changes that occur when critical thresholds are exceeded. Under these conditions, changes are rapid and the immediate destruction of existing natural systems makes way for processes that create new systems. Major earthquakes, 1,000-year floods, immense volcanic eruptions, and massive landslides can completely eliminate the previous conditions in brief periods of time. History and folklore document the nonbenevolence of Mother Nature in a multitude of natural disasters ranging from the Noachian Flood of ancient times to the contemporary volcanic eruptions of Mt St Helens and Mt Pinatubo.

Land Degradation: its Definition

The term "land degradation" is a relatively new addition to our scientific vocabulary. Even today it is not recognized as a category within the Library of Congress (USA) reference system. As a result, there is a lack of universal agreement on its definition. Is it but a synonym for environmental and resource deterioration?

Examination of its usage in the literature and in the *Journal of Land Degradation and Rehabilitation* (first published in 1989) implies that there is general agreement concerning two critical aspects of land degradation. First, there must be a substantial decrease in the biological productivity of a land system; and, second, this decrease is the result of processes resulting from human activities rather than natural events. Thus the results of exogenic forces such as geologic erosion and climatic change, as well as natural catastrophic events, such as earthquakes, volcanic eruptions, and flooding – unless exacerbated by human activities – lie outside the realm of land degradation, even though areas can become less productive biologically due to these natural changes.

On the basis of how the term is currently being used, we define land degradation as *the substantial decrease in either or both of an area's biological productivity or usefulness due to human interference*. Because biological productivity is determined not only by the attributes of the land resource but also water properties, land degradation incorporates those aspects of the hydrologic domain that are significant in a given area. Usefulness is also a crucial attribute of land degradation. The biomass productivity in an area could remain constant, yet land degradation might still occur. For example, overgrazing could result in a decrease in an area's soil fertility and a diminished ability of the environment to support the growth of palatable forage plants. When the vegetation cover reestablishes itself after grazing pressure is reduced, even though the grass cover (biomass) is similar in density to the initial cover, the species that replace the original vegetation often are not nearly as palatable. As a result, a smaller number of livestock can now survive on these lands. Thus, after this change, the area is no longer as valuable to its inhabitants. This decrease in an area's biological productivity and usefulness (land degradation) may be either reversible in the short term or nonreversible in the long term.

With our definition, not all resource and environmental deterioration falls within the rubric of land degradation. For example, if desertification – the impoverishment of arid, semiarid, and subhumid ecosystems (Dregne 1976) – is the result of increasing aridity due to climatic change, it is not a manifestation of land degradation. However, if the increased desert-like conditions are the result of overgrazing, water well drilling, or other human activities, then the desertification of the area is an example of land degradation.

Any area, regardless of its climatic, topographic, or other environmental characteristics, can undergo degradation. Cape Cod, Massachusetts, represents a humid coastal setting where lands are degrading, sometimes due to natural changes and at other times due to human activities. The Cape's origin is due to Pleistocene glaciation. During this epoch large quantities of nonconsolidated glacial materials (outwash and till) were deposited. This deposition occurred in a nonmaritime environment as sea level was approximately 100 m lower than today (figure 1.1). Since the end of the Ice Age, large areas of the Cape have been modified or lost to the sea, due both to a rise in sea level and coastal erosion primarily resulting from storm waves. Wave action continues to erode the Cape, especially along the eastern margins of the outer Cape that face the open sea. It is inevitable that the lands located in zones of wave energy concentration shall continue to be lost to the sea, regardless of human attempts to stop this natural trend (plate 1.1).

In contrast to this natural erosion of the Cape, along some of coastal zones, especially along private beach areas, some human activities have accelerated erosional processes that are irreversible under current environmental conditions. Because private ownership of some beaches exists, individuals owning beach frontage often attempt to stop the natural migration of "their" sand beach. Previously, when legal, the most popular method was to build groins. While groins can stabilize beach location for the short term, they actually increase sand loss to deep water and thereby decrease the sand supply available for beaches. Thus, the long-term net effect of groins is to accelerate beach erosion. The first example of coastal erosion typifies the dynamic nature of the Earth's lands due to natural processes, while the latter example of accelerated beach erosion represents an example of land degradation.

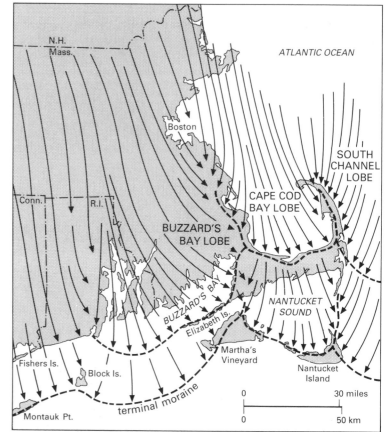

Figure 1.1
Cape Cod's present shoreline and Pleistocene coast (after Strahler 1966)

Plate 1.1
Coastal erosion due to wave action on Cape Cod

Long-term (Nonreversible) and Short-term (Reversible) Land Degradation

Stability and resilience

Stability is the ability of a system to return to its previous equilibrium state after a temporary disturbance (Holling 1973). A critical aspect of stability is the speed with which a system will return to its previous state after a disturbance. Like a cupie doll, a carnival game figurine with a low center of gravity that is difficult to knock over, the more rapidly the system returns to its original state, the more stable is the system. A tropical forest that regenerates quickly after being cleared by shifting cultivators is more stable than a semiarid grassland that takes a long time to regenerate after its grass cover has been cleared for cropland. The difference in the behavior of the two systems is linked to the amount, frequency, spatial distribution, and variability of available moisture, the basic energy component in the system.

Resilience, on the other hand, is the capacity of a system to absorb change without significantly altering the relationship between the relative importance and numbers of individuals and species of which the community is composed. This characteristic of a system is analogous to a boxer whose opponent delivers numerous blows to body and head without apparent effect. The more blows the boxer can absorb without his ability to continue being impaired, the more resilient he is. However, continuing to box may mask sustained damage to his internal organs and central nervous system. Sudden and unexpected collapse can occur from even a light blow when the resilience limit is exceeded. Camels have a legendary capability to recognize when their master can no longer continue to add to the burden that they bear. Of course, if the camel's owner does not recognize this limit of resilience, the proverbial piece of straw will break the camel's back! An environmental example of the same phenomenon is the use of rivers, lakes, seas, and oceans as dumping grounds for agricultural and urban wastes. The limit beyond which the water body can no longer absorb the pollutants is often suddenly reached, with consequent catastrophic collapses in the fish and other aquatic populations that depend on decent water quality for life. Resilience limits are not easy to identify, but once they are exceeded, the system rapidly seeks equilibrium at a different level of production.

When an area is disturbed, resulting in a degraded state, two conditions must be met if it is to have the ability to return to a nondegraded state; that is, to the stability and resilience characteristics that the area formerly possessed. First, the disturbance must be only temporary. Second, it also must be of a low enough magnitude and duration so that the changes resulting from the disturbance have not set in motion changes that alter the fundamental environmental conditions of the area. If these two conditions are not met, the resulting land degradation type is considered to be nonreversible and long-term.

If the catalyst causing the land degradation is temporary and of low enough intensity that critical thresholds have not been reached, once the stimulus is removed, the area will evolve back to its previous nondegraded state. Under these conditions, the land degradation would be classified as both short-term and reversible. If the catalyst results in system changes that are too extreme for the area to absorb, a new balance will emerge at a totally different level of productivity. When productivity is lower, the system has degraded. For any land system, there are tradeoffs in the use of the system. Actions that ignore the stability and resilience limits of the area, and that discount the differing impacts that variable technologies can have on the system, are likely to initiate serious and long-term degradation. Also, they are likely to create cultural–ecological systems that are precariously poised on the brink of disaster.

Time

The time dimension, that is to say the period required to ameliorate any damage to the productivity of the land, is a crucial factor in land degradation. The feasibility of a particular rehabilitation ecology, applied to any land facet, is largely determined by both the economic costs and the time period required for improving the environmental situation. In most cases, no matter how severe a degradational process is, given enough time, natural processes will repair the damage without human assistance. An exception to this observation is the case of synthetic materials, some of which are so stable that, in many cases, the time factor is ever increasing as technology creates more and more new materials that are less and less susceptible to decomposition. However, in general the time scale implied in nature's rehabilitation ecology, when damage is excessive or when it occurs in environmental situations in which many natural processes are slow (i.e., cold and arid areas), can be geologic in duration. Likewise, to reverse degradation may be so costly economically – such as importing soil into Haiti to replace the eroded soil from many of its hills and plateaus (plate 1.2) – that the restoration processes are meaningless from the perspective of human needs.

Within the context of this book, because of the duration of the human life cycle and the demands of livelihood systems, any degradational process that is set in motion by human actions and cannot be reversed in a time period of less than 50 years is considered as long-term degradation. For practical purposes, this type of severe land degradation is a permanent change from a human point of view, even though it is a relatively inconsequential period from a geologic perspective.

Short-term degradation occurs in any cultural–ecological system. For example, the cultivation of most crops results in at least a short-term decrease in soil fertility and accelerated soil erosion rates. If

Plate 1.2
Severe soil erosion in Haiti

these trends remain unchecked, yields will decrease. Under extreme circumstances farming will no longer be viable and land will have to be abandoned. In this situation, short-term degradation becomes long-term. Short-term degradation can be relatively easily countered in most agricultural systems from a technical perspective (i.e., conservation practices). It would appear to be in the land manager's self-interest to do so, because usually it is less costly to invest in short-term, *status quo* maintenance rehabilitation than it is to abandon land altogether and begin anew elsewhere.

Short-term rehabilitation ecology is achieved by practices such as manuring and the spreading of chemical fertilizers, contour plowing, mulching, terracing, crop alteration, and a host of other strategies (OTA 1988). Yet because of a multitude of reasons, ranging from accounting procedures to limited capital, these remedial technological interventions are often not implemented. For example, under corporate farming, the farm manager and stockholders are often primarily concerned with maximizing annual profit (the manager's annual bonus and even his job often depend on this). Technological interventions require investments. Most accounting analyses ignore environmental concerns. In particular, they rarely incorporate a declining soil resource as a decreasing asset. Thus, under current accounting practices, it is often in the immediate interest of the corporate farm to minimize investments associated with environmental quality unless it shows up in the "bottom line."

Individual farmers, if they are short of capital, may wish to invest in conservation, but their economic constraints may be so severe that it is impossible. This is the general case in many parts of the "developing world" and is also widespread for many small farmers in the United States. When these land rehabilitation mechanisms are either not implemented or are not sufficient, more expensive interventions are required to rehabilitate an area. In extreme cases, such as many rural areas in northwestern Haiti and Madagascar where the abuse of the land has been great and the damage severe, rehabilitation of the land is not a likely, viable option from an economic cost/productivity improvement perspective.

In arid areas Dregne (1977b) suggests that rehabilitation interventions should focus on high-productivity ecotypes and agricultural systems (i.e., irrigation), rather than on low-productivity and extensive systems (i.e., livestock grazing). If this philosophy becomes widely accepted, most extensive and low-productivity areas would continue to be exploited and degraded. Not only does this viewpoint have serious ethical implications, but it ignores the areal geomorphic and hydrologic linkages that exist in natural systems. This type of cost-effectiveness approach to rehabilitation ecology makes many zones capable of rehabilitation ineligible for governmental concern. It suggests that degradation may become functionally irreversible long before it is beyond rehabilitation.

In theory it is easy to define the properties of long- and short-term degradation. In practice, it is often difficult to classify which type of degradation is occurring within a specific area. Sudan's Kordofan area (figure 1.2) illustrates some aspects of this classification problem.

Desertification in Kordofan – Short-term or long-term land degradation?
Beginning around 1955 and continuing until the middle 1970s, the population of the Kordofan area increased at an annual rate of about 2.7 percent (Stern 1985). Up to the early 1970s, the cultivated area simultaneously increased at approximately the same rate as the population. Generally, throughout this period precipitation was greater than normal. Beginning in 1968, and continuing through the early 1970s, drier conditions returned to the Sahel. With the arrival of drier conditions in Kordofan, areas under cultivation decreased (Olsson 1985). One explanation for this decrease in cultivated area is that many lands were degraded and had to be abandoned. According to Horowitz (1981), in many parts of this area, social factors led to the desertification of lands due to over-cultivation, grazing, and firewood demands. These pressures were often phrased in terms of population growth requiring that more land be developed and brought into production. Other social changes, such as an increase in urban population, which required more charcoal for heating and cooking, were also seen as contributing to increased pressure on rural landscapes to generate more fuel wood for urban consumption.

All of these pressures were met during the 1950s and 1960s when rainfall amounts were generally higher than the long-term average. With the environment producing more, farmers and herders were able to cultivate more land and raise more sheep with greater success than ever before. However, these increases in productivity could not be sustained when environmental fluctuation produced a

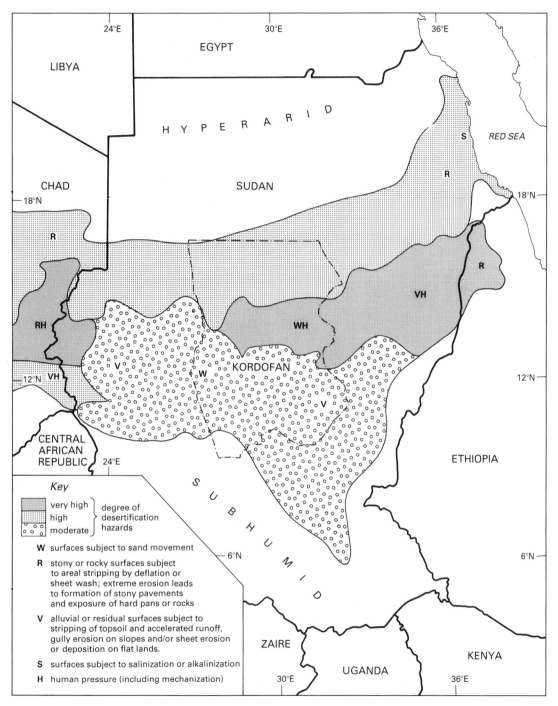

Figure 1.2
The desertification risk in Sudan's Kordofan Province

return to drier conditions. Historically, these environmental fluctuations could be accommodated because the pressure of human resource use was less and could be sustained by existing cultural–ecological adaptations. Now, this perspective contended, good environmental conditions masked basic changes in nature–society relations. Because of the intense social pressures, serious land degradation occurred during this period of precipitation shortfalls (Horowitz 1981). The implication was that a long-term decrease of the land's biological productivity occurred throughout this semiarid area.

This is the general impression of what happened throughout the Sahel during the drought of the late 1960s and early 1970s. Lamprey (1975) concluded that the desert has expanded by between 90 and 100 km in Kordofan during the past 17 years. Eckholm and Brown (1977) and other publications of the Worldwatch Institute likewise describe continuing desertification throughout the Sahel. Other reports, both popular and professional, agree with these findings (Dregne 1977a; Grainger 1990). They contend that in response to human activities and climatic stress (deficient rainfall), formerly productive farmland is evolving into nonproductive farm and grazing lands (i.e., long-term degradation: Ibrahim 1984). These observers view the trend as one that is rapidly approaching a level of long-term degradation from which the prospects of recovery are bleak.

Contradicting the perception of widespread land degradation throughout the Sudanese Sahelian areas, Olsson (1985: 147), through the analysis of Landsat data, simulation, and GIS techniques, concludes that "... reports on desertification have been very exaggerated." Most of the pessimistic studies have based their conclusions on the comparison of data sets collected *prior* to the Sahelian drought and *during* or shortly after the drought. He concludes that "not surprisingly, the environmental conditions degraded from the first (wet) to the second (dry) situation" (Olsson 1985: 147).

Olsson found that desertification exists only during the extremely dry periods. When rainfall returns to the wetter phase of the climate, both natural vegetation and agricultural crops return to more typical production levels. These findings are based on an array of data gathered for conditions before, during, and after drought conditions. These include: analysis of the area's albedo values; examination of changes in the textural properties of the study area; and statistical analysis of crop production and climatic factors. Thus the decrease in biomass production noted during the droughts is largely in response to natural conditions, and hence is not illustrative of even short-term land degradation.

This is not to imply that desertification is a myth nor to underestimate the likelihood of desertification occurring in some parts of the Sahel. However, this example illustrates the difficulty in determining whether land degradation is occurring or whether decreases in biomass production just represent normal adjustments to the variable but normal natural rhythms of meteorological phenomena. To substantiate desertification within the Sahel, data must be gathered over a time period of sufficient duration to include the range of climatological events that occur within this zone. Unfortunately, it is the norm for this semiarid area to have highly variable climatic conditions, both temporally and spatially (Nicholson, Kim, and Hoopingarner 1988). Drawing conclusions about land degradation based on time-limited data can result in serious misinterpretations of both status and trend.

The American Dust Bowl – Sixty years of land degradation?
In the most historically accurate context, the American Dust Bowl is restricted to portions of the states of Colorado, Kansas, New Mexico, Oklahoma, and Texas (figure 1.3). However, through years of general usage, the Dust Bowl has been extended areally to include all of the Great Plains. In either context, it represents the prototype example of ecological failure resulting from drought in areas where rainfed agriculture has been extended beyond the climatological limits of humid areas (Worster 1979; Braeman 1986) into semiarid areas.

While precipitation variability in all climatic types is normal, as climates become increasingly deficient in moisture, variability

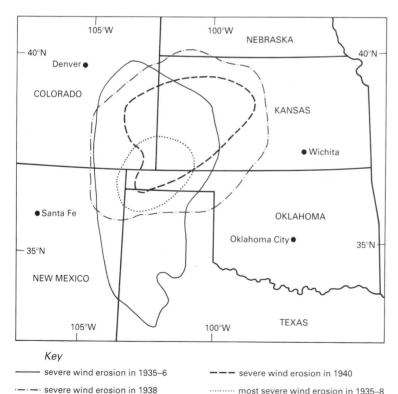

Key
—— severe wind erosion in 1935–6   – – – severe wind erosion in 1940
·—·—· severe wind erosion in 1938   ·········· most severe wind erosion in 1935–8

Figure 1.3
Severe wind erosion in the US Dust Bowl, 1935–40 (after Worster 1979)

increases. In semiarid areas, except in selected favorable environmental settings such as floodplains, most crops are exceedingly sensitive to these moisture fluctuations. During more humid periods, the moisture deficiency is either extremely small or a small surplus may exist during a growing season in semiarid settings. During these wet episodes, agricultural yields are good and erosional problems, especially those related to wind, are minimal. This is especially true in the American Great Plains where a significant proportion of the moisture supply falls as snow. Under typical conditions in the Great Plains, at the advent of the spring growing season soil moisture content is high, since it is the late winter and early spring snowmelt that recharges the soil moisture. Under these conditions, adequate soil moisture along with the sprouting vegetation minimize wind erosion.

All semiarid and subhumid areas, including the Great Plains, experience dry cycles during which precipitation is notably less than average. Under pristine natural conditions, the continuous sod cover of these areas protects the underlying soil from extraordinary erosion during these periods of moisture stress. The environmental situation remains within an equilibrium condition even though drought conditions exist and the grasses do not attain their normal growth. The sod cover and permanent vegetation protect the soil from the ravages of wind and water. With the introduction of agriculture, the natural conditions of semiarid and subhumid lands are disturbed. The ability of these areas to absorb the environmental stresses resulting from drought conditions is reduced. In the Great Plains during the 1930s dry period, sodbusting, required for agriculture, exposed the silt and clay components of the soil to deflation processes. Hence wind erosion, not significant in prior and similar dry periods when a complete sod cover existed, became effective once desiccation of the exposed cultivated soil occurred. During the dry years of the 1930s, windstorms were now effective in removing clay and silt size soil particles from the bare ground and transporting huge quantities of them in suspension as dust. Under similar climatic conditions in the 1750s, 1820s, 1860s, and 1890s, dust storms did not occur, as the soil remained protected by the area's complete sod cover (Stockton and Meko 1983). The inference is that it was the exposure of the bare soils from agricultural activities, and not climatic change, that created the American Dust Bowl conditions of the 1930s.

The first great dust storm began on May 11, 1934. As dust from this storm travelled all the way to Washington, DC and beyond, out into the Atlantic, the nation was clearly alerted to the need to protect agricultural lands. During the great windstorms of March 1935, when clouds of suspended dust reached heights in excess of 3,000 m (9,843ft) and estimates of the eolian load in central Kansas were in excess of 35×10^6 kg/km^2 (Skinner and Porter 1987), clearly massive human-induced wind erosion took place. During the 1930s, in selected areas, wind erosion removed over 1 m of soil. This is in comparison to long-term average regional estimates of only a few centimeters per thousand years under natural ground-cover conditions (Skinner and Porter 1987).

In the early 1990s soil erosion and land degradation remain a problem within the Great Plains. Despite modern conservation practices, wind damage to land in the Great Plains has increased sharply in recent years. This is mainly because of lingering drought in North Dakota and the exposure of soil resulting from agricultural activities. The US Soil Conservation Service has reported that 2×10^6 hectares (4.93×10^6 acres) were damaged between November 1989 and February 1990, while about 1.9×10^6 hectares (4.7×10^6 acres) suffered damaged by wind in the same four months of 1988–9. However, statements that contend that erosion is worse under contemporary conditions than during the Dust Bowl may be questioned (SWCS 1984). Plains climatic and crop data indicate that the scattered dry periods in the 1970s had fewer environmental impacts than those of the 1930s and 1950s (Riebsame 1986).

Unlike contemporary "dust bowls" in many other semiarid areas, such as the Sahel, the Thar, and the former USSR's virgin lands, where major capital constraints, educational limitations, nonprivate land ownership, poor governmental infrastructures, high population pressures, and a multitude of other economic, governmental, and social factors are cited as both the cause of and rationale for the lack of progress in arresting the problem, this is not the case for the American Great Plains. For example, not only have governmental policies been developed and implemented, but a host of forms of high-technology intervention has been developed and made widely available to the farmers of this area for decades (Harris, Habiger, and Carpenter 1989). Despite all of this intervention, evidence exists that the resource base is still declining in the Great Plains. Most national farm bills still do not make natural resource conservation and environmental protection central features of American agriculture and agricultural policy (Cook 1989).

Since the early 1970s, policy and conservation practices emphasizing the prevention of future major dust bowls and extreme soil erosion resulting from channelized runoff have been favored by most US federal farm legislation. This policy deals with those aspects of the soil resource base where degradation is clearly manifested. This perspective officially became dominant with the implementation of the Soil and Water Resources Conservation Act of 1977. The selected goal of this act was to minimize soil loss from cropland. It ignored most other environmental aspects that have long been considered critical in the ecological sciences (Richards 1984). One result has been that the low-intensity, slowly accruing, chronic forms of land degradation have not been addressed for the Great Plains and other areas. The relative productive potential of any soil is determined not solely by erosion rates, but by a soil's ability to provide a good environment for root growth and development. These conditions are determined by a multitude of soil properties such as an adequate pH, available water capacity, nutrient characteristics, organic matter, and soil tilth (Lawson et al. 1981).

To date, the slowly accumulating environmental impacts of modern farming have been offset by huge energy inputs (chemical fertilizers), new hybrid crops, and the use of irrigation. But environmental danger signals associated with agricultural intensification – the use of large inputs of chemicals and pesticides to achieve high yields – are becoming evident. As but one example, soils in the rich French agricultural areas of Beauce, Brie, and Somme exhibit significant declines in their organic content. This has resulted in their soils becoming "a neutral medium to which everything must be added: fertilizers, mineral elements and, because the crops grown there are so fragile, huge amounts of pesticides" (James 1993: 17). The question remains as to whether continuous technological advancements will remain economically viable as the soil resource slowly degrades and associated pollution problems increase in intensity.

Human Causes of Degradation

Introduction

Although environmental change and degradation have their origins in natural processes, the main contemporary factor in environmental deterioration is the impact of human activity. Identifying the level of human responsibility in any given situation is often a complex task. The difficulty arises because humankind and nature are linked in an interactive system in which cause and effect, and process and response often blur. Unexpected natural events such as a prolonged drought, a flood, or a shift in the location of an ocean current (e.g., the Peruvian Current – El Niño) may overwhelm the adaptive capabilities of a particular resource use system. What appears to fishermen and government agencies to be a safe level of fish catch may, in the face of an environmental fluctuation of an unexpected magnitude, initiate a sudden collapse of the fishery. In this context, people often are less the purposeful initiator of degradation and change than they are accidentally in the way of environmental fluctuations that are (given our present state of knowledge) essentially random, and for which they are unprepared.

An example of this situation is found in many traditional pastoral nomadic communities, whose characteristic initial response to drought is to move their families and herds to the best known, most reliable local water source (Johnson 1973). The concentration of animals in this restricted zone initiates overgrazing and local degradation. As long as the dry spell falls within normal expectations, long-term degradation is not a threat either to the biomass production of the local habitat or to the survival of the herder and herd. Howerver, if the drought is an extremely low-frequency event, occurring once or twice in a century, it may be of sufficient magnitude that the local "reliable" and other local water supplies dry up. Under these extreme conditions which, despite their severity, are within the range of normal climatic variability, the herding community faces serious problems.

Unless an early decision is made to move to better pasture and water conditions elsewhere, significant herd losses are likely to occur. Long before animals die in sufficient numbers to balance local forage and water availability, severe, but highly localized, overgrazing has likely taken place (plate 1.3). This places a heavy responsibility on community leaders to reduce grazing pressure at an early point in the drought/degradation cycle in order to avoid long-term deterioration of the resource base. Yet group experience, based on traditional ethnoscience knowledge, demonstrates that prolonged droughts are rare (100-year frequency) and that local resources usually can meet the demands placed upon them. For this reason, herders are reluctant to move too early, since they can never be entirely certain of grazing and security conditions elsewhere. The knowledge of local limits and how to prevent overuse is an integral part of resource use in many dryland pastoral systems (Draz 1974; Hobbs 1989; Galaty and Johnson 1990; Johnson 1993a). Of course, there are some critical problems: How do you determine the magnitude of a dry period as it is occurring? What are the appropriate measures that should be taken once an area enters a dry period? Unlike the Biblical story in which it was foreseen that seven good years would be followed by seven bad ones, it remains difficult, if impossible, to assess the duration of a drought until after it has ended. Unable to react with sufficient rapidity as moisture conditions deteriorate, human-initiated degradation appears as more the unforeseen consequence of

Plate 1.3
Overgrazing near a traditional well at Mazrub, Kordofan, Sudan

how a normally well-adapted livelihood is unable to adjust to unexpected environmental fluctuations than a ruthless and willful misuse of resources.

The ecological transition

Nature no longer is an external force, separated from human culture. It is increasingly incorporated into the matrix of human experience and action (Bennett 1976). This trend reflects the degree to which human societies now influence and alter natural systems, and is dubbed the "ecological transition" by John Bennett. Ever since humankind constructed its first simple tools, environmental alteration was possible, although at first these changes were on a microscale. Domestication of fire permitted the first major areal modification of nature. Coevolved ecosystems were the result, in which grasses and thick-barked trees were favored over more fire-sensitive species. In this process of altering the ecological balance between people and the environment, two types of societies emerge: those that, for longer or shorter periods, are able to create sustainable agroecologies, and those that are not.

The ecological transition is a product of both humankind's increasing technological capabilities and its emergent anthropocentric value systems. These changes are increasingly viewed as producing an estrangement from the natural world that is reflected in an aimless art, a bankrupt philosophy, and a boring literature (Trussell 1989). A more powerful technology makes it possible to overcome the capability constraints (Carlstein 1982) that limit society to the exploitation of the local habitat. Resources, to satisfy a wide range of needs, come from an ever increasing distance to meet local site requirements for manufacture and consumption. Mines in Chile produce the copper that covers church roofs or is drawn into wire in industrial communities in the United States. Coal mining and thermal-powered electricity generation in northwestern New Mexico sends clean electrical energy to southern California. Tall local smokestacks in Ohio and Pennsylvania lift pollutants high enough into the atmosphere to be carried out of the local habitat; the acid rain resulting from these airborne pollutants, which corrodes buildings and poisons lakes further to the east in Upper New York, New England, and eastern Canada, was an impact that was unforeseen originally. High chimneys were built to minimize local pollution: the intent was never to transfer this problem eastward – the assumption was that the unwanted elements would disperse in the vast reservoir of the atmosphere without unwanted impacts.

The environmental change produced in the mining district is remote from the conscience and knowledge of the manufacturing populace. Likewise, the atmospheric pollution transported great distances from plants is not of immediate concern to plant managers and local residents. Yet the ability to cause environmental change, destroy vegetation, alter wild animal habitats, acidify lakes, remove unwanted overburden to reach mineral ores, and affect groundwater distribution and quality is great. Distanced from the activity site, with the costs of environmental change and degradation exported onto other peoples and places, the initiating population ignores the impact. Basking in the warm glow of self-satisfied security generated by elevated material well-being and by apparent mastery over the natural world, it is easy to view environmental degradation as someone else's problem. This rampant consumerism and exaltation of economic growth and materialistic gain has been gained at the expense of the spiritual values that sustain concern for the well-being of the biosphere.

Values are an integral part of the causal mechanism in degradation. The animistic attitudes toward nature characteristic of "primitive" peoples promote a less aggressive exploitation of environmental resources. Respect for the spirits inhabiting bush and beast does not prevent the hunter, gatherer, or cultivator from killing, collecting or cutting. However, it does promote an ethic that recognizes limits to useful extraction, and it does undergird a sharing value system that is hostile to excessive accumulation and hoarding (Lee and DeVore 1968). This value system has been in retreat ever since the invention of agriculture. With the manipulation of domesticated plant and animal species both surplus accumulation and storage become possible. The surplus sustains greater local populations on a sedentary basis and concentrates environmental impacts that a mobile hunting and gathering livelihood system spreads more evenly over a greater territory. Control over plants and animals helps to assure humankind of its mastery over the natural world, and generates the imagery found in *Genesis* whereby nature is placed at the service of *Homo sapiens*. The shift in value orientation places humankind at the center of the Universe, created in the image of, and as Earthly surrogate for, the Divine. A type of species loyalty emerges that envisages the human species as the ecological dominant. This encourages the direction of energy toward our species to the detriment of other denizens of the biosphere. While this tendency probably existed in the first hominid, tool-using cultures, it was mitigated against by both the animistic ethic characteristic of early hunting and gathering communities and by the relatively limited power of early technology.

The emergence of first the agricultural revolution and, more recently, the industrial revolution destroyed or pushed into more remote and marginal habitats the more reverential animistic ethic. In its place was substituted the more anthropocentric value systems characteristic of modern civilization. These value structures favor an I/it orientation toward nature rather than an I/thou attitude. The significance of this shift was noted by numerous observers, many of whom follow Lynn White (1967) in attributing the change to the influence of the Judeo-Christian heritage. This perspective is manifested in the conflict between "...the idea of stewardship of land [being] pitted against the belief in soil exploitation for personal gain and that soil is merely an economic commodity in the marketplace" (Jenny 1984: 161).

Yet causation is more complex than this neat association between religious and cultural heritage appears to suggest. As White himself notes, reverential attitudes can be found within the Judeo-Christian tradition. Moreover, reverential images of nature in the literature of non-Western cultures often mask the reality of ruthless exploitation and degradation. Tuan (1970) points out that China has a very checkered record of deforestation and soil erosion, despite its literary and artistic motifs of reverence for and harmony with the natural world. It was the environmental degradation of the Greeks and Romans, long before Christianity ever appeared on the scene, that inspired Marsh (1864) to examine environmental change in the Mediterranean, and in so doing serve as the founding father of the modern environmental movement. Jacobsen and Adams (1958) have shown that salinization in the fields of ancient Sumer in 2000 BC was responsible for significant decreases in yields and in a shift from salt-sensitive wheat to more resistent barley. When presented with modern technology, few hunting societies are immune to the value shifts that encourage a more wasteful and exploitative use of existing resources. Kemp (1971) demonstrated this with Baffin Island Eskimos who, after gaining access to the rifle, the outboard motor, and the snowmobile, clearly became more wasteful with regard to the wildlife in their hunting area.

Ideological blindness can also promote an absence of ecological concern, as the recently acknowledged ecological impacts of the smokestack industries of the former socialist/communistic societies of Eastern Europe graphically illustrate. Parts of Poland and the Czech Republic have perhaps some of the most environmentally devastated lands in Europe, a situation that developed under their former state socialist regimes. The economic, utilitarian, anthropocentric values that underpin degradation transcend political, cultural, and religious ideologies, and occur at all points in the time and space record of our species.

Intentional and unintentional change

The human dimensions of degradation are further complicated by two additional factors. First, environmental change is both intentional and unintentional in its causation. Frequently, people set out to initiate change knowing full well that some of the consequences will constitute degradation. For instance, strip mining inevitably produces blighted landscapes and degraded land and water systems. Only recently has a feedback loop been created in the United States that requires the restoration of such degraded areas. In most countries, this remedial measure is sacrificed to immediately perceived economic gains. On a smaller scale, in some semiarid areas (see chapter 2 for a detailed description) runoff farmers degraded the hillslope zones in small river basins to increase surface flow toward their valley-bottom fields. In so doing they sacrificed the productivity of the more marginal facets of their habitat, the hillslopes, in order to enhance the useful productivity of the higher potential elements in their land use system, the valley bottoms.

African pastoralists often use fire as an environment-modifying tool, both to prevent the spread of trees and bushes, thereby increasing the amount of grass available for their animals, and to remove habitat that harbors the tsetse fly, the intermediate vector responsible for spreading trypanosomiasis (sleeping sickness) to cattle and humans. These strategies have one thing in common: they reflect a conscious commitment to environmental change even when those changes represent a biological degradation in part or all of their habitat.

Despite these instances of change that accept the possibility of degradation with some equanimity, most of the degradation that occurs is of the unintentional variety. In most cases when humans promote environmental change, they envisage the change as being one for the better. They expect that the land-use system that they create will be more productive for their purposes than the unaltered natural system that is replaced. The increase in productivity that occurs is usually true in the short term, but often is less true in the long term. A classic example of this is the use of irrigation in order to improve the productivity of drier landscapes that are risky or impossible for rainfed agriculture. No one intends to promote soil degradation through the spread of salinization and waterlogging. The intention is to make the desert bloom, to create a paradise in the midst of the wilderness, to intensify existing agricultural systems.

The Aswan Dam in Egypt is a controversial example of this problem, for its success in promoting multiple cropping, desert reclamation, and hydroelectric power is also associated with initiating serious land degradation problems (Kishk 1986). Today, on a global scale, irrigation agriculture is locked into a treadmill effect whereby every new land unit brought into production is virtually offset by a unit lost somewhere else (and often not very far away) to salinization and waterlogging. Farmers who open up a new field in hilly terrain by plowing a field with straight furrows perpendicular to the contour follow cultural tradition and plan to make their farms productive. They do not intend to promote soil erosion. Planners who insert deep borewells into pastoral zones have no intention of encouraging overgrazing and wind-aided soil erosion. The frequent negative consequences of these changes come as a surprise to the perpetrators, and are often expensive, difficult, or impossible to reverse. None of these degradational effects are necessarily the product of malicious intent.

A second factor contributing to degradation is that causal variables may be both internal and external to the production system and human actors most immediately implicated in change. In the Negev, ancient agricultural systems were created that were extremely sensitive to the nature and rhythm of the local arid ecosystem.

These farmers deliberately encouraged hillslope degradation in order to promote transport of soil and water into adjacent valley bottoms where these resources could be trapped and concentrated behind stone check dams (Evenari 1977). The terraces that were created were highly productive and illustrate three useful principles. First, small-scale systems that are sensitive to the local ecological setting are most likely to achieve success. Second, sacrifice zones (the hillslopes) were created at the microscale. One portion of the environment (the sacrifice zone) was deliberately degraded in order to increase the productivity of another. Because both the benefits and the costs of this action were localized, it was possible to erect a sustainable agroecology. This long-term sustainable system illustrates the third principle, that of creative destruction. Creating production systems useful to humankind often requires the destruction – or at least massive alteration – of nature, with consequent changes in species diversity and habitat quality. The critical variable is the erection of a system that can withstand the test of time.

The Nabatean run-on farming system, developed in an extremely arid setting, was able to survive for hundreds of years, and its inherent viability is indicated by the fact that archaeological recreations of the ancient field systems work under contemporary arid rainfall regimes (Evenari 1977). A similar example in a very wet environment is the pond field agricultural terraces of central Luzon in the Philippines (Conklin 1980). These fields are masterpieces of ethnoscientific engineering that transform incredibly steep slopes into productive fields by destroying tropical rainforest. By creating a totally artificial habitat with great stability characteristics, an extremely productive agroecosystem is established.

While creative destruction can establish long-term sustainable agroecologies, continuous maintenance of the new system is required. When upkeep ceases, the system declines. The Negev system depended for its success on subsidies from outside the immediate environment. When trade routes shifted, decreasing the base of the Nabatean economy, and the central authority of the area shifted to elsewhere in the Mediterranean basin, Negev agriculture also went into eclipse. Unlike the Negev run-on farms, the central Luzon terrace farms rely entirely on local labor, management, and initiative. Yet, the terrace systems of central Luzon are equally vulnerable to forces operating outside their immediate habitat. Maintenance of the terrace fields is intimately linked to an adequate labor supply to repair terrace walls, to level fields, to transplant rice, to channel water, and to perform a host of other tasks in the agricultural system. Withdraw male labor by migration to coastal cities and the elaborate terrace systems, the product of half a millennium of effort, could quickly collapse.

The limitations of economic evaluation

If land degradation could be solved by technological interventions alone, by this date it should be declining in areal extent. Even with

the multitude of international agencies, national government agricultural departments, and private nongovernmental agencies help-
ing to facilitate and implement conservation practices worldwide,
the processes of land degradation continue to spread their havoc
throughout the world (Blaikie and Brookfield 1987). Along with
other human factors, before land degradation will be solved, it will
be necessary to develop widely accepted economic analyses that
realistically evaluate the various costs of environmental degradation, including both aesthetic as well as production factors.

One reason for widespread "mining" of renewable natural
resources is that accounting procedures rarely evaluate the deleterious environmental effects of these actions. In the short term, at the
individual or institutional level, it is often economically rational to
pollute and destroy resources. Per-Olov Johansson (1990) believes
there are three critical reasons why adequate environmental costs
must be incorporated into any activity that impacts on resources,
such as land quality. First, all of humanity can benefit from the
direct effects of many environmental resources. For example,
breathing clean instead of dust-laden air (e.g., the Dust Bowl) has
both health and economic ramifications for the population at large.
Thus the direct costs of soil erosion go beyond the immediate user
of the resource, the farmer, and the consumer of the crops. Second,
citizens may value the opportunity of using a resource in the future
more than the value they would gain from using it in the present.
For example, many people are in favor of protecting the gorilla,
rhinoceros, or American buffalo and its natural habitat for the
future possibility of seeing them. This is true even if there is a
high likelihood that the individuals will never actually see them
directly: that is, there is a perceived value for their existence. This
is also true for preserving unique landscapes for their aesthetic
values without regard to wildlife. The perceived value of the
Canyonlands in their natural state has been a critical reason why
major dam construction in the American Southwest ceased in 1963
with the construction of the Glen Canyon Dam on the Colorado
River (Martin 1989). The successful campaigns to prevent further
major dam construction on the Colorado, despite the region's
strong traditional economic development forces, gives credence to
Johansson's second reason. His third and final point, and this is
especially a crucial component in the Green movement, is the desire
to preserve the Earth's environmental heritage for future generations. Society grants us the right to exist, regardless of whether
we are useful. If for no other reason, the same privilege should be
extended to both plant and animal species.

Today, techniques that assign costs to environmental impacts
are being incorporated in an ever increasing proportion of cost–
benefit analyses. However, the counter to this positive trend is that
there is no widespread acceptance of any single technique as the
most appropriate way to evaluate environmental costs. With the
multitude of techniques available, in most cases it is possible to
justify diametrically opposite outcomes when using identical data.

Thus it is still difficult to incorporate – let alone resolve – environmental conflicts using numerical accounting procedures.

In today's domestic and international markets, competition is a deterrent to linking environmental impacts with resource utilization. Some arguments against the creation of a Canadian–United States–Mexican free market are based on the different environmental laws in these three countries. Weaker laws are perceived to give an economic advantage for attracting industry from one country to another. Without incorporating the environmental value into every product's or service's cost, the presence of externalities indicates that, in attempting to maximize their utility or profits, households or businesses will not generally make socially optimal decisions (Fuchs 1986). Most consumers are not willing to pay for additional production expenses associated with environmental preservation when similar products are available, and are less expensive because they are produced with minimal concern with environmental degradation.

If the state of Iowa passes legislation that requires its farmers to minimize soil loss while the farmers in the state of Indiana are not governed by similar regulations, the cost of Indiana corn (maize) might be less because of lower conservation costs in their state than those of Iowa. The result could be that land degradation is arrested in Iowa and the long-term benefits for Iowa are clearly greater; but Iowa farmers could be out of business, or at least might earn less money than their Indiana counterparts. The grocery chains and food manufacturing companies would likely purchase the less expensive Indiana corn at the expense of the Iowa producers.

Most politicians are not willing to resolve environmental conflicts and environmental regulation under the present limitations inherent in current environmental and costing evaluations. The debate over acid rain in both Europe and North America during the 1980s and 1990s reflects the latter point (Torrens 1984; Bunyard 1986). While almost all parties agree that acid rain is causing environmental damage, to date it has been difficult to obtain a consensus on the environmental or economic ramifications of this problem. One reason why political accountability is minimal in this problem is the shortcomings in environmental and costing techniques. In the United States, the coal-producing areas, such as West Virginia, are minimally affected by the acid rain problem; but it is the burning of this fuel that is a major cause of acid rain. Under current practices, if controls are placed on air pollution emissions, which in terms of the acid rain problem largely originate in the Midwest, in order to protect New England's land, water, and forest resources, the Appalachian and Midwest areas will be burdened with higher production costs and yet receive minimal environmental benefits. This is clearly a high-risk situation for any Appalachian or Midwestern politician.

Governmental contradictions

In the United States, as in most other countries, reversing land degradation (as well as other environmental problems) is further complicated by contradictory policies among governmental agencies, including the federal (national), state (regional), and local levels. For example, the federal government both subsidizes the exploitation and the conservation of land and water resources. In the Food Security Act of 1985, the soil and water conservation behavior of farmers is leveraged against crop price supports, credits, and availability of crop insurance. This is clearly an attempt to encourage conservation. Yet at the same time, federal tax laws, in reality subsidies, have promoted the over-pumping of groundwater reserves in the Southwest. In this dry region, this has encouraged excessive irrigation and in most cases wasteful use of water. A consequence of this policy has been the mining of the region's groundwater reserves. A renewable resource is becoming nonrenewable. Widespread deterioration in the quality and quantity of groundwater is one result of these subsidies. The resulting shortage as well as poorer quality of the groundwater are two factors contributing to the increasing salinity of soils in these areas. Many of the region's existing irrigated lands likely will go out of production in the future due to the resulting soil degradation encouraged, at least indirectly, by government policy.

Summary

Land degradation is not a new problem; nor is it limited to any particular political or cultural ideology. Its "uniqueness" is that it exists virtually everywhere to one degree or another. One estimate is that worldwide 26 billion tons of soil are being eroded from crop and grazing land annually (Brown and Wolf 1984). Besides soil loss, an equal if not more important aspect of land degradation is the deterioration of soil fertility, including chemical, biological, and structural properties. Estimates on the magnitude of this component of soil degradation are not available.

The causes of land degradation are complex. It is a response to a multitude of complex interacting physical processes along with human values and constraints. From a purely technical perspective, the technology exists today to prevent, arrest, and rehabilitate the majority of land degradation situations. Yet, all indications are that the problem is getting worse. The land resource remains continuously under attack. In widely diverse settings, ranging from the forests of Western Europe to the Amazon Basin, from the irrigated lands of the American Imperial Valley to Rajastan in India, and from the chemical poisoning of the land in Poland to the Valley of Mexico, land degradation continues to occur, and in many places it is accelerating.

In this introductory chapter we have introduced the basic dimensions of this phenomenon. In the remainder of this book we examine the problem of land degradation with the intent of developing the understanding necessary to create a sustainable land resource.

TWO

Land Use and Degradation in Historical Perspective

Human Causation in Land Degradation

It is a widely accepted truism that the greater the density of the human population, the greater are its impacts on the surrounding environment. This principle holds true generally also in nature outside human-managed systems if concentrations of animals are encouraged by the distribution of scarce resources. A water hole is an example of this phenomenon. Large numbers of animals coming to drink in a specific locale will inevitably trample the vegetation around a water site. Limits on degradation are set by nature, because too extensive overgrazing near the water source will result in an absence of fodder for the animals. Lack of fodder makes the water resource unusable, and sets in motion processes of stress and movement that operate to conserve the basic land resource. Nonetheless, the vicinity of any water hole is always more degraded in nature than is the adjacent territory, without having to invoke human intervention as a causal agent to explain the deterioration.

However, human impact is the most frequent explanation for negative environmental change. One does not have to search hard for evidence in the historical record. Even in small numbers and at low population densities, humankind initiated significant altera- tions in the natural environment. The most primitive tools, such as digging sticks, crude stone crushing and chopping axes, and butchering blades, gave small human societies increasing control over their local environment. The higher levels of caloric energy directed toward *Homo habilis* and away from other species gave early hominids and their descendants a comparative advantage. However primitive these tool technologies may appear from a twentieth century perspective, they were of tremendous ecological significance. The better levels of nutrition and higher rates of survival made possible by tool technology, combined with superior cognitive abilities, cooperative patterns of behavior, and great adaptability, enabled the descendants of *H. habilis* to gradually spread throughout the Earth's major ecological habitats. These attributes also enabled increased human populations to be sustained in local concentrations.

While these early technologies gave humankind a comparative advantage over other organisms in the quest for food and shelter, they were not sufficient to transform the Earth beyond a purely local scale. The domestication of fire altered this relationship (Sauer 1956; Pyne 1993). Full control of fire and its conscious use as a tool to modify the environment is at least several hundred thousand years old, and may extend significantly further back in time (Stewart 1956). First used to provide domestic warmth, protection from predators, and a device – via cooking – by which to kill parasites and harmful bacteria, fire increasingly was used in hunting. By driving game into settings in which hunters could more easily dispatch their prey (mired in bogs, driven over cliffs, or directed into constrained "shooting galleries"), fire was an important part of the hunter's tool kit. In the process of setting fire drives, habitat was modified. The frequent use of fire discouraged the survival of most woody perennials, and promoted the growth of annuals and selected woody individuals that were genetically adapted to resisting or evading fire. In the dry tropics and subtropics, this meant that grasses were promoted over trees. Measured in terms of standing biomass, the result is a degraded ecosystem from an ecological standpoint, even though it was a more desired system from the standpoint of the human population inhabiting the space. This is the case because the spread of grassland at the expense of forest also meant that herbivores were at a comparative advantage. Since these animals were often the same species as preferred by the hunters, the result was the development of a coevolved ecological system of hunter, and later herder, grazing animals, and grassland. The land degradation that occurred was desirable from the perspective of the hunter and herder because it expanded the grazing habitat. In Africa, the reduction of woody biomass additionally reduced habitat favorable to tsetse infestation. Thus lands that in their natural state had limited utility for humankind through repeated use of fire evolved into settings able to support more people. From the viewpoint of forest economies, whether of hunting or farming, this loss of habitat was distinctly undesirable. In environmental terms, changes introduced by the widespread use of fire were negative insofar as they reduced the diversity of plant and animal species supported in a given area. However, as long as the burning was not so extensive that it favored annuals over perennials, little long-term loss in basic soil productivity through exposure of extensive surfaces to the erosive effects of wind and water took place.

If tools and fire gave prehistoric communities greater control over the natural world, the impacts that were the result were minuscule compared to the changes wrought by the agricultural revolution. The beginnings of plant and animal domestication some 10,000 years ago fundamentally altered the way in which humans were able to affect their environment. Increased mastery over the location and reproduction of favored food-producing species of plants and animals improved the quantity, quality, and availability of food. Small bands of hunter–gatherers were no longer required to move seasonally in search of sustenance. Instead, increasingly permanent settlement became possible and occurred.

More people in denser concentrations resulted in greater impacts on the local environment. In wooded tropical and mid-latitude environments, slash and burn agricultural systems developed that were extensive in nature and used long forest fallow periods to regenerate soil fertility. Whenever these shifting agricultural communities had to reduce the fallow cycle, whether due to internal population growth or the constraints placed on mobility by neighboring communities, land degradation in the form of grassland expansion at the expense of forest was the result (Conklin 1954). This process of agricultural impact by native societies on surrounding forest land was the cause of the extensive grasslands found in the midst of the hardwood forests of eastern North America when European explorers and settlers first encountered the New World (Rostlund 1957). A similar process is occurring today in the Amazon Basin, where pioneer peasant farmers are unable to sustain a viable agriculture once the forest is cleared, and the outcome is more extensive rearing of cattle on the grasslands that develop. In extreme cases, land degradation is of sufficient magnitude that even cattle rearing is impossible on the cleared lands (Denevan 1981).

Through history, with each major change in technology, the number of people who can be supported increases. Concomitantly, the ability of those populations to have an adverse impact on their environment increases. Graphing human population dynamics arithmetically gives a false picture of the importance of each revolution. A logarithmic graph of the same population data (Deevey 1960) suggests the powerful impetus provided to population growth by each major change in ecotechnology (figure 2.1). Eventually the rate of growth slows as a rough balance between nature and society is achieved. At the same time, this pattern is only true for the overall population at a global scale. At more regional levels of analysis, local population–resource crises can produce dramatic declines in population (Whitmore et al. 1990: 26). In some places the combination of altered social and environmental conditions can result in degradation so severe that permanent production losses are produced. Under these circumstances not only are fewer individuals supported but also, on a human time scale, rehabilitation becomes impossible.

Up to the beginning of the industrial revolution, the pace of change fostered by new technologies was relatively slow. The less

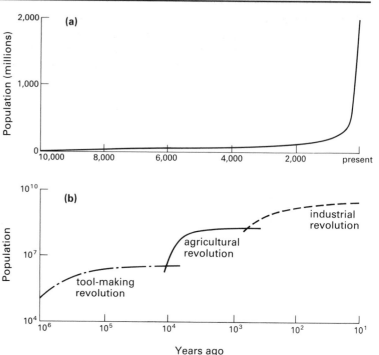

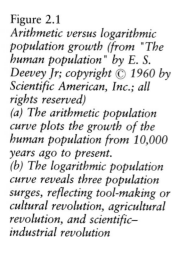

Figure 2.1
Arithmetic versus logarithmic
population growth (from "The
human population" by E. S.
Deevey Jr; copyright © 1960 by
Scientific American, Inc.; all
rights reserved)
(a) The arithmetic population
curve plots the growth of the
human population from 10,000
years ago to present.
(b) The logarithmic population
curve reveals three population
surges, reflecting tool-making or
cultural revolution, agricultural
revolution, and scientific–
industrial revolution

powerful the technology, the greater the amount of time needed for the practitioners of that technological tradition to have a negative impact on their habitat. Also, less powerful technologies were generally not able to affect distant areas. The combination of a slow pace of change and a local scale of operation meant that negative impacts could be seen and reacted to soon enough for remedial measures to be taken. As a consequence, small-scale, slow-paced, low-technology human interventions tend to be self-regulating. They coevolve with the increasingly humanized landscape that they create and the ecotechnology that they employ. Such societies are not stable and static: rather, they slowly develop in ways that progressively enhance productive capacity and the ability to support more people.

This process of the gradual development of a sustainable ecotechnology is based on the fundamental principle of *creative destruction*. The transformation of the agroecology that takes place inevitably involves changes that destroy and degrade some parts of the habitat. As long as these changes do not diminish the overall productivity of the habitat, the net result is beneficial. However, often even successful ecotechnologies are unable to maintain themselves indefinitely. Changes in society and economy outside the control of local communities may undermine the viability of an apparently successful ecotechnology. Thus, in assessing the durability of a given agroecology, careful attention to the stability and resilience parameters of the system is essential. Equally important is a time trajectory that enables the observer to place a given eco-technology into its historical setting and to predict where trends in that adaptive system are heading in the future. Fortunately, there are several historical examples that elucidate the complexity of the concept of creative destruction. From the array of potential studies, we examine two well documented historical cases, the agropastor-alists of the Negev and the irrigation farmers of Mesopotamia, each practicing a contrasting ecotechnology. Other historical and con-temporary cases are explored in later chapters.

Run-on Farming in the Negev

The Negev is an arid region occupying the southern half of Israel (figure 2.2). It is a small segment of a much larger dry zone that

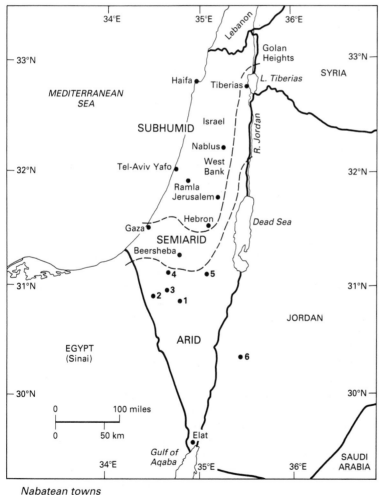

Nabatean towns

1	Avdat	3	Shivta	5	Kurhub
2	Nitzana	4	Khalutza	6	Petra

Figure 2.2
Nabatean settlement in the Negev in its regional context

stretches from the Sahara to Central Asia. Of the region's 12,500 km^2 (4,826 sq. miles), only 350,000 hectares (864,859 acres) in the northern Negev ever supported ancient settlement (Evenari et al. 1961). Within this area's foothill and upland zones, only a small fraction of the total surface contains sufficient soil and water resources to provide foci for settlement. Lack of reliable rainfall is the primary reason for this paucity of resource use opportunities.

The Negev lies at the southern end of the Mediterranean climate zone, and therefore shares in the winter–spring rainfall and summer–fall drought climate pattern typical of that region. Although a Mediterranean rhythm occurs, precipitation totals decrease southward and eastward from the sea. Furthermore, rainfall variability increases toward the south and east too. In the northern Negev just south of Beersheva, the annual average rainfall is between 200 and 350 mm (7.9–13.8 in). South of this transition zone, rainfall diminishes rapidly, vegetation becomes ever more sparse, and few agricultural and pastoral resources exist to attract human settlement. Even in the northern Negev, rainfall is an irregular resource; for as rainfall totals diminish, variability in interannual precipitation increases. Years with practically no rain are juxtaposed with years that exceed the normal by considerable amounts. This variability in precipitation is the norm for almost all arid areas. In 1962–3, at Shivta, Evenari and his colleagues (1971: 146) recorded a total precipitation of 28 mm (1.1 in). The next two years produced 153 mm (6 in) and 165 mm (6.5 in) respectively at a station that, over seven years, averaged 93 mm (3.7 in) of annual rainfall. The consequence of this small and unreliable water supply and highly variable environment with widely dispersed resources is a very hazardous habitat that most communities, especially agricultural ones, would shun.

These conditions of water resource poverty place severe constraints on those human groups required to live in the area. In order to survive in the region, inhabitants must first have a compelling reason to locate in what by any objective measure is an inhospitable and hazardous area. Second, they must develop sustainable patterns of resource use that are resilient and stable for long periods. Characteristically, those livelihood systems that are most successful are those that are able to adjust their activities to the natural ecological rhythms of the district (the *genius loci principle*) and to convert apparent negative features of the environment into positive contributors to human sustenance.

Both pastoral nomads and sedentary agricultural communities have managed to accomplish this feat. Nomadic pastoralists survive in arid environments by frequently moving from poor conditions to locales with better resources. This movement is dictated by an ecological rhythm that governs the general availability of grass and water. By using domesticated animals as an intermediate converter of pastoral resources, and by shifting these animals to seasonally favored sites, nomads are able to capture spatially dispersed resources over wide areas. The Negev has a history of successful use of its resources for long periods by nomadic animal-herding groups. Because these nomadic groups are characteristic of this, and many other, dry regions, they are of only passing interest to us at the moment.

Of far greater interest are the settled agricultural communities that also occupied the northern Negev for substantial periods in the past. On several occasions, farmers and urbanites lived in the desert for periods of several centuries if not millennia (Evenari, Shanan, and Tadmor 1971). Because the earliest epochs are not well represented archaeologically, if only because so much of their settlement remains have been reworked by subsequent settlers, our attention is focused upon the period from 300 BC to AD 650 when settled agriculture flourished most extensively.

The settlement history of this epoch is closely associated with a group of people called the Nabateans. They were an Aramaic-speaking Semitic people who occupied a broad swath of the pre-desert zones from the northern Negev near Beersheva through Jordan to the Jebel Druze just south of Damascus. The Nabatean economy originally was based on two complementary activities: animal husbandry and caravan trading. The two activities supported each other, since many subsistence needs were met by animal products while animals provided beasts of burden for the mercantile operations. At the same time, consumption goods not readily provided locally could be purchased in more favored areas near the coast and carried into the Nabatean heartland by the caravans passing through the area. A strategic location controlling the overland caravan trade provided the Nabateans with considerable wealth, and over time much of this wealth was invested in a settled infrastructure. Heavily influenced by the urban Greek culture of the coastal districts, the Nabateans created a spectacular capital at Petra in present-day Jordan and controlled a number of other urban centers, such as Avdat (Israel), of varying size throughout the region.

The growth of urbanization apparently was the product of a conscious policy on the part of the Nabatean monarchy throughout the first century AD (Bowersock 1983: 64ff). A deliberate effort to promote sedentarization and the expansion of settled farming was apparently part of this policy. All of the northern Negev urban centers exhibit considerable growth in this period. Bowersock (1983: 72ff) believes that a motivating factor in the process of agricultural expansion was the gradual decline in the viability of Nabatean overland caravan trading in the face of competition from more northerly routes through Palmyra and more southerly ocean traffic along the Red Sea. A succession of farsighted kings invested heavily in the development of a special form of agriculture that was ideally adapted to the local arid, topographic and pedological environmental conditions.

This development process involved both special circumstances and an application of the *principle of creative destruction*. Most societies living in drylands with low agricultural potential do not possess significant resources to invest in agricultural development. This was not the case with the Nabateans. Their mercantile activities generated the monetary resources needed to invest in agriculture. Roads, farmhouses, terraces, and urban centers all were made possible by the profits generated by the caravan trade. This same cash flow made it possible to obtain the labor resources needed for heavy construction. While some of the labor, especially in skilled categories, undoubtedly was paid, much of the manpower was forced to work. This uncompensated labor not only included slaves, who were obtained through the caravan trade and war, but also comprised the labor power of the military. After AD 106, the Nabatean Kingdom was incorporated into the Roman Empire as the province of Arabia. From this point on, Roman legions not only provided security in the pre-desert frontier zones, but they also provided much of the labor that constructed roads, forts, and settlement centers throughout the area. The traces of these infrastructural investments are widely visible today (Kennedy and Riley 1990). The increased security provided by imperial protection also encouraged the settlement of farmers in the frontier districts of the northern Negev. Indeed, it was a conscious policy of the Roman Empire and its Byzantine successor to promote the growth of a defense in depth zone by settling retired soldiers in potentially insecure districts where they functioned as a frontier militia. Thus, during the entire period in which agriculture flourished in the northern Negev, government investment in infrastructure, labor subsidies, and security were substantial and essential to the success of the agricultural system.

The agricultural system that the Nabateans created was based on the principle of creative destruction. Long experiential knowledge of the desert environment enabled the Nabateans to develop a simple system that was rooted in a sophisticated understanding of their environment. The secrets of this resource management system have been unraveled only in the past 30 years (Evenari, Shanan, and Tadmor 1971; Negev 1986; Rubin 1991). This system was based on the observation that, under natural conditions, the rainfall occurring in the region was too sporadic spatially and temporally, as well as of insufficient magnitude, to be of much use for agriculture. Although contemporary meteorological observation shows that most of the region's rains fall in small amounts of less than 10 mm (0.4 in) (Evenari, Shanan, and Tadmor 1982) with only limited runoff, larger storms do occur. The surface flow that accompanies these storms can produce destructive flash flooding that causes serious erosion. The essential problem was to figure out a way to concentrate the water that did fall into places where it could be useful, while minimizing the destructive potential of flash flooding. From this standpoint, most of the foothills and uplands of the northern Negev were quite unsuitable for agriculture, since their slopes were too steep, too stony, too thin soiled, and too lacking in water retention capacity to be of any use in crop production (figure 2.3).

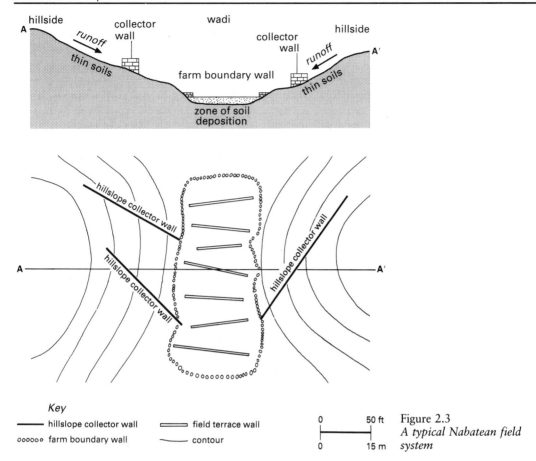

Key

— hillslope collector wall   ▭ field terrace wall

ooooo farm boundary wall   — contour

Figure 2.3
A typical Nabatean field system

The system that the Nabateans evolved concentrated agricultural production in the valley bottoms of the Negev's sporadically flowing streams. These valley bottoms were the *critical zones* upon which the viability of the entire Nabatean agricultural system depended. The valley bottoms were critical zones in the Nabatean system for two reasons. First, they were the places where soil deposition took place naturally. As a consequence, these lowland sites were the locations where the soils with the greatest agricultural potential existed. Second, under natural conditions, the valley bottoms had the greatest water storage capability in the area. These were the places where runoff collected and, therefore, where perennial vegetation was most densely concentrated. Focusing agriculture in these critical zones reflected a sophisticated application of the *genius loci principle* that was based on a rich understanding of local environmental processes, rhythms, and potential possibilities. Instead of fighting nature and trying to conquer it, the Nabateans devised a resource use system that operated within the constraints of the environment and was aligned with natural forces. By linking their agricultural system to natural processes, they were able to get nature to work for and with them. What the Nabateans did was to direct, channel, and enhance the natural processes in ways that worked to their benefit. They were able to do this because they had lived in the area for a long time, and had gradually evolved a technology that was based on a rich experiential knowledge of the region. Thus, individual elements of technology might be quite simple, such as the roughly coursed check dams used to impede water and sediment in the valley bottoms, but the technology was rooted in a very sophisticated grasp of hydrology and other operational principles of the world in which they lived.

Much of their cultivation was located in the smaller tributary wadis ("wadi" is an Arabic word meaning an ephemeral flowing stream channel; the identical features are called "arroyos" in the American Southwest) of the region. In these areas, the available quantities of water and soil were manageable. Likewise, these small watercourses were also the places where concentrated runoff would first occur in the modest rainfall events that characterized the Negev. They were the sites that would respond quickest to management with the most limited labor inputs. In these wadi bottoms the Nabateans built a succession of low stone walls that acted as leaky dams, extending across the valley at right angles to the stream flow. When runoff occurred, the floodwater ponded up behind the crude stone barrier. This had three important consequences: (1) it reduced the velocity of the water flow, thus decreasing the risk of soil erosion along the valley bottoms where crop planting took place; (2) soil particles eroded and carried by the runoff from the surrounding hills settled out of the ponded water with its lower velocity and deposited on the valley floor, and over time a terrace of richer, agriculturally useful soil developed behind the check dam; and (3) water trapped behind the wall increased the time available for the ponded water to infiltrate into the soil behind the dam. Because of this increase in the time available for infiltration, the soil moisture storage in the soils on the terraces often reached its capacity. This increase in soil moisture storage was usually sufficient to sustain a crop even if only minimal rainfall occurred during the remainder of the growing season.

The wadi terrace check dams seldom extended more than 30 cm (11.8 in) above the terrace surface. This height proved sufficient to retain enough water to provide the terrace with its soil moisture needs. Holding back more runoff water on a particular terrace with a higher wall served little purpose, since the oversupply of water would not increase the potential for crop utilization. The 30 cm (11.8 in) dam height provided sufficient soil moisture recharge needed by available crops. Allowing surplus water to flow to a lower terrace once the base soil moisture needs of upstream terraces had been met distributed available water over the maximum of potentially productive land. Once sedimentation raised a terrace surface so high that inadequate water was being retained to replenish the soil moisture store, the wall was raised by adding another layer of stones. In this way, terraces grew upward without changing the basic relationship of the agricultural land surface to the available water.

The runoff water produced by rainfall events to nourish the terraces came from the surrounding stream catchment. These hilltops and slopes that make up the land within the upstream areas were unsuited for agriculture due to their thin soils. However they played a critical role in the Nabatean farming system because it was water derived from these areas, once transferred to the adjacent lowlands, that maximized the benefit from the region's limited rainfall. Extensive surveys of the ancient farmsteads in the northern Negev by Evenari and his colleagues (1961; 1971: 104, 109) reveal that there was a consistent average relationship of 20:1 between the uncultivated water-producing catchment area and the cultivated fields. The importance of this relationship is revealed by considering the water made available for the cultivated fields. If only one-fifth of the precipitation produced by a 10 mm (0.4 in) rain shower appeared as runoff, a 20 hectare catchment would generate 40 mm (1.5 in) of runoff. Combined with the 10 mm (0.4 in) of rain that the shower directly produced on the 1 hectare (2.47 acres) of field to which the catchment area was attached, the water available on the valley-bottom field was the equivalent of having received 50 mm (2 in) of direct rainfall. Since the areas around Avdat receive approximately 100 mm (4 in) of precipitation annually, the accumulated and concentrated runoff from the surrounding catchment could potentially produce available moisture equivalent to a rainfall regime that was five times as abundant. As long as the crops cultivated could withstand the rigors of a hot, desiccating climate, the field environment created by the Nabateans reflected the soil moisture conditions of a much less arid habitat. For agricultural purposes, therefore, it was essential to view the area's two land-use components as part of an integrated system, and the Nabateans did so. Moreover, wherever possible, the Nabateans were interested in increasing the flow of runoff to their fields.

They did this by treating the hilltop and slope component as a *sacrifice zone*, the productivity of which was best exploited by the transfer of water and soil to higher potential land facets downslope. On the hillsides adjacent to their terraced fields they invested considerable labor to collect into large mounds the stones found on the surface. At one time it was thought that these mounds were intended to serve as dew collectors. Vines planted at the base of the stone mound and trained over the rocky surface were thought to have benefitted from the increased soil moisture content of the soil beneath the mound. However, careful analysis of the amount of moisture concentrated by dew condensation on and beneath the mounds indicated that vineyard cultivation based on stone mound dew collection was not possible due to the insufficient quantities obtained (Evenari, Shanan, and Tadmor 1971: 134). Since documentary evidence indicates that vines were cultivated, the region's vineyards must have received their supplies from more conventional runoff sources. The real purpose of the stone mounds was to reduce

surface roughness for the slope areas as a whole, diminish the
amount of infiltration that occurred on the slope, and encourage
greater runoff to the valley bottoms where the water could be
utilized in agriculture.

In this effort the Nabateans were aided by the characteristics of
the local soil type. Many of the northern Negev's soils are fine-
grained loess soils with limited pore space and a tendency for
individual grains to swell when wet. As a result, the soil reacts to
wetting by permitting relatively limited infiltration before it forms a
sealed surface layer. Once this clogged surface develops, downward
water infiltration into the soil proceeds slowly, and most of the
subsequent rainfall flows away as runoff. By removing the stones
from the soil surface, large pore spaces between the stones are
eliminated, surface roughness that encourages retention of water
for longer periods on the slope is reduced, and larger amounts of
bare soil are exposed to rapid swelling.

What the Nabateans did was to encourage water export and soil
erosion from the slopes for the benefit of the valley bottoms. Soil
erosion was not the prime purpose of the slope denudation; it was
an ancillary benefit of the main objective of increasing water capture. Once soil removed from the slopes by the increased runoff
reached the nearby wadis, it was trapped behind the valley check
dams. Captured in relatively tranquil pools, the floodwater was
able to infiltrate slowly into the terrace soil. The suspended soil
particles carried in the water settled on the soil surface as a soil
improvement addition. When all the captured water had infiltrated
into the soil or evaporated from the surface, the resulting soil crust
reduced the removal of water from the soil moisture store by evaporation and capillary action, thus locking up the runoff as available soil moisture for plant utilization.

Other water collection devices essential to the Nabatean agroecology also were developed. An important element in the water
management system was the slope water channel. These water
transportation devices were built on hillsides at an angle to the
slope gradient. Their purpose was to collect runoff from a particular portion of the slope or watershed and conduct that water to a
designated terrace field in the valley bottom (Evenari, Shanan, and
Tadmor 1971: 182ff). The main wadis, with larger and deeper
channels than the tributaries, were more difficult environments to
manage. Extracting water from them required the use of diversion
dams in the stream channel in order to raise the water level to the
height of the adjacent fields. The water diverted from the flood was
directed into conduits that led the water to the terraces. In this way
fields had three water sources: floodwater extracted from the main
wadi channel, the spillage of surplus water from upstream terraces,
and the specific slope runoff from the adjacent watershed dedicated
to a designated field. Within the field system, a network of canals
and weirs insured the distribution of water to the individual terrace
fields, and made it possible to shift excess water from field units
saturated with moisture to deficit units. This cascading quality in

the water distribution system guaranteed that upstream fields would receive the moisture needed to grow crops before any surplus was passed on to fields further down the farm and the wadi system. However, it also held on upstream terraces only the water needed for crops grown on that field, thereby insuring that downstream terraces received a fair share of the available supply. Only the size of the individual flood event determined the number of terraces in the system that would receive water at any one time.

The more elaborate slope collection channels and terrace field distribution systems that were typical of main wadi agriculture required a much larger investment of capital and labor than did the simple check dam systems of the smaller tributary wadis. The more complicated field systems developed on the main wadi floodplains where much larger sized fields were possible, and therefore larger yields could be produced to justify the added capital investment. These main wadi fields also needed much larger catchment areas, often measuring dozens of square kilometers, in order to generate sufficient water to supply the terraces. Because of their greater scale and larger capital requirements, the main wadi agricultural systems developed later than did the smaller-scale system of the tributary wadis. These more elaborate agricultural enterprises represent a form of intensification within the Nabatean dryland farming system. It is not surprising that the major examples of these more elaborate farms are concentrated near the region's main towns. The more intricate infrastructure involved, and the larger amounts of floodwater with which they had to cope, made the main wadi agricultural systems more vulnerable to the destructive effects of episodic extreme floods. These floods required both more elaborate preventive protection devices and more costly reconstruction efforts than was needed in the subsidiary wadis. Eventually, these main wadi systems could not be maintained, and they were abandoned. In contrast, many of the smaller-scale systems continued to be exploited by pastoral nomads after the settled agricultural system collapsed. In fact, many are still used by Israeli Bedouin who farm these relics of the Nabatean period.

The denizens of the northern Negev were equally adept at developing all available water sources for drinking water. These included wells, both shallow and deep, to tap aquifers and the gravel beds of the regions ephemeral streams (wadis) as well as an elaborate array of cisterns. There is no evidence in the northern Negev for the existence of the *qanat* systems (underground water collection galleries) that are found in many parts of the Old World drylands, although one "chain of wells" has been located at Yotvata in the Arava rift valley (Evenari, Shanan, and Tadmor 1971: 173ff). The cistern systems were invariably linked to centers of human habitation, since no household could hope to survive without adequate drinking water. The flat roofs of houses, cleared areas near the farmhouses, the streets in the villages and towns, the roofs and courtyards of public buildings – all were connected to cisterns that stored the precious water collected after each rainfall. Slope

collector channels were also used to conduct water to nearby cisterns from small catchment areas that were not linked to agricultural terraces. Often equipped with sediment tanks, and protected
from evaporation by their underground location, even small cistern
collection systems could produce and store substantial amounts of
high-quality water. Over a seven-year period, for example, a small
1.2 hectare (3 acre), catchment area near Shivta produced an annual
average of 150 m^3 (5,297 ft^3) of water for its associated cistern
(Evenari, Shanan, and Tadmor 1971: 146; 166). A half-dozen pastoral nomadic families and their animals could survive for a year on
this amount of water.

Relatively little is known about the specific crops grown by the
Nabateans and their Roman and Byzantine successors. The few
local written records from the period are fragmentary, but they
suggest that the crops used were the wheat, barley, vine, and olive
complex commonly employed throughout the Mediterranean
(Rubin 1991: 200–201). None of these crops, as well as minor
associated species, could survive in the northern Negev on natural
precipitation. All required the supplemental irrigation developed
by the Nabateans. Decay in the farming system rather than land
degradation or climatic change is the reason why these crops are
so infrequently encountered today. In sum, the crops cultivated in
the northern Negev reflect an adaptation of the basic
Mediterranean system to the more marginal and extreme conditions of the desert fringe.

The basis for the success of this system was its diversity. Cereal
crops, tree crops, and animals were a stable trilogy upon which the
settled farming community rested. Each of the three elements was
capable of withstanding or escaping drought to a considerable
degree. Wheat and barley are relatively hardy grasses, and many
of the traditional varieties in the region possess waxy surfaces that
reduce evapotranspiration loss. Olives, vines, and other dryland-
adapted trees, such as almonds, are deeply enough rooted so that
they can reach groundwater and survive even in a year when rainfall and run-on water supplies are inadequate to grow cereals. By
using microcatchments (Evenari 1977: 87–8; 1981: 11–13) that
concentrate runoff from a small zone around each individual tree,
modern experiments have demonstrated that sufficient water can be
generated to nourish fruit trees in very desolate sites (figure 2.4). *In
fact, this very strategy is used in Israel to permit isolated roadside
sites of tree patches to exist along their Negev highways.*

Animals, particularly goats, are the survival insurance *par excellence*, because they can be moved from areas of water and fodder
deficit to districts unaffected by precipitation shortfalls. Many of
the animals – especially sheep, who have a high water consumption
demand – must have been based around the cisterns that were
located at a distance from the main zones of wadi farming. Here
they could have found more than adequate water and just adequate
rough forage, supplemented with post-harvest residues from the
cereal fields, to survive. The camels raised by the Nabateans

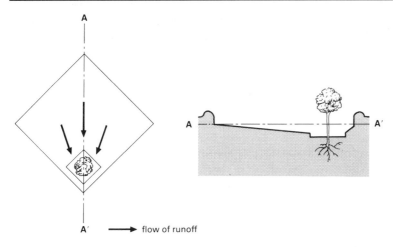

flow of runoff

Figure 2.4
The plan and cross-section of a
microcatchment plot

(Negev 1986: 40) were an essential part of their economy, for not only could they contribute hair and milk to domestic subsistence but also the camel was the mainstay of their overland caravans. When caravan trading declined in importance in the Nabatean economy, the wealth of stables and votive horse figurines found in the Negev indicates that horse breeding became a principle source of local prosperity for the next six centuries (Negev 1986: 106). Whether or not cultivation of fodder grasses was common in the period is unknown, although it is clear that well-adapted perennial species do exist, including indigenous wild oats, that give high yields and successfully resist severe drought. Thus an ecological trilogy of cereals, tree crops, and animal products could be produced in the desert fringe because a suitable set of water generation and conservation techniques existed.

A spectacular efflorescence of settled life based on the Nabatean system occurred in the fifth and sixth centuries AD in the Negev. This high point in the development of the region featured both rural agricultural and urban development (Negev 1986; Rubin 1991). What emerged was a *limes* (frontier) zone that featured a defense in depth concept rather than a rigid fortified line of walls and forts. The idea behind the intensified settlement system was to provide protection for the higher productivity coastal districts from raids conducted by more nomadic groups from the desert interior. Settlers during this period, in turn, depended on logistical support from the Byzantine Empire. For it was the imperial system of the Byzantine center that promoted the steady expansion of the northern Negev periphery as a buffer. The ability of the frontier settler militia to act as a sponge, soaking up the aggressive attacks of tribal groups from outside the imperial system, was directly linked to the degree of support and security that the imperial military and logistical system could supply. When this imperial center came under increasing pressure from the forces of an expansionist Islam in the middle of the seventh century, the conditions that were conducive to settled life in the Negev ended.

The termination of settled life did not occur as a cataclysmic collapse, but rather as a gradual decline in the viability of sedentary existence (Rubin 1991: 204). People gradually decreased their investment in field and tree crops, and increased their emphasis on animal husbandry. Towns and farmhouses were abandoned as the site of permanent residence, and a more mobile nomadic existence became the norm for the next 13 centuries. In this shift away from settled exploitation of the region in favor of a less intensive livelihood mode, the post-Byzantine era mirrored the historic rhythm of alternating settlement and mobility. Nonetheless, even in the more nomadic episodes, the most manageable of the agricultural terraces were cultivated and the functional cisterns were still utilized. The same fixed points in the settlement system attracted the region's inhabitants. Only the manner and intensity with which these resources were utilized changed.

The Nabatean settlement system functioned successfully for over nine centuries because it was rooted in the *genius loci principle*. By fitting into the rhythm of the local ecosystem and by building on specific features of the landscape, the Nabateans were able to promote developments that constituted *creative destruction*. Erosion encouraged on the slopes (sacrifice zone) increased runoff, which then ran on to adjacent lowland areas. The collection of water and soil in these lowland sites impoverished the steeper slopes, encouraging soil erosion and vegetation degradation in the least productive parts of their habitat. The Nabatean system possessed both *stability* and *resilience*, and as such it was able to exist for centuries. As long as a central government in a higher potential zone – whether Roman or Byzantine – was willing to subsidize activities in the Negev, recovery from any perturbation was possible and usually was accomplished quickly. A well-conceived, diverse agricultural system, based on a system of water collectors, terraces, and cisterns, provided the resilience needed to absorb the impact of drought as well as to make the most of limited available moisture. Nabatean agricultural activities concentrated on *critical zones*, the wadi bottoms, that had high potential in nature as traps for energy, soil, and water. The system was also based on *capturing wastes*, in this case the rainfall that fell on steep slopes and the thin soils that covered them, and transferring those wastes to sites where they could be concentrated and used more effectively. Water that was surplus on one terrace, rather than being wasted by being retained there, became an input as run-on moisture for the next terrace in the system. Managed systems that use the "waste" from one sector as the "input" for another sector are extremely efficient and stable enterprises.

All of this was not accomplished without *exporting costs* onto other parts of the larger ecumene. Some of these costs were borne by the central administration, whether Nabatean, Roman, or Byzantine. In each instance, these costs were apparently regarded as justified, since without them the frontier could not be effectively

controlled, caravans could not be operated, border guards could
not be supported, and defense buffers could not be maintained.
Other costs were inflicted on nomadic pastoralists, whose best
pastures were converted to cereal crops and whose rough hillside
pastures were denuded (where anything other than ephemeral vegetation existed) to promote the movement of water and soil to lowland sites. The response of these more mobile elements of the
population to these losses and costs was a factor in the instability
following the Arab conquest that undermined the viability of settled
life in the northern Negev. But the denouement of this remarkable
settlement system should not obscure its importance as an example
of *creative destruction*. The slope areas were sacrificed (degraded)
for the increased productivity of selected valley bottoms.

Irrigation in the Mesopotamian Lowland

Mesopotamia is the alluvial lowland of Iraq that lies between the
Tigris and Euphrates Rivers (figure 2.5). It is an exceedingly flat,
generally featureless expanse of land that is crossed by two large
rivers, the Tigris and Euphrates, which drain the rugged highlands
of Anatolia and the Zagros Mountains of Iran. The hydrologic
regime of each of the twin rivers is quite different, a product of
their respective source areas. The Tigris is the shorter of the two

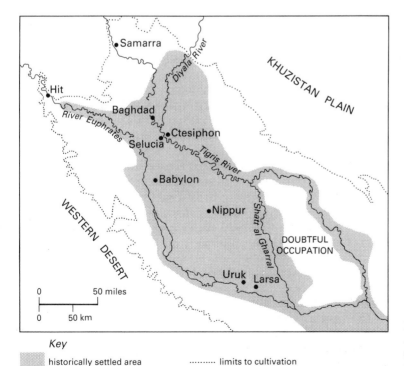

Key

▓ historically settled area   ·········· limits to cultivation

Figure 2.5
*Mesopotamia and the Tigris
and Euphrates floodplain
(after Turner et al. 1990)*

rivers but, because it is joined by a series of tributary streams that drain high elevations in western Iran and eastern Iraq with significant winter precipitation resources, its discharge is close to twice that of the Euphrates (al-Khashab, 1958: 41). Rising in the less elevated central and eastern districts of Turkey, the Euphrates travels a much longer distance to the Gulf, has a lower river discharge, is joined by far fewer tributaries – none of which are located within contemporary Iraq – and floods earlier in the year than the Tigris. Both rivers share two common characteristics: in their lower courses neither river receives significant additional moisture, and both rivers are characterized by highly variable interannual flow regimes. This extreme variability in the flow behavior of the Tigris and Euphrates introduced an element of unpredictability into the Mesopotamian environment that was quite different from the orderly behavior of the Nile in Egypt (Frankfort 1956: 53). In Mesopotamia, people were aware of the uncertainty of life and of their profound inability to control many of the events that had enormous power over them.

Flowing across a broad and – to the outsider – largely featureless, nearly tabletop flat plain in southern Iraq, each river has a low gradient. Meandering stream courses characterize both rivers, and elongated meanders are prone to being cut off. In these circumstances, frequent changes of river course occur (Adams 1981: 9), with consequent complications for the human populations located along the cutoff stream channels. The old levee banks of these abandoned channels provide a microrelief that is not readily observable to the naked eye. Nonetheless, these meander scars play a significant role in the hydraulic regime of the floodplain. The barrier effect that these minute, convoluted, sinuous topographic ridges play in the region's surface drainage is reflected in the more moist soils and denser seasonal and perennial plant cover found on their upslope side.

Coping with the complications of a highly variable, unpredictable, and complexly interrelated river regime was essential to successful human use of the floodplain. Moreover, the alluvial floodplain of the two rivers south of contemporary Baghdad receives less than 200 mm (7.9 in) of precipitation. While the hills of northern and eastern Iraq receive enough precipitation to support field crops, the insufficient rainfall in the south is unable to sustain dry farming. Without irrigation, settlement would be confined to the immediate banks of the two rivers and only small population totals could be sustained. It is ironic that one of the Earth's greatest early civilizations arose in an inhospitable desert plain with few indigenous resources other than the water of the two great rivers that crossed it, the muddy alluvial soils that the streams deposited, and the date palms and reeds that grew in abundance along the region's constantly shifting water courses.

This riverine, desertic lowland is the setting for the second of our historical examples of land degradation. In contrast to the Negev, Mesopotamia illustrates the operation of *destructive creation*. The manifold accomplishments of the Sumerians, who developed the world's first urban civilization over 5,000 years ago (Kramer 1963; Adams 1966; Redman 1978), and their successor cultures not withstanding, irrigated agriculture in Mesopotamia failed to maintain human populations and environmental resources in a sustained fashion. Two and a half major cycles of population growth and decline have occurred in Mesopotamia (figure 2.6), an oscillating pattern that is due in large part to the internal inconsistencies and instabilities that are deeply embedded in the structure of the region's irrigation system (Adams 1965, 1981; Adams and Nissen 1972; Johnson and Whitmore 1987; Whitmore et al. 1990).

By *destructive creation* we mean a resource-use system that contains critical flaws internal to the system that undermine its viability. The flaws that promote destructive creation are often insidious, masked by shorter-term surficial signs of success. However, these flaws operate either to erode the system's sustainability in a direct fashion – for example, the deposit of sediments that clog water delivery canals – or to create such a delicately posed system that it is vulnerable to sudden, often irreversible changes of state when confronted with unexpected events. Irrigation systems are vulnerable to such catastrophic collapses whenever nature, managerial mistakes, or military conflict lead to an alteration in the main stream courses that provide water to the agricultural operation. This situation is different from the *creative destruction* that produced the run-on farming systems of the Negev because the factors that promote collapse in Mesopotamia are internal to the agricultural regime. In the Negev, external support was essential to the full flourishing of the Nabatean system, and its loss after the Arab conquest resulted in the decay of the Nabatean agricultural enterprise. However, 13 centuries later the same infrastructure, operating on the same management principles worked as well as it ever had when it was restored. The same cannot be said for Mesopotamia.

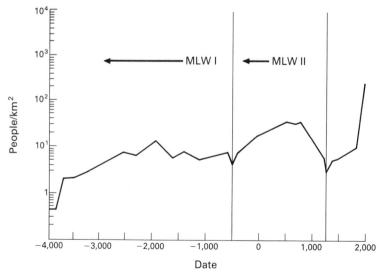

Figure 2.6
Cycles of population growth and decline in Mesopotamia (after Turner et al. 1990)

The irrigation system of the Mesopotamian lowland exists historically in two states, disaggregated and integrated. The disaggregated system is the oldest irrigation complex, and its appearance is associated with the spread of agriculture in the lowland. The hunter–gatherer communities of the floodplain were primarily confined to the vicinity of the region's major rivers, their multitudinous branches and subsidiary channels, and associated swamps and marshes. The bulk of the floodplain was too dry to be of any real importance in terms of food production. In the earliest stages of settlement, harvesting of wild plants and animals was the dominant mode of production. However, agriculture proved to be a superior food production system in two ways: it provided larger yields for a growing (albeit at a very slow rate) population, and it produced surpluses that could be stored in order to make it easier for people to survive the hazards of the natural environment.

Agriculture itself could be made more secure, and the yields from cultivated plots along the river levees and levee backslopes could be increased by digging small ditches through the levee. This permitted the earliest farmers to withdraw water for their fields for a longer period and reduced their total dependence on the seasonal flood to deliver all their irrigation water. These small, exceedingly primitive delivery ditches depended for their construction and maintenance only on the labor of local kinship groups. This meant that small communities of several hundred individuals could generate the labor and leadership needed to operate the system. The region's meandering natural streams were the main components of the water delivery system. Provision of drainage by the farming community was not considered, and excess water flowed from the fields down the floodplain's gentle gradient until it had fully infiltrated into the soil outside the cultivated zone or encountered the barrier of a nearby levee. Here excess moisture accumulations favored the development of swamps.

Because the disaggregated system existed at a small, local scale, such swamps were limited in size. Individual communities remained vulnerable to large fluctuations in the environment. Floods might wipe out an individual community, or changes in a river course due to a meander cutoff might destroy the productive habitat of a local population unit, but the survival of the floodplain population as a whole was never in doubt. Few, if any, of the driving forces that promoted change in local nature–society relations were under the direct control of the local population, and the system exhibited great stability over time.

These conditions began to change dramatically when the irrigation system evolved from a disaggregated state, in which isolated settlement beads on a riverine string were scattered along the region's winding natural watercourses, into an integrated system of water distribution and management. This change is associated with the rise of larger urban agglomerations and the development of a series of small city–states (Wooley 1965a,b; Adams 1966; Redman 1978; Nissen 1988). The change is reflected in the

architecture of the region's canals, which no longer meander ran-
domly across the countryside but rather slash in straight lines
through the landscape, delivering water from stream bed to farm
fields. Evidence also exists in the form of documentary data recovered from an ancient irrigation archive dating to the Isin-Larsa
period (ca.1890 BC) nearly a century before the famous ruler,
Hammurabi of Babylon, who united the entire southern alluvial
lowland into the first large-scale state structure (Walters 1970).
This archive contains documents that make possible a reconstruction of the irrigation bureaucracy of the period, as well as time and
effort calculations for canal construction. For example, 1,800
working days were needed to dig a one and one half mile (2.4
km) long canal that was five feet (1.5 m) deep and five feet wide
(Walters 1970: xix). In short, by 5000 BC there was in existence in
Mesopotamia a centrally managed irrigation system, the operations
of which appear to be analogous to those of a contemporary Corps
of Engineers!

The ancient irrigation system that developed in Mesopotamia was
an exceedingly sophisticated structure in its aggregated state. For
long periods this system was able to maintain large populations in
a condition of relative prosperity and security. However, this surficial success in maintaining irrigation system functions masked latent
flaws that either emerged episodically, after lurking beneath the
veneer of prosperity only to appear when conditions were appropriate, or imposed increasingly greater burdens on the long-term
productivity of the system. One set of long-term factors was a consequence of the way in which the Mesopotamian irrigation system
dealt with its *sacrifice zones*.

These *sacrifice zones* were two in number; the forested uplands
and the lowland swamps. Lowland Mesopotamia is a singularly
one-dimensional environment from a resource standpoint. Aside
from its rich alluvial soils, it possesses few natural resources. The
petroleum riches that have fueled modern Iraqi development could
not be exploited until the twentieth century. Sumerians, and all
subsequent agriculture-based civilizations in Mesopotamia, lacked
basic timber and stone materials for construction. Except for soil,
they were completely without mineral resources. These items could
be found only in the adjacent highlands; and their acquisition figured from an early date in the economic and imperial designs of the
lowland states. The Gilgamesh epic (Kramer 1959: 184; 1963: 191–
3) describes the expedition of a mythical ruler to a distant place
(Lebanon or Turkey?) where cedar trees, the abode of a fearsome
guardian, are cut down. There is no indication that the timber was
brought back to Sumeria, yet it is not impossible that the epic
adventure of this mythical Sumerian hero records a more prosaic
functional activity. Although glazed mud bricks and palm logs were
the prime lowland construction material, there was no substitute for
the hardwoods of the upland forest in any major construction.

Economic penetration of the highlands for mineral products sti-
mulated both settlement expansion in the uplands and the removal

of forest cover. The result was a millennia-long process of deforestation in the highlands that contributed to soil loss from these environments. In these developments the acquisitive demands of lowland irrigation communities were not the only culprit, for the expansion of rainfed agriculture in the uplands played the major role in opening up clearings in the highland forests and exposing bare soil to the weather (Wagstaff 1985: 46–7). In a real sense, the health of many upland areas was sacrificed for the overall benefit of lowland civilizations. The result of this process of highland landscape change was increased sediment generated by erosion and moved into the upstream portions of the river systems. This resulted in increased siltation rates in the low-gradient floodplain districts of Mesopotamia. This increase in siltation imposed a heavy maintenance burden upon the irrigation system, a factor that will be discussed in more detail below.

The conversion of the highland forest habitat into more open environments dominated by seasonal grasses created new opportunities for the region's nomadic pastoralists, who found enhanced grazing for their flocks in their seasonal movements to summer pastures in the highlands. This demonstrates the degree to which almost no land degradation occurs without some human community finding a way to benefit from the negative development. The swamps of the alluvial lowlands were the second major sacrifice zone. In large part, the extensive swamps of the alluvial plain are natural habitats. They are produced by a combination of low gradient close to sea level, abundant seasonal stream flow, impeded drainage caused by the old levee banks and meander scars of abandoned stream channels, and the steady deposition of sediments by the region's rivers into the head of the Gulf. The majority of these swamps are permanent (Larsen and Evans 1978), but about one quarter are seasonal (al-Khashab 1958: 62), the product of the annual flood regimes of the Tigris and Euphrates. The irrigated areas in the settled zones are characterized by the existence of significant swamps that are responsible for considerable water loss. Evapotranspiration by the vegetation and evaporation from the surface waters found in these swamps as well as infiltration into the groundwater and return seepage into canals and stream channels represents the major feedbacks from swampland into the hydrologic cycle (figure 2.7).

These swampy areas archetypically develop at the tails of the irrigation system. As an Akkadian poet depicted the relationship:

> After Anu had created heaven,
>   Heaven had created earth,
>   Earth had created rivers,
>   Rivers had created canals,
>   Canals had created marsh,
> Marsh had created worm ——
>   (Kramer 1961: 123)

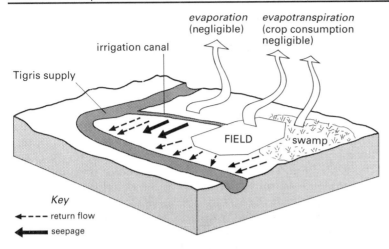

Figure 2.7
Water allocations of a hypothetical irrigation system unit

Swamps are encouraged by the microtopography of the irrigated portion of the floodplain and by irrigation practice (Adams 1965: 8). The region's irrigation systems are gravity fed, with the high point in the distribution system being found along the levees of the natural river courses and main distribution canals. Flow is down slope and away from the main water sources across the backslopes of the levee system. Excess water accumulates at the tails of the distribution system at the lowest points in the irrigated landscape. Usually, these areas also are associated with old meander scars that impede further lateral surface water movement.

The growth of initially ephemeral, but over time more permanent, wetlands is aided by the tendency of all farmers in all epochs to over-irrigate if water is available. The Sumerian Farmer's Almanac (Kramer 1963: 341), for example, urges its readers to water their barley at least three times after seed begins to germinate, but to throw on a fourth watering if possible in the interest of even better yields. In an environment such as southern Mesopotamia, where the water table is close to the surface, frequent over-irrigation inevitably encourages waterlogged soils, especially in microtopographic settings that promote water accumulation. This same tendency to over-irrigate whenever water supplies are available is intensified by the difficulty of being sure of adequate water for the planting season, which occurs at the same time as the low-water stage in the twin rivers. Historically, this is an important contributor to the spread of land degradation in this area due to salinization (Adams 1978: 330) and swamp encroachment (Adams and Nissen 1972). These zones become sacrificed, from an agricultural standpoint, in favor of the better-drained soils in the prime farming zones. This is particularly true when, as is typical of ancient irrigation systems in Mesopotamia, no provision is made for draining either the irrigation plots or the swamps that develop at the tails of the system. That these swamps can also create ideal habitats for the snails and mosquitos that serve as transmitters of schistosomiasis and malaria also indicates that they can become important indirect negative impacts on

the integrated irrigation systems that are their primary creators
(McNeill 1976).

In south and southwest Mesopotamia, an important factor in
swamp development was a failure to maintain crucial infrastructure. Late in the fifth century AD and early in the seventh century,
catastrophic floods broke through dikes the maintenance of which
had been neglected for many years. These floods innundated vast
tracts of lowlying land. Although vigorous reclamation efforts partially rehabilitated lands flooded in the first disaster, the second epic
flood proved more permanent in its impact. Despite very strenuous
efforts on the part of the Sassanian administration to reestablish the
broken flood-control works and drain the waterlogged lands, little
progress was made (Le Strange 1905). The lowered productivity of
land resources damaged by both natural (floods) and human
(neglect) impacts undoubtedly contributed to the inability of the
Sassanian state to resist the Muslim invasion and conquest in the
second quarter of the seventh century AD (Lapidus 1981). And while
Arab administrators used much of the surviving Sassanian water
distribution and flood-control infrastructure (Moroney 1984), agricultural recovery was not uniform in the Mesopotamian floodplain,
particularly in the salinized and swampy south.

It is important to recognize that the impact of swamp expansion
has positive implications. In the far south of Mesopotamia, centered
in the vicinity of contemporary Basra, entire cultures emerged that
were based on fishing in the fresh, brackish, and saline swamps of
the region (Adams 1981: 16). Analogous adaptations to the swamps
are found in contemporary southern Iraq among the Marsh Arabs
(Thesiger 1964). Although there are indications that fishing declined
in importance after 2300 BC, if the number of fish species mentioned
in ancient texts is any guide (Kramer 1963: 110), the existing evidence could also indicate a reduction in species diversity under the
pressure of an increasingly saline environment. At the same time, an
expansion of swamp vegetation would increase the habitat available
for water birds of all types, and more than four dozen variety of
edible birds are noted in the ancient texts (Kramer 1963: 110).
Moreover, swamp grasses constitute a valuable fodder source with
which to support substantial domesticated animal populations. This
resource constitutes a valuable grazing niche for the migratory herds
of nomadic pastoralists, whose animals could combine freshwater
wetland grazing in the winter with postharvest stubble resources and
the semiarid steppe grazing of the nonirrigated floodplain districts.
In sum, while the expansion of swamps was a negative consequence
of irrigation system integration and expansion, for certain ethnic
and economic groups the degraded wetland environments became
a positive resource.

The *critical zones* for the functioning of floodplain irrigation
systems are the perennial irrigation zone along the natural river
channels and the high potential floodplain soils commanded by
the artificial canals of the integrated, centrally managed irrigation
bureaucracy. The best drained soils on the river and main canal

levee banks are a vitally important resource facet, since these districts are the only ones in the Mesopotamian floodplain that can be cultivated on a year-round basis. The use of simple lift devices to raise water to the fields, the application of manure (often provided by nomadic herds grazing on post-harvest stubble), and the application of considerable labor is all that is needed to keep these soils in constant production. Alluvial floodplain soils require the application of a local *genius loci principle* in order to remain productive. This principle is based on two factors: alternate-year fallow of cultivated fields (Gibson 1974: 15–16) and the survival of salt-tolerant perennial bushes in the cultivated fields (Adams 1965: 5, 18). A related factor in maintaining the critical, high-potential zones is a competent managerial system.

Alternate-year fallow is an integral part of the Mesopotamian agricultural regime. Only the limited, multicropped, perennial production areas along the rivers and distribution canals where low-flow water is available are an exception to the alternate-year fallow principle. Leaving alluvial soils fallow every other year is essential to productivity because drainage is not a part of the Mesopotamian irrigation system. Without the wetting provided by irrigation, soils drained adequately and dried in their surface layers. The water table declined to sufficient depth to enable the next production year's irrigation to wash any accumulated salts out of the surface soil and away from the root zone of cultivated plants, thus preventing salinization. Reduction in the frequency of fallow encouraged a rise in the water table and a movement of salts toward the soil surface. Any increase in salt content in the root zone is immediately reflected in a lowering of yields for the cultivated crops. Historically, salinization appeared suddenly near the end of the third millennium in Mesopotamia, at a time when irrigation systems were aggressively expanded by both local and regional administrations. These increases in soil salinity are reflected initially in a shift away from cultivating wheat in favor of the more salt tolerant barley, a three-fold decrease in soil fertility over a seven-century period as measured by barley yields, and the appearance of surface patches of salt in temple fields (Jacobsen and Adams 1958; Helbaek 1960: 194–6; Walters 1970: 160–1). Explanations for the appearance of salinization are linked to the availability of more irrigation water, as local rulers dug canals that tapped previously unused Tigris River water (Jacobsen and Adams 1958: 2) to supplement the much more extensively developed canal distribution network based on Euphrates River water (Jacobsen 1960). The presence of greater water supplies that were available for longer periods throughout the year had the potential to encourage more extensive irrigation of the critical central perennial cultivated areas than could be attained using simple lift devices. Thus, indirectly, the integrated irrigation system had a high potential to become an inadvertent "engineered disaster" (Gibson 1974: 15). Also implicated are the demands of rulers for a greater return via taxes from the peasant farmers under their control (Gibson 1974: 16). In many cases the immediate returns to the central government were greater from large landholders, with whom autocratic governments were often linked politically and socially and who they naturally tended to favor, than from small land-owners organized in a more tribal system. Favoritism of a landed rural elite, often resident in urban areas, debased the peasantry, reduced the availability of rural labor as sharecroppers were driven out of operation and fled, and initiated a downward spiral in rural productivity and irrigation system maintenance that was difficult to arrest (Gibson 1974: 16–17). Whenever taxes increased as a means of concentrating rural surpluses in cities in support of centralized military and managerial bureaucracies, the farmer had little recourse other than to violate the alternate-year fallow cycle. Once initiated, only a short time (often less than a decade) was needed to bring the water table sufficiently close to the surface to begin to affect production negatively.

Perennial shrubs play an important role in support of the alternate-year fallow practice. Two shrubs in particular, *shauk* (*Prosopis farcta*) and *agul* (*Alhagi maurorum*, camel thorn), are legumes that have a major role in the traditional agricultural system (Adams 1965: 5, 18). Each plant has a tap root that reaches deep into the ground. Evapotranspiration from their leaves, in the absence of irrigation, has a positive effect on the depth of the water table. In effect, *shauk* and *agul* function as natural pumps, lowering the water table in a natural way without requiring the expenditure of human or animal energy. Both plants are legumes, and thus contribute to nitrogen fixation. Their deep roots help to loosen the soil and encourage greater water infiltration. Their leaves constitute a nutritious source of fodder once residual, post-evapotranspiration salt has been washed off, and their branches can be collected for firewood. For all of these reasons, these fallow plants were a critical component contributing to the well-being of the traditional irrigated farming system. Until modern times, farmers left these "weeds" unhindered in their fields, a coexistence that was favored by the limited soil-turning ability of the traditional local plow, which was unable to root out these shrubs.

Rather than competing with the shallow-rooted local cereal crops for water and nutrients, *shauk* and *agul* tapped deeper resources and recycled these to the surface. In an irrigation system in which systematic planned drainage ditches did not exist as a way to lower water tables and reduce soil salinity, these deep-rooted natural pumps were vital ingredients in maintaining overall system stability. While none of the primary or secondary sources suggest that removal of these useful shrubs was carried out systematically, it is not difficult to imagine pressures for increased perennial production contributing to their removal. The reduced presence or absence of these useful plants from irrigated fields would have created a much more delicately balanced field system that demanded increasingly sophisticated micromanagement in order to avoid disaster.

As the irrigation system became increasingly integrated, it depended more and more on the managerial skills and competence of the irrigation bureaucracy. In the middle of the third millennium BC apparent incision of streams and shifts in the course of the Euphrates (Adams and Nissen 1972; Nissen 1988: 129–64) instituted stress that placed even greater demands on the stability and organizational skills of the irrigation managers as they competed, often in military conflicts, for access to secure water sources. City–states that were unable to control reliable water supplies declined in size and prosperity, and the center of gravity of political, military, and economic power shifted northward in the floodplain.

Episodes of civil war over water made an early appearance historically in Mesopotamia (Kramer 1959: 35–44), as did efforts to reform abuses of power and taxation (Kramer 1959: 45–50). Whenever these conflicts broke out for any prolonged period the result was a decline in the viability of the irrigation system. The need for stability was substantial because the irrigation system required organized labor to counter the threat of siltation. Only a central administration could organize sufficient labor to keep the entire system clear of silt. Silt loads throughout the floodplain were always high, but environmental changes in the adjacent uplands increased the sediment loads of the river system. Only constant vigilance could keep the canals clear of silt and insure a secure supply of water for the fields (Jacobsen and Adams 1958: 7). Weaken this central administration, and rapid decline would follow.

However interested local communities might be in keeping their segment of canal in functioning order, their success did not depend on their efforts alone. What happened upstream dramatically affected their prospects for prosperity. Without an effective administration to keep all units functioning in a coordinated fashion, the populations at the tails of the system experienced immediate negative impacts. Decline tended to spread from the insecure margins of the irrigation system back toward the centers of power in the more advantageously placed major cities. Greedy rulers promoted taxation and land management policies that precipitated mismanagement and a steady decline in population and productivity. Efforts to bring in slave labor in the Abbasid period, in order to replace a pauperized peasantry in carrying out land reclamation projects in the salinized and swampy south of the floodplain, produced a servile rebellion that nearly capsized the imperial regime. A neglect of investment in irrigation infrastructure, the loss of authority to an exploitative rural land-owning elite, the institution of rampant and corrupt tax farming, civil wars and conflicts with hostile neighbors, and the military inability of a weak central administration to provide security against the incursions of nomadic peoples led to the centuries-long decline of the Abbasid Caliphate (Lapidus 1988: 135–6).

Three centuries of progressive internal decline culminated in the Mongol *coup de grâce* when the sack of Baghdad in AD 1258 pushed the second major population growth phase in Mesopotamia into total collapse and reduced the irrigation system to a disaggregated state. This story of an apparently successful system self-destructing is repeated episodically in Mesopotamian history, and is the bleak side of the destructive creation that generated the spectacular, albeit instable, efflorescences.

That Mesopotamian irrigation was able to cope with a host of factors over which it had little direct control suggests that the system, finely poised though it was, possessed considerable powers of resilience. Tectonic changes that controlled the advance and retreat of the head of the Gulf and shifts in stream channels were coped with through location shifts in the population, the development of new cultural adaptations to deal with a swampier environment, and the construction of major artificial canals that linked much larger regional units together in one irrigation system. Efforts to expand into alluvial areas previously unaffected by salinization were attempted, often successfully. Changes in crops in favor of more salt-tolerant varieties, such as barley and the date palm, made it possible to deal more effectively with an increasingly saline environment. More use of previously underdeveloped Tigris River water through a complex system of regional canals characterized the Sassanian (Persian) pre-Islamic agricultural and population expansion. Yet none of these coping strategies could overcome the basic vulnerability of the Mesopotamian system to salinization, siltation, and mismanagement.

Summary

Creative destruction and *destructive creation* are opposite sides of the same coin. Each has operated throughout human existence, as individuals and communities have struggled to extract a living from the resources at their disposal. Whether by accident or by design, successful systems are those that are able to achieve sustainability for more than the transient lifetime of an individual. In the human use of environment, change, even destruction, in natural systems is inevitable. The slower that change, the more likely it is that adaptations can evolve that recover rapidly from disturbance and absorb change without a drastic alteration in system state. The process of creative destruction is able to accomplish this feat, replacing a less humanized natural system with a new structure that has long-term staying power. Destructive creation is the creative process gone awry in ways that are inviable.

Human efforts to survive in the Negev and in the Tigris and Euphrates illustrate these two contradictory processes. Both resource management systems developed in harsh, arid environments. Each system is built on a particular characteristic feature of the local resource base. In the Nabatean instance, it is the relative impermeability of local soils to infiltration that is at the basis of their ecotechnology. In Mesopotamia, it is the potential abundance of two exotic rivers that, if harnessed successfully, promises prosperity. In both instances, rainfall and river flow regimes are highly variable in time and space, creating great hazards of drought and flood with which the resource user must cope.

Integral to a successful ecotechnology is the selection of sacrifice zones that can be destroyed in the interests of protecting and enhancing the zones that are critical to sustained use. In the Negev, the Nabateans deliberately sacrificed the crests and hillside habitats in order to encourage soil erosion, water movement, and water harvesting. Aided by the tendency of the region's soils to form a plastic crust that reduced infiltration, these upland water and soil losses were impounded behind low check dams and terrace walls in the nearby valley bottoms. Recognition of the soil characteristics of their habitat, as well as conceptualization of how water and soil when concentrated in the valley bottoms that were critical zones for agricultural production, could overcome the limitation of local climate, was a positive application of the *genius loci principle*. That the system created by the Nabateans eventually declined when regional political conditions changed in the seventh century AD does not diminish the magnitude of their accomplishment. The same infrastructure, albeit on a reduced scale, continued to function right down to the present. By constructively using the waste products of one part of their habitat as an input for their critical agricultural zone, the Nabateans turned a negative development, soil erosion, into a positive asset, a truly creative act. Contemporary Bedouin still reap benefits from the relics of the Nabatean system through the planting of crops on the former terraced fields.

In Mesopotamia, the Sumerians and their successors also developed a creative solution to coping with a desert environment. Unfortunately, their ecotechnology contained internal inconsistencies that undermined effective long-term system viability. Land facets at the tails of the irrigation system and in depressions were sacrificed because provision of drainage, which would have been difficult given the low gradient of the area, was not part of the overall design. Although some use of these habitats could be made by both domestic livestock and wild animals, their overall contribution was negative. Disease vectors flourished in the swamps and wetlands created as an adjunct to irrigation, evaporation and transpiration contributed to an increasingly saline environment, and waterlogging encouraged a rise in the regional water table. With each downstream water transfer, water quality declined as surface water and groundwater returns to stream and canal flow contained an elevated salt content, thus making the situation of downstream users more precarious. This failure to use wastes as a constructive input diminished the effectiveness of the Mesopotamian irrigation adaptation.

Over time, Mesopotamian irrigation managers, for all their evident sophistication, were unable to protect the levee backslope and rich alluvial soils from salinization. Violation of alternate-year fallow, the *genius loci principle* of the region, episodically promoted salinization and initiated a spiral of land degradation that was sudden in onset and rapid in producing negative impacts. Excessive siltation from the upland sacrifice zones created managerial problems for canal clearance that reached crisis proportions whenever civil conflict or invasions from neighboring states occurred. In sum, the Mesopotamian irrigation adaptation was a finely poised creation that operated close to the stability and resilience parameters that defined its existence. Destructive tendencies embedded within its social, political, and agronomic spheres promoted episodic land degradation and ultimate collapse.

THREE

The Physical Domain and Land Degradation

Introduction

With the diversity of physical settings found on our planet, not all areas of the Earth are equally stable or resilient for given activities. Environmental conditions prone to land degradation already were identified by the mid-1970s (Farnworth and Golley 1974). One of the earliest publications, a Food and Agriculture Organization (FAO) report (Rauschkolb 1971), identified numerous factors contributing to agricultural land degradation. The most ubiquitous cause was soil erosion. Terrains having the highest potential for deterioration under agricultural, silvicultural, and pastoral use systems began to be called "fragile lands" (DESFIL/PID 1985).

In these cases, the term "fragile" is used to designate areas that, once disturbed by human activities, have a low resistance (resilience) to high-frequency (common – not exceptional) natural events such as heavy rains, strong winds, and low to moderate magnitude earthquakes. A wide range of ecological zones are fragile. However, in particular – with regard to *low-resource* rural land-use systems – the principal areas of interest for many international donor and governmental agencies became semiarid areas (e.g., the Sahel), humid tropical lowlands (e.g., the Amazon Basin), and zones of steep slopes (e.g., Nepal). (Low-resource rural land-use systems are the form of economic activities practiced by farmers, pastoralists, and other individuals that are based primarily on the use of local natural resources. Today, these systems usually also make limited use of external inputs, including technology.)

A primary reason for emphasis in these areas was that degradation was destroying a significant percentage of the national territory. If this trend were to continue unabated, the economic viability of large regions would be threatened. In many cases, lands were not just evolving into less productive conditions, but were becoming worthless for the production of foodstuffs and other agricultural commodities. Land abandonment was accelerating.

It must be emphasized that land degradation on fragile lands is not restricted to either poor countries or low-resource systems. Vulnerable lands in developed, industrialized nations also experience similar environmental problems. Moreover, there is no compelling evidence to suggest that land degradation is necessarily of greater magnitude in low-resource systems when compared to industrialized countries. Both in intensity and in spatial distribution, land degradation affects rich and poor peoples and industrialized and industrializing countries in similar and serious ways. There is strong evidence that documents land degradation in industrialized countries. For example, in Australia, widespread decreasing soil fertility, increasing salinization, and deteriorating water quality were clearly documented as early as the 1960s on semiarid grazing lands, and by the mid-1970s in humid areas of the country (Chisholm and Dumsday 1987). In the United Kingdom, severe erosion resulting from human activities has occurred since Romano-British times and has become more widespread over the past 250 years as agriculture has expanded (Evans 1993). In the United States, excessive erosion, increasing groundwater salinity, and salinization of irrigated lands continues in many farming areas despite numerous governmental incentives and educational programs (Sheridan 1981; Cook 1984).

Furthermore, land degradation is neither limited to fragile environmental settings nor to rural activities. Significant land degradation resulting from industrial pollution in the eastern United States was already under investigation and documented by the middle 1950s (Schumm 1956). Recent reports indicate that industrial pollution is widespread in Eastern Europe (Edwards 1993). In fact, land degradation in varying degrees of magnitude exists on all of the continents excluding Antarctica.

In areas of low resource use systems, the resilience of the physical (natural) environmental situation plays a large role in determining whether a human activity will result in land degradation. In stark contrast, while not minimizing the importance of the natural environmental conditions, in areas of high resource systems often it is the magnitude and duration of the human activity that determines whether deleterious changes result. Under the onslaught of powerful modern technological interventions, even the most resilient and stable area will degrade if the human activities are not managed in an environmentally sound manner.

Regardless of the environmental situation, when a Chernobyl or Three Mile Island type nuclear accident occurs, the areas affected by radioactive releases will clearly undergo land degradation. While the majority of human activities today are not as critical as those associated with radioactive disasters, the increase in the magnitude of many activities using contemporary technologies has increased the risk for severe land degradation. For example, the immense size of contemporary supertankers can result in oil spills that create significant deterioration in regional environmental systems and affect thousands of square kilometers. This scale of impact is a sharp contrast to the more local impacts of former days when oil tankers were much smaller. Today, there are a multitude of by-products from intermediate and high resource activities that result in land degradation when these activities are not practiced under environmentally sound conditions. Intensification of agricultural and pastoral lands, increasing use of chemical insecticides, disposal of urban and industrial wastes, expansion of irrigation practices, and transport of petroleum products are examples of potential hazards that could result in land degradation depending on the environmental setting and management practices utilized.

The natural environmental setting, including its energy properties and earth materials, interacts with any given human activity. These energy and material factors incorporate both gravitational and biochemical attributes, climatological peculiarities, tectonic characteristics, and the soil, regolith, and rock types, as well as their structural arrangement. Depending on the combination of the properties of these factors and the intensity of the processes found in an area, there will be a specific interaction between the natural and the human systems. For most activities, this interaction determines whether an area's response to the human utilization results in land degradation.

In this chapter, we examine why diverse physical settings have different susceptibilities (sensitivity) to degradation. To accomplish this, we shall identify and explore many of the critical aspects of the climatological, geomorphological, and vegetational domains that significantly determine a given area's response to human activities. While we shall explore attributes of these three domains separately, in reality they are interdependent and generally interact simultaneously in determining the land's response to human activities. For example, erosion rates resulting from water are higher when:

- rainfall is intense and of a long duration (climatic)
- runoff is large and rapid (climatic, geomorphic, and vegetational)
- soils are easily eroded (geomorphic)
- slopes are steep (geomorphic)
- vegetation cover is poor (climatic, geomorphic, and vegetational)

Climatological Factors

Climate is the integration of an area's meteorological phenomena over a period of years. The expected ranges and the variability of the meteorological elements are critical components of climate. Average values of temperature and precipitation often are stressed in describing an area's climate. However, with regard to land degradation, average climatic values are insufficient predictors: that is, they are not sufficiently sensitive indicators by which to determine the susceptibility or likelihood of an area to degrade under different human land-use systems. With relevance to land degradation, average values along with other temperature and moisture attributes (e.g., the number of months having temperatures less than 0°C, the number of months with moisture deficits, and the interannual variability of rainfall) are needed to describe an area's climate.

A multitude of climatic parameters exist for any area. Cloud cover, annual ranges in the length of daylight, absolute humidity, pollen count, wind intensity, evapotranspiration, dust, and daily insolation represent just a sample of the variables that together determine an area's climatic characteristics. For the purposes of this book, the question that must be answered is: What climatic attributes determine the vulnerability of an area to land degradation processes? It is these variables that we must both identify and examine.

Three main geomorphic processes that have direct impacts on the land degradation problem are erosion, deposition, and mass movements. Therefore, we shall explore those climatic attributes that affect these processes and for which data are also widely available from national meteorological services or the scientific literature. In particular, we shall emphasize those aspects of precipitation, temperature, and moisture balance that affect biological productivity.

Precipitation factors

Both Earth science and agricultural studies agree that the precipitation properties of an area are one critical set of environmental conditions affecting an area's stability, especially once it is disturbed by human activities. The amount, intensity, duration, and variability of precipitation under a wide range of environmental and livelihood systems are some of the most important determinants affecting the land resource (Douglas and Spencer 1985; El-Swaify, Moldenhauer, and Lo 1985).

Erosivity
Two separate but interrelated processes are required for soil erosion to result from precipitation. First, soil aggregates must be detached from the ground (initial movement). Second, the detached soil must then be transported in flowing water to a new location. The erosional force associated with precipitation that initiates both soil detachment and transport is called erosivity. The potential erosivity of a rainstorm is determined by the storm's *kinetic energy*, a value largely determined by the interaction between the storm's rainfall intensity (v) and the total rainfall of a rain event (m). Kinetic energy $(KE) = \frac{1}{2}mv^2$, where m is mass (here the amount of rainfall) and v is the velocity. In general, it would be expected that as the total kinetic energy of a rain event increases the potential

for erosion would increase. However, this is not necessarily true. In a very low intense rainstorm (drizzle) (less than 4 mm/hr, or 0.15 min/hr) with a duration of 45 hours could have the same total kinetic energy as a summer thundershower lasting only 15 minutes but having a maximum intensity of 150 mm/hr (5.9 in/hr). For the drizzle, minimal soil detachment would be expected as the effect of raindrop impact would be minimal. Also, the precipitation would probably result in minimal runoff, since its intensity would likely never greatly exceed the rate of soil infiltration. For the thunderstorm, both soil detachment and large but brief surface water runoff would be expected. Thus, while the storms' total kinetic energy could be similar, the erosional potential of the two systems would be unequal. For this reason, a number of empirically derived erosivity indexes have been developed to estimate the erosional potential of different areas in terms of their regional rainfall characteristics.

No single index of erosivity universely results in accurate predictions of erosional potential, because of the diverse physical conditions found on our planet (Lal 1988). This reflects the highly variable nature of worldwide precipitation patterns found in different regions. In western Nigeria, typical of many lowland tropical areas, where many storms are of a high intensity but of short duration, the $AI_{7\,1/2}$ erosivity index is a good predictor of soil loss (Lal 1976; Lewis 1981). The AI index is defined as the product of the $7_{1/2}$ minute maximum rainfall intensity of the rain event and the total rainfall of the event. The $AI_{7\,1/2}$ index increases in magnitude as the erosional potential of the storm increases.

In the United States and many other middle-latitude areas, where precipitation is largely associated with the passing of extratropical cyclones and their associated warm and cold fronts, studies indicate that the erosional potential of rain events correlates highly with a storm's maximum 30 minute rainfall intensity. This empirical information was incorporated into the R rainfall and runoff factor that is used widely in middle-latitude locations to access the effect of rainfall on soil loss (Wischmeier and Smith 1978; Renard et al. 1994).

In Zimbabwe and Sri Lanka, another empirically derived measure to evaluate soil loss potential was developed to reflect the local conditions found in these areas. Here the $KE > 25$ index obtains the best correlations with soil loss, where KE (kinetic energy) equals $29.8 - (127.5/I)$ and I is the rainfall intensity for rain events having an intensity of at least 25 mm/hr (0.98 in/hr)(Hudson 1976; Joshua 1977). Because of the diverse nature of precipitation patterns from region to region, and since most erosivity indexes are empirically derived for specific areas, they should be first evaluated for local conditions before they are used to estimate soil loss potential in terms of erosivity. Otherwise large errors in the estimates are likely.

Globally, the areas with both high amounts and intense rainfall have the highest erosional potential for water-related erosion. Compared to other major ecological units, humid tropical rainforest areas (figure 3.1) have the largest annual precipitation.

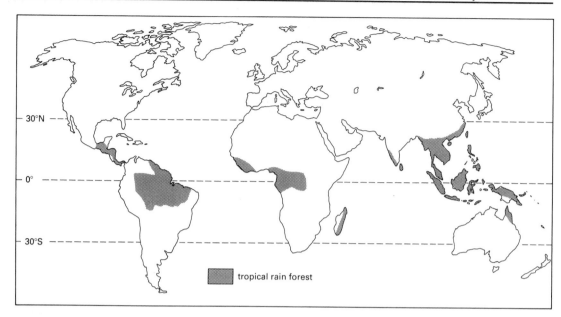

Figure 3.1
The global distribution of tropical lowland forest (after Northington and Goodin 1984)

Additionally, as a very large proportion of rainfall in the tropics results from convective systems (e.g., thunderstorms), most rain events are generally intense. The result is that tropical lowland areas are associated with some of the highest annual erosivity values. Their rainfall characteristics are a significant reason why these areas usually fall within the fragile land category. Once the natural vegetation cover protecting the soils in these areas is destroyed by human activities, the high-erosivity rains can directly strike the soils and accelerated soil losses usually result. If a perennial vegetative cover, such as banana plants or rubber trees, that exhibits many of the properties of the natural tropical forests is the replacement, soil losses should not differentiate greatly from the natural state. Hence, even when major changes are introduced into tropical fragile settings, land degradation need not occur if the *genius loci principle* is integrated into the transformation.

The trend from humid to arid climates and from warmer to cooler areas is characterized by decreasing annual erosivity. These general relations are illustrated using data from Hawaii (tropical) and the other 48 contiguous states (figures 3.2a and b). As rainfall increases with elevation on the windward (eastern) slopes of the five islands of Hawaii, the R-values increase; on the drier western lowlands (rain shadow), the R-values are lower than those found on the more humid eastern coastal areas. The overall pattern of decreasing R-values in humid areas as annual temperatures become cooler is depicted by the decrease in R-values in the northward direction from the Gulf of Mexico to the Canadian border on the map of mainland United States. The decrease in R-values reflects both the general decrease in precipitation as temperature drops as well as the increasing proportion of the precipitation that occurs as snowfall.

(a)

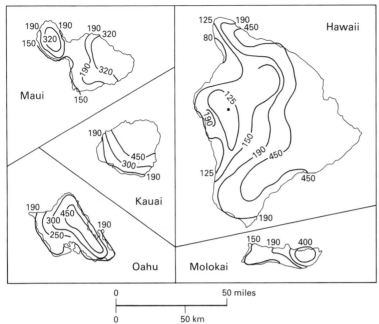

Figure 3.2
Approximate average annual rainfall index values for (a) Hawaii and (b) the continental United States

(b)

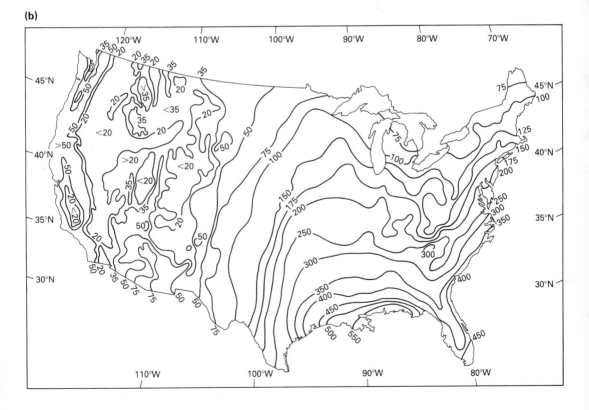

While arid areas have lower annual erosivity values than humid areas, individual – albeit infrequent – rainstorms in drylands often are associated with intense periods of rainfall. As a consequence, these intense showers often are of sufficient erosivity to cause significant erosion. Individual severe rain events in arid areas are associated with huge quantities of materials being removed and deposited elsewhere (Schick, Lekach, and Hassan 1987). Torrents in desert areas reflect this characteristic in that high concentrations of suspended and bedload materials (eroded materials) are carried in most arid and semiarid ephemeral streams during their brief periods of flow.

Thus, while average annual erosivity values are one indicator of an area's overall erosional potential, it is the actual erosivity of individual storms that are the catalyst for soil loss and hence land degradation. Even though annual erosivity rates are low in the Negev, the Nabateans took advantage of the high erosivity of individual rain events and used this property to accelerate soil loss on the hillslopes (see chapter 2). Erosion caused by water is a discrete process in both humid and arid areas. Furthermore, it occurs only when critical thresholds are surpassed. The result is that most rain events, even in humid tropical areas, are associated with minimal soil losses (Lewis 1981). Usually only a few intense storms with sufficient rainfall each year result in significant erosion in humid areas. As aridity increases, the frequency of erosional events decreases under natural conditions. Yet when these thresholds are surpassed, change can be rapid. When these critical high-energy conditions occur, management practices must be in place to minimize erosional losses if land degradation is to be controlled.

Precipitation variability
There are four major dimensions of precipitation variability; (1) temporal, (2) spatial, (3) type, and (4) intensity. Each of these dimensions affects the environmental conditions found at any location. In the previous section we explored some characteristics of intensity within the context of erosivity. In the following two sections we shall focus on the temporal and spatial aspects of precipitation variability related to the problem of land degradation. In the process of this discussion, the importance of precipitation type (snow or rain) will also emerge.

Temporal Precipitation at any location is temporally variable. On some days it rains and on some days it does not; some months are wet and some are dry; and from year to year, a site is either wetter or drier than the previous year. It is normal for most meteorological elements, which together give a location its climatic character, to vary. Hence it is expected that an area's precipitation will experience a range of values on a daily, monthly, and annual basis.

How a site's precipitation values are distributed around a measure of central tendency (mean, median, and mode) represents its variability characteristics. Every set of precipitation values is associated with a frequency of occurrence. For example, at Ibadan, Nigeria, 900 mm (35 in) of annual rainfall has about a 13 percent frequency of occurrence (figure 3.3). Also, Ibadan has a mean annual rainfall of 1,230 mm (48 in) and a mode of 1,200 mm (47 in). Its measured annual precipitation ranges from a low of 715 mm (28 in) to a high of 1,925 mm (75.8 in) (figure 3.3): that is, over the period of time during which rainfall data have been collected at Ibadan, its absolute range in precipitation is 1,210 mm (47.6 in). Clearly, the environmental conditions at Ibadan differ greatly during dry years when moisture deficiencies are severe compared to conditions during wetter years.

Two common statistical measures used to assess precipitation variability are the standard deviation and the coefficient of variation. (Detailed explanations of both the standard deviation, $s = [(1/n)(x_i - \bar{x})^2]^{1/2}$, and the coefficient of variation, $C.V. = s/x$, may be found in any elementary statistics book.) As either of these statistical parameters increase, a location's distribution of precipitation becomes more variable: that is, as variability increases, the measure of central tendency (e.g., the mean) becomes a poorer description of the site's likely (expected) precipitation for any given week, month, year, or longer period. With regard to precipitation, as these statistics increase for an area, a greater range of moisture properties are experienced (normal). The interaction of various magnitudes of precipitation with other environmental factors, such as the nature of the ground cover or water runoff, often make areas having wide ranges in moisture more difficult to manage than more consistent zones. To a large degree, it is the variability of moisture that makes semiarid and subhumid areas especially prone to land degradation. Management strategies developed to maximize agricultural outputs during the wetter periods often place these lands under stress during drier phases.

Temporal variability tends to increase as regional precipitation decreases. In humid areas, it is relatively low; for subhumid and semiarid areas, it increases in magnitude; and, in arid areas, it reaches maximum values. These characteristics are illustrated in the comparison of four locations along the north–south moisture gradient of West Africa (figure 3.4). At Calabar, at the southern end of the gradient, where the climate is very humid (mean annual precipitation 2,962 mm, or 117 in), the coefficient of variation (C.V.) is 15 percent; the standard deviation is 469 mm (18.5 in). Annual precipitation in the Calabar area for 68 out of 100 years should range between 2,493 mm (98 in) and 3,431 mm (135 in) (2,962 ± 469). This precipitation range is less than a factor of 1.38.

At Ibadan, a humid location close to the subhumid boundary (Lewis and Berry 1988), the C.V. equals 21 percent, and the standard deviation of annual precipitation is 263 mm (10.3 in) (figure

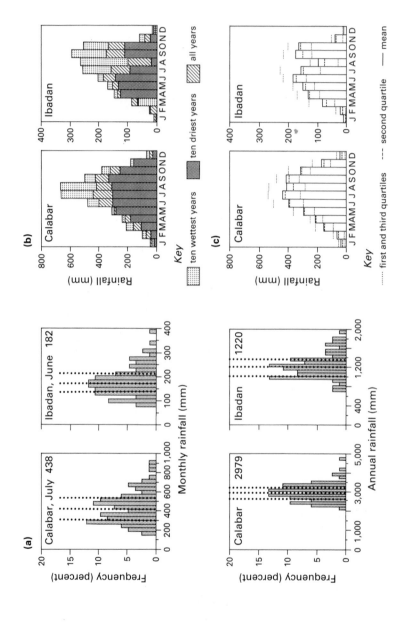

Figure 3.3
The rainfall characteristics of Ibadan and Calabar, Nigeria (from Nicholson, Kim, and Hoopingarner 1988).
(a) Frequency distribution of rainfall: dotted lines indicate the first, second, and third quartiles – the mean (in mm) is given in the upper right-hand corner
(b) Monthly mean rainfall
(c) Monthly rainfall statistics

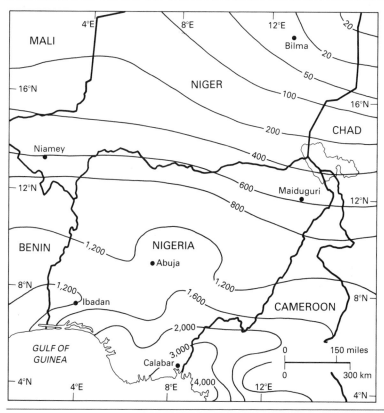

Figure 3.4
*Temporal variability in rainfall
along a West African moisture
gradient*

	Jan.	Feb.	Mar.	Apr.	May	Jun.	Jul.	Aug.	Sep.	Oct.	Nov.	Dec.	Annual
Mean monthly precipitation (mm)													
Calabar	38	65	156	213	297	394	437	410	414	318	173	45	2,962
Ibadan	10	23	84	133	151	184	158	98	176	166	40	10	1,230
Maiduguri	0	0	0	8	34	69	168	213	100	15	0	0	609
Bilma	0	0	0	0	0	1	3	9	3	1	0	0	17
Standard deviation of monthly precipitation (mm)													
Calabar	45	56	80	73	100	153	165	191	125	101	78	48	469
Ibadan	23	26	54	54	50	62	85	101	89	73	32	15	263
Maiduguri	1	1	1	15	25	42	68	83	49	19	1	0	153
Bilma	1	0	0	0	2	4	4	13	6	7	0	1	16
Coefficient of variation of monthly precipitation (%)													
Calabar	118	86	51	34	33	38	37	46	30	31	45	106	15
Ibadan	229	113	64	40	33	33	53	103	50	43	79	150	21
Maiduguri	—	—	—	187	73	60	40	38	48	126	—	—	25
Bilma	—	—	—	—	—	400	133	144	200	700	—	—	94

3.3). From these statistics it is inferred that for about 68 out of 100 years Ibadan's annual precipitation (mean annual precipitation 1,230 mm, or 48 in) will range between 967 mm (38 in) and 1,493 mm (58.8 in). This is a factor of 1.54.

At Maiduguri, a semiarid location (mean annual precipitation 609 mm, or 24 in), the C.V. is 25 percent and the standard deviation is 153 mm (6 in). Here the expected range in precipitation for

68 out of 100 years is between 456 mm (18 in) and 762 mm (30 in) or a factor of 1.67.

At Bilma, Niger, a very arid location, the mean annual rainfall is 17 mm (0.66 in); the C.V. equals 94 percent and the standard deviation is 16 mm (0.62 in). In this arid zone, where the precipitation is very variable, 68 out of 100 years should receive between 1 mm (0.03 in) and 33 mm (1.3 in), a factor of 33.

The increasing variability of precipitation with increasing aridity is an important natural phenomenon with regard to land degradation. As moisture deficiency becomes a limiting variable, variability increases. As a result, land-use strategies in the transition areas (subhumid and semiarid) must be able to cope with a wider range of moisture-related conditions than in more humid locales. Plowing the soil for field preparation in a humid area will likely be followed shortly by rainfall and plant germination. However, plowing the soil in a semiarid area is risky. If the rains do not come when expected, the loose and exposed soil will continue to become drier and make the area increasingly susceptible to wind erosion.

In humid areas, precipitation variability results in different degrees of moisture surplus from year to year; but most humid locations remain humid during their expected ranges. Likewise, in arid areas, even though the variability in precipitation is at a maximum, the range results in only a change in the severity of the moisture scarcity. Again, as in humid zones, no critical moisture climatic boundary is crossed due to the ranges of precipitation encountered in these zones. Arid locations remain arid even under "wetter" conditions, just as humid zones remain humid during their "dry" years.

In contrast, precipitation variability is a particularly sensitive climatic attribute for subhumid and semiarid areas. Unlike humid and arid areas, the expected ranges in precipitation in these climatic zones often cross critical moisture boundaries. Because climatic conditions usually remain persistent over brief periods of time (dry and wet periods usually occur in cycles of a few years in duration), human systems must be sensitive to these fluctuations, and it must not be assumed that either the wetter or drier conditions will persist: that is, neither is the norm. This is a critical difference between humid/arid and subhumid/semiarid lands with regard to our inquiry into land degradation. Even during drier periods, humid areas usually do not experience major climatic stress with regard to moisture. Moisture stress conditions in arid areas remain high, even during their periods of higher rainfall. Therefore, with regard to land utilization: (1) strategies developed for humid conditions remain valid in humid areas for most if not all expected moisture conditions; (2) tactics for arid land management remain valid in arid areas for the expected ranges in precipitation found in these zones; but (3) in subhumid and semiarid zones strategies need to be developed that can successfully cope with the different environmental conditions experienced under both moisture deficit and surplus states.

Subhumid and semiarid areas are utilized to a far greater degree than arid areas. During dry periods, when these areas experience low precipitation, the record indicates that many aspects of their environment undergo stress due to human activities. As a result, they are particularly vulnerable to degradation for a large percentage of the time. Furthermore, even during humid annual cycles, it is normal for these two climatic types to experience dry periods of several weeks to months in duration. Thus, there is a need continuously to utilize strategies in these two zones that result in environmental stability both under humid and arid conditions.

Spatial   Spatial variability of rainfall is apparent in the irregular distribution of precipitation on short time scales (daily and monthly) and in longer time frames. Spatial variability tends to be greater in low-latitude regions, where precipitation is largely convectional in origin, than in middle and higher latitudes, where a significant proportion of rainfall is associated with air mass fronts. At all latitudes, spatial variability usually increases as dryness increases. Hence, humid regions are generally less variable than arid ones.

The high degree of spatial variability (areal discontinuity) of precipitation in tropical areas is related to the randomness of the convective process, the dominant cause of tropical rainfall. Spatial variability is particularly great in arid warm climates (Sharon 1972). Not only do most rains in these zones occur in isolated convective cells, but also usually the precipitation cells are less than 10 km in diameter. Because there is a very limited atmospheric moisture supply in most dry areas, rain-producing cells generally cover only between 10 and 15 km^2 per 100 km^2 even under favorable conditions. This means that rainfall is highly variable from place to place during a single storm. A heavy downpour can occur within one small river basin while all of the surrounding catchments remain absolutely dry. Because of the isolated nature of the convection cells in these areas, rainfall is highly variable both during individual rain events and from year to year. Therefore, daily, monthly, and annual precipitation conditions at nearby locations often differ greatly, even when long-term precipitation averages throughout the area are comparable. Land-use strategies must be developed to cope with the environmental constraints of this extreme local spatial variability, which often results in juxtaposed areas having markedly different moisture conditions for any given time period – but especially so for daily, weekly, and monthly periods.

Data, collected in areas of high spatial variability, that are to be used in developing planning strategies must be analyzed carefully, since inter-site correlations will probably be low, even though they come from a region of homogeneous temporal precipitation variability. Therefore, risks are especially high for inferring areal rainfall conditions from data gathered at individual weather stations in regions of high spatial rainfall variability. A dense network of weather stations is ideal for this type of setting but, in reality, these are generally areas of sparse data collection, as drier areas with their generally lower population densities usually have fewer weather stations.

Spatial variability is not nearly as great in humid tropical areas as it is in drier low latitude climes, even though in both cases convection is the dominant cause of precipitation. One reason for this difference is that there is a greater reservoir of moisture in the atmosphere under humid conditions. Thus a smaller volume of air can supply sufficient moisture for precipitation in humid compared to drier areas (figure 3.5). The result is that neighboring precipitation cells in humid low-latitude areas are closer together (figure 3.5). Furthermore, as there is a greater atmospheric moisture supply in the humid areas, rain cells persist longer in humid areas and allow a greater area to experience rainfall. Also, because a zone is humid, rain events occur more often in these areas than in drier zones. Thus, if a specific area receives no rainfall from today's storm, unlike in drier areas the likelihood is that another rain-producing system will arrive shortly thereafter. With the greater frequency of rain events, the probability of any area experiencing a large moisture deficiency over any length of time is slight in the humid zones, even given the spatial variability of individual storms.

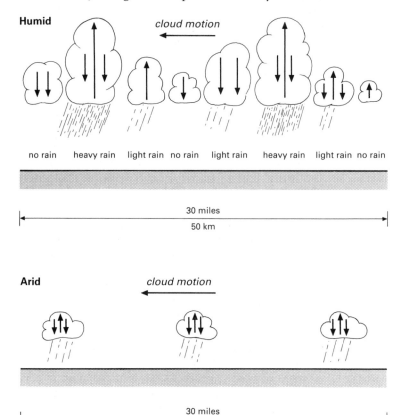

Figure 3.5
*Spatial variability in rainfall
between humid and arid areas*

In contrast to low-latitude areas, middle latitudes experience a significant proportion of their precipitation along air mass boundaries. These fronts, being hundreds of kilometers in length, result in contiguous precipitation zones of relatively similar conditions covering hundreds of square kilometers (figure 3.6).

Furthermore, in most of the middle latitudes, a proportion of precipitation occurs as snowfall. In the colder middle latitudes where snowfall represents a significant proportion of annual precipitation, winter snow accumulation offsets to a degree the ramifications of the spatial variability of precipitation. Spring moisture conditions in these areas are determined significantly by winter seasonal totals rather than by individual snow events. The magnitude of the spring snow pack determines numerous hydrologic conditions in these areas. The ground moisture conditions at the start of the growing season, the period when annual crops are particularly vulnerable, can be estimated by measuring the moisture content of the snow cover. Likewise, the probability and the magnitude of stream flows (e.g., floods, and water available for storage in reservoirs) can be accurately predicted on a regional basis.

During the summer months, the spatial attributes of precipitation in the middle latitudes approximate some of the conditions found in the lower latitudes. With the warmer air masses existing during the summer months, an important proportion of the rains result from convection processes, the dominant processes of the low latitudes. This results in a percentage of the mid-latitude summer rainfall being distributed in a disaggregated spatial pattern. This phenomenon is reflected in weather forecasts during warm muggy summer conditions: it is on these days that widely scattered afternoon showers are predicted. Nevertheless, because frontal precipitation also occurs during the summer, spatial variability of precipitation remains less variable in the middle compared to the low latitudes.

Temperature factors

One of the most crucial ecological factors limiting biological activity is temperature. Both high and low temperatures are critical environmental attributes. For example, the potential length of an area's growing season is the continuous number of days free of freezing temperatures. Moisture deficiencies in arid areas can be offset by irrigation. However, when severe cold exists, options to negate it for large areas do not exist: plants either become dormant or die.

Low temperatures also affect the inorganic world. Underlying approximately 20 percent of the land area of the world is permafrost (figure 3.7). Areas of permafrost are found in locations that have annual mean temperatures less than 0°C. Permafrost is perennially frozen ground; by definition, sediment, soil or bedrock that remains frozen for two years or longer. Permafrost lands present a host of environmental demands on human utilization. Many zones

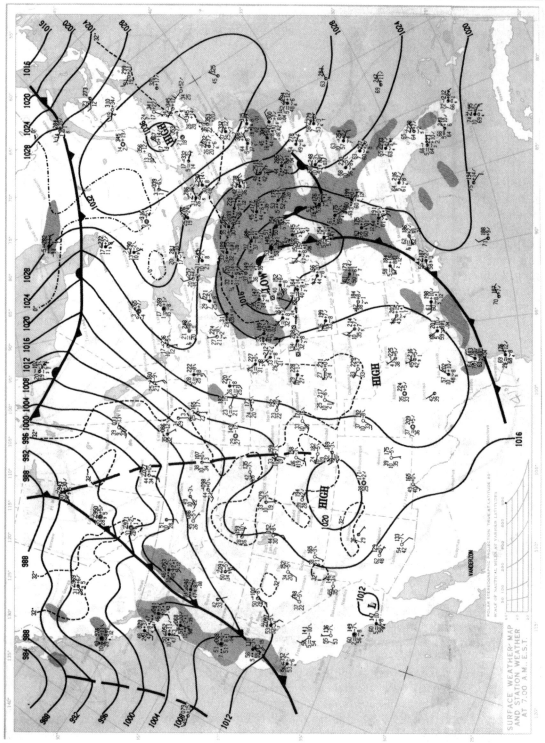

Figure 3.6
Spatial characteristics of frontal rainfall in North America: Tuesday, March 23, 1993

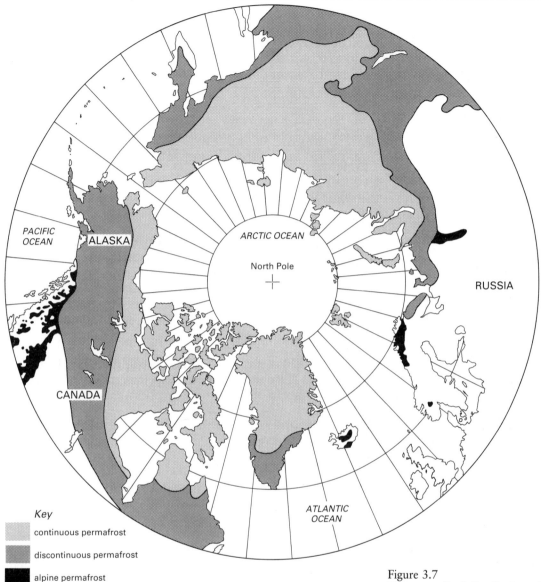

Key

continuous permafrost

discontinuous permafrost

alpine permafrost

Figure 3.7
The generalized distribution of permafrost in the Northern Hemisphere (after T. L. Péwé, Department of Geology, Arizona State University)

within the permafrost areas clearly fit into the definition of a fragile land. Particularly with the quest for mineral wealth, many of these subpolar and polar lands are experiencing ever increasing modifications. During the subpolar summer, when the surface thaws but the subsurface remains frozen, disturbances to the surface may damage the vegetation mat. These surface disruptions take years to repair themselves – if, indeed, they ever recover. Because of the fragile vegetation mat, land degradation is increasing as these areas are exposed to increased economic activities. In the remainder of this section, we examine aspects of temperature that have important relations with the environment.

Chemical activity

Temperature is a critical factor affecting the intensity of chemical reactions. Every student recalls heating a mixture of chemicals during a chemistry laboratory experiment in order to initiate a chemical reaction. Van't Hof's law asserts that, statistically, for every temperature increase of 1°C (1.8°F) the intensity of a chemical reaction increases by a factor of 0.25. The worldwide distribution of weathering rates, the breakdown of earth materials by chemical and physical processes, reflects this temperature–chemical relation. Thus weathering rates are rapid in warm areas. This is an important relationship, since without weathering most earth materials would be too resistant for erosion to occur. Granites, which break down slowly in cool humid climates, are resistant rocks in these climatic settings. However, in the tropical warm moist areas where chemical weathering is more intense, granites are eroded relatively easily. The different rates at which granite disintegrates illustrate how a region's sensible temperature properties are a contributing variable in determining the susceptibility of an area to erosion, an important factor in determining the likelihood of land degradation.

Biological activity

Temperature also affects almost all biological activity. Both plants and animals function at optimum temperatures, which vary according to species. For photosynthesizing plants, the optimum temperature is normally between 20 and 25°C (68–77°F) (Tricart 1965). This is reflected in the creation of biomass. Overall, maximum production of biomass occurs in regions where monthly temperatures range between 20 and 29°C (68–84°F)(Tricart 1965). These temperature ranges prevail in humid tropical and subtropical lowlands throughout the whole year, and during the summer months in humid middle-latitude areas.

The opposite of biomass production, decomposition of organic matter, also is greatly influenced by temperature. The species of bacteria that destroy organic matter are very active at temperatures around 28°C (82°F), and many reach their thermal optimum between 30 and 35°C (86–95°F) or greater. Thus, a tropical rainforest that produces three to four times more vegetal matter per area than a mid-latitude forest ends up with less humus on the surface and in the soil than its cooler counterpart. Despite its high production, microorganisms decompose the dead matter at about the same rate as it is produced in the warm tropics (Lewis and Berry 1988). With any increase in ground temperatures, a likely outcome when forests are cleared, microorganisms potentially become more active, and organic material on the ground and in the soils decreases. This results in both less fertile soils and conditions that increase erosional potential. The sparseness of decaying matter on the surface affords the soil minimal protection from the heavy rains that prevail in the warm tropics, and from their high erosivity.

Physical activity

In locations at which temperatures fluctuate above and below the freezing point of water during a portion of the year, freeze–thaw processes occur in all but the driest environments. Water expands as it changes from its liquid to its solid phase; conversely, it contracts in its conversion from ice to water. It is this property of water and its interaction on earth materials that has important environmental ramifications. Under some conditions, frost action is very beneficial. Frost action associated with the numerous freeze–thaw cycles experienced in mid-latitude soils helps to prevent soil compaction in these areas. In contrast, warm tropical soils, once compressed by heavy machinery, often remain compacted. As a result they become less permeable, runoff increases and erosion accelerates (Lewis and Berry 1988). Thus, plowing methods that, in mid-latitude areas, do not degrade soils could increase the likelihood of accelerated soil loss in tropical areas, due to the absence of freeze–thaw processes that combat soil compaction. This environmental difference illustrates how the transfer of technology that evolved in one physical setting can act as a catalyst for land degradation in a different climatic landscape.

Freeze–thaw processes are clearly important geomorphic processes that are temperature driven. In regions that experience temperatures both above and below freezing during their annual cycle, management strategies need to be sensitive to the problems that arise due to the expansion and contraction of water during freeze–thaw cycles. Particularly in the areas in which permafrost exists, major environmental problems result from the interaction between these processes and human activities. It has been estimated that when water is frozen to a temperature of $-22°C$ ($-7.6°F$) its conversion to crystalline ice and its associated expansion will produce a theoretical maximum force of 2,100 kg/cm^2 (4,630 lb/in^2) (Embleton and King 1975). The forces involved in the freezing and thawing of water have important effects on the land resource in permafrost zones. The alternate thawing and freezing of arctic soils, caused by heat lost from the pipeline, has made it essential to install oil pipelines above ground in order to minimize the risk of disruption, breakage, and leakage, which could have catastrophic impacts on the relatively unresilient tundra ecology. Disruption of the natural environment almost always upsets the existing local heat balance in these zones. This results in changes within the permafrost, often creating severely degraded environmental conditions. Some of these changes are examined in this chapter's section dealing with geomorphological factors.

Evapotranspiration

Potential evapotranspiration (PET) is a measure of the maximum possible evaporative and transpiration water losses due to climatic factors. It assumes that for the location being evaluated there are no periods of moisture constraints. PET is largely dependent on temperature and to a lesser extent on the duration of sunlight, wind, and humidity conditions. PET increases with higher temperatures, clear skies, longer hours of sunlight, greater wind velocities, and lower humidity. When precipitation is greater than PET, a moisture surplus exists, and humid conditions prevail; when precipitation is less than PET, arid situations are found; and when neither a moisture surplus nor a deficit exists, that is PET equals precipitation, the environment is subhumid.

As discussed previously, for any given year or time period, locations classified as subhumid rarely experience a moisture balance – the term subhumid represents an average condition. Because of precipitation and temperature variability, a site usually experiences surplus or moisture deficits during any specific short period such as a week, month, season, or year. It is only over longer time intervals, such as decades, that regions actually are subhumid with their moisture and evapotranspiration in balance.

Wheat is a common crop in many mid-latitude and some tropical highland subhumid areas. Bumper years for wheat occur during periods when the growing season has ample moisture (humid). Under these optimal conditions, groundcover is usually good and minimal wind erosion is likely. Less than maximum yields occur during dry conditions. In fact, during very dry conditions, crop failures occur. When annual crops are grown in subhumid areas, during arid periods the soil is often exposed. The potential for wind erosion is large and land degradation may occur. This is especially true since the wind only removes the finer components of the soil. After eolian erosion occurs, the remaining soils are coarser. This results in a generally less fertile soil with a lower capacity to hold moisture, two characteristics of a degraded soil. Cropping strategies must cope with diverse moisture conditions in subhumid areas if land degradation is to be prevented. However, today, in most areas, the attempt is to maximize production during the wet years. One result is that the lands are poorly protected by a vegetation cover during dry years. Under the management strategy of attempting to maximize production during the humid cycles, these lands become stressed when "unexpected" dry cycles occur. This has resulted in an increase in both wind and water erosional activity during these dry periods. With land poorly protected by a crop that has failed due to the paucity of moisture, the erosivity of whatever rain does occur is often sufficient to initiate accelerated soil loss.

Many environmental situations are associated with the status of an area's moisture balance. In areas with negative balances (precipitation less than PET), river flows are often intermittent or ephemeral, grasses are a dominant vegetation, vegetation cover does not always result in a complete coverage of the soil, and wind is a potentially effective erosional agent. Conversely, in areas with positive moisture balances, permanent river flows prevail, trees are the major vegetation type, and exposed soil is rarely found under natural conditions. In these conditions, wind as a geomorphic agent is nominal.

Moisture balance attributes are one of the critical determinants of an area's environmental status. Various land degradation problems are associated with moisture balance conditions. Salinization problems related to irrigation projects and wind erosion due to over-grazing cluster in moisture-deficient areas; problems related to acid rain are predominant in many regions that have a moisture surplus. Given the importance of water balance relations, the aridity index – the ratio between net radiation and precipitation at the Earth's surface or the Budyko ratio – was developed to integrate a location's PET and precipitation traits (Henning and Flohn 1977). (The aridity index equals R_{net}/LP, where R_{net} is net radiation, L is the latent heat of vaporization, and P is precipitation.)

The aridity index compares the average amount of energy available for evaporation with the amount required to evaporate all the precipitation at the site: that is, this ratio quantifies how many times (the multiple) the radiation at a location could evaporate the site's annual precipitation. A value of 0.5 indicates that one-half of an area's precipitation is surplus. Unity indicates that precipitation and evaporation are in balance. A value of 2.0 implies that precipitation would have to be doubled before the moisture deficiency would be removed.

The annual average worldwide distribution of the aridity index is illustrated in figure 3.8. Values less than one indicate humid areas. The 1.5 estimate delimits the drier margin of the transition zone between humid and dry lands (subhumid), areas in which rainfed agriculture becomes increasingly risky. The 2.0 isoline designates semiarid regions. Values greater than 4 represent arid conditions, the "humid" boundary of deserts.

One advantage of this index is that annual moisture conditions in different climates can be compared directly in a manner that is physically relevant. Regardless of location and terrain, regions, with aridity values between 1 and 4 are found to be prone to desertification processes (Henning and Flohn 1977). The 2.0 isoline running north–south from Mexico to Canada across the United States indicates that the severity of the moisture deficit in this portion of the Great Plains is equal in magnitude. The large difference, almost double, in annual precipitation between the cooler northern parts (400 mm, 15.7 in) and warmer southern portions (750 mm, 29.5 in) is countered by the lower available radiational energy (i.e., the potential evaporative power) in the northern areas. The increase in the index value in the westward direction indicates a trend toward greater moisture deficits.

When supplemental water is added to drylands as the aridity index becomes larger, the probability of degradation associated with water shortfalls increases. Salinization problems will increase unless sufficient irrigated water is added to insure the flushing of soluble minerals from the soil in dry areas. Therefore, with regard to the information provided by the aridity index, there is an indication that the potential for salinization problems increases in a westward direction within the Great Plains. Thus the range of values for the aridity index within the Great Plains indicates that this region, perceived to have a homogeneous climate (semiarid), requires different management strategies in order to minimize salinization throughout the region.

Because precipitation's temporal variability increases with increasing aridity, during dry years there is a good chance that water available for irrigation will be in short supply in areas such as the Great Plains. Unless there is a reduction in the area under irrigation, there is a good chance that the inadequate water supply will not be able to meet both crop needs and the prevention of the accumulation of salts in the upper portions of the soils. In the United States, as urban demands for the limited water supply

Figure 3.8
The aridity index of North America (after Henning and Flohn 1977)

increase, the increasing scarcity of this resource could raise its cost. This could increase salinization in irrigated areas if farmers cut back on the flushing of salts out of their soils in order to maximize short-term profits.

In general, the risk of problems associated with moisture shortages increases as the aridity index values become larger. To prevent both the waste of the limited water supply as well as degradation in drylands, strategies of land management must be sensitive to the ranges in the aridity index that occur from year to year and season to season. Failure to integrate the variability of aridity from month to month and year to year into the management of many irrigation projects is a major contributing factor in the widespread growth of waterlogging and salinization problems on irrigated lands.

Salinization

In humid areas (aridity index less than 1) where precipitation is greater than evapotranspiration, there is a surplus of moisture. Within the soil/groundwater system this results in a net removal of soluble minerals from soils by the leaching processes. With sufficient moisture, soluble minerals found in soils, such as salts and carbonates, will slowly dissolve in the water, percolating through the soil and the underlying earth materials until the mineral-enriched water reaches groundwater. As moisture surplus increases, the likelihood that water will remain undersaturated within soils increases, and minerals from soils will slowly continue to be leached from the upper portions of the soil. Under humid conditions, the dissolved minerals slowly migrate downward through the soil and eventually are removed from the local area in both groundwater and stream flows. The net removal of soluble minerals is one of the significant processes that contributes to the general acidity of soils (pH < 7.0) found in humid areas: that is, the outflows of water are accompanied by the removal of the soluble minerals. For this reason salt and other soluble mineral accumulations rarely occur in soils found in humid settings. In fact, to counter their removal, lime is often added to soils in order to counter the acidity problems encountered in many soils of humid areas due to the leaching of the soluble minerals.

In contrast, drier areas are characterized by smaller outflows of runoff and groundwater. It is largely because of this that many soils found in arid areas are characterized by active processes of salt and alkali accumulation. This is particularly true in arid lowland settings, where intensive evaporation of shallow surface water and groundwater annually brings new salts and other soluble minerals into the soil. In these arid hydromorphic settings, highly saline or alkaline soils have naturally evolved (Kovda 1980: 74). Because of their high pH properties, soils in these arid settings are not potentially useful for irrigated agriculture. The arid soils that have a high potential for irrigation are those situated in geomorphic settings that are not favorable for salt accumulation; namely, locations in which surface waters do not accumulate, and the groundwater table is deep or nonexistent.

Salinization is the process that leads to an excessive increase in the salinity of the soil, due to agricultural practices, such that plant growth is prevented. Most salinization processes result from poor agricultural practices associated with irrigation. Contemporary processes of salt accumulation in irrigated areas are largely determined by:

- the salinity of the water used in an irrigated area
- the groundwater balance of the area

The best waters used for irrigation usually contain at least 200–500 parts per million of salt (Nebel 1990: 225–41). Most utilized irrigation waters are even more saline. As irrigation water is evaporated or transpired, the salts that were in solution are largely left behind and accumulate in the soil. Gradually these salts will become sufficiently concentrated to preclude plant growth unless surplus water is added to flush the salts out of the soil. Since water is a scarce commodity in most areas where irrigated agriculture is practiced, often short-term strategies of maximizing agricultural output with the limited water supply take precedence over the utilization of scarce waters for flushing salts from the soils and increase the long-term susceptibility of the irrigated lands to salinization. This is especially true during dry years when short-falls in water supply could require all farms within an irrigation project either to reduce acreage under cultivation and hence lower the year's crop outputs (short-term), or "waste" some of the water on the reduced acreage as a means of insuring sufficient flushing of the salts from the remaining fields under cultivation (long-term).

Most irrigated lands were situated originally in areas in which the groundwater table was sufficiently deep to preclude any capillary rise of moisture from the water table to the root zone. Once a water table rises to a height sufficient enough to permit the capillary rise of groundwater moisture to the root zone, the dissolved minerals in the groundwater will have the potential to begin to be deposited in this critical sector, and the soil's pH will begin to rise. Water tables rise when inflow (precipitation and irrigated water) exceeds consumption (evaporation and transpiration).

Because of bad design and/or poor management, water inputs on irrigated lands often are less than optimal (Warren and Maizels 1977). Leaking irrigation canals and too much water placed on lands are common problems in many irrigated areas. Thus, local water tables are usually very low at the start of most irrigation schemes. However, where excessive inflows resulting from canal seepage and/or poor management occur, this will result in a net annual upward movement of the water table. Eventually the groundwater comes close enough to the surface for capillary rise of the water to occur and excessive salts begin to be precipitated in the root zone. If the water table continues to rise into the root zone, this is called waterlogging. Under this condition, plants die from excessive moisture.

Both salinization and waterlogging can be prevented through good irrigation design and practice. Since most irrigated waters contain a high percentage of dissolved minerals, to prevent salinization often it is necessary to place more water on the land than is required to meet a plant's water needs. The strategy of applying surplus water to irrigated lands results in creating a net downward movement of water through the soil. This prevents the salts in the saline irrigated waters from accumulating in the root zone. However, unless this surplus water is removed from the irrigated lands before it reaches the groundwater, it will result in a rising water table. Thus this strategy of preventing salinization can contribute to a waterlogging problem if, simultaneous with this application of surplus waters, no strategy (or an inadequate one) is undertaken to prevent this surplus from reaching the water table.

Drainage canals that capture the incremental, salt-flushing waters from the irrigated soils prior to reaching the water table, and remove these waters from the area, are as critical in managing a sustainable irrigation scheme as the irrigation canals that deliver the water to the fields and permit plant growth in the first place (Withers and Vipond 1980). Yet, a negative off-site by-product, inherent in this drainage-flushing strategy, is that the waters in the drainage canals possess more dissolved solids than the original waters used for irrigation. When this waste irrigated water is returned into a stream system, it means that the downstream users of these waters will receive ever more saline water. Thus, not only does the water resource degrade for all future users, but it becomes increasingly difficult to prevent salinization on irrigated lands in the downstream direction. For example, in the Colorado River Basin approximately 75 percent of the waters applied to irrigated fields are utilized in evapotranspiration as part of plant growth (Burns, Billard, and Matsui 1990). Thus, on average, 25 percent of all irrigated waters are returned to the water system after utilization and are available for recycling downstream on other irrigated farms. However, these waters became more saline than during their previous use. One implication of this change in water salinity is that when the waters are returned to the river system, they increase the river's salinity and make the prevention of salinization increasingly difficult toward the river's delta (see chapter 7).

The flushing of salts from irrigated lands is not the only strategy for preventing salinization. Traditionally, in parts of the plain of the Tigris and Euphrates Rivers, farmers would fallow their irrigated fields on a biannual basis (see chapter 2 and Gibson 1974). This fallow system not only inhibited excessive salt accumulation, but it also allowed any rise in the local water table that occurred during the irrigated period to fall during the fallow. This curtailed both the likelihood of salinization as well as waterlogging. However, a major shortcoming of this tactic under contemporary conditions is that it limits the area under irrigation at any one time. It thus restricts agricultural output in an area that is deficient in these products. For this reason, it has decreased in importance as a method for controlling salinization.

The linkages between salinization and irrigation are understood and strategies exist to prevent it. Nevertheless, salinization/irrigation continues to be a major cause of land degradation. Over 30 percent of all lands irrigated already have been salinized (table 3.1). Furthermore, between $1-1.5 \times 10^6$ hectares ($2-4 \times 10^6$ acres) continue to be salinized each year (Nebel 1990: 178). Shortages of sufficient water to irrigate lands, especially during dry years, and poor management are two major causes of this phenomenon. The reality is that poorly designed and managed irrigation projects are a major contributor to land degradation worldwide.

Western Australia and salinization

Irrigation is not the only cause of salinization. Today, in southwestern Australia, over 400,000 hectares (988,410 acres) of rainfed agricultural farmlands are experiencing secondary soil salinization. Over 260,000 hectares (642,467 acres) of these lands have been degraded by salinization to such a degree that yields have dropped, and in some places agriculture is no longer possible (Barrow 1991: 188). While salts have naturally accumulated in this area's soils and groundwater for thousands of years, salinization was not widespread until the natural vegetation cover was altered by farming during the twentieth century. The origin of a large proportion of the salts that today are causing salinization problems in this area of Australia are associated with the region's rainfall. The marine winds flowing inland from the coast are laden with airborne suspended salts. When precipitation occurs, the salts are deposited along with the rainfall throughout the southeastern area (Hingston and Galaitis 1976).

The lands being affected by secondary salinization today were originally forest (figure 3.9). Only during this century were they

Table 3.1   Salinization of irrigated lands

Region	Area (acres (hectares) $\times 10^6$)		Percentage degraded
Africa	15	(6)	32
China and Mongolia	25	(10)	30
Asia (other)	188	(72)	34
Australia	5	(2)	19
Europe	15	(6)	25
Latin America	30	(12)	33
United States and Canada	50	(20)	20
Total	327.5	(132)	30

Modified from *World Resources 1986*, World Resources Institute, Washington, DC.

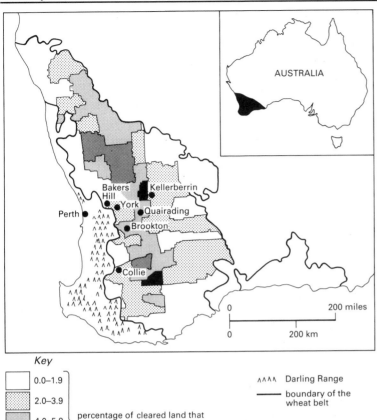

Figure 3.9
Areas impacted by salinization in Western Australia (after Guy Foster, Department of Geography, University of Western Australia). Saline, regional groundwaters beneath the confining hardpan are under pressure and leak salty water to near-surface locations through old root channels and other macropores.
Throughflow, which accounts for the greater proportion of water moving in the landscape, is relatively fresh. However, the rising saline groundwaters mix with throughflow, which then spreads the mixture downhill. The salty water accumulates in aquifers perched on top of impermeable hardpans in valley floors. Surface runoff then spreads the salts further down the valley. Thus, farmers and others may be affected by salinity problems originating in distant watersheds

cleared for farming and livestock production: "The transformation from natural vegetation to agriculture has disturbed the soils and the hydrologic cycle and has redistributed the salts (that have been deposited through the millennium) closer to the soil surface" (Conacher 1990: 6). The change from forest to cropland was the original explanation for rising water tables in this area. Because wheat and other annuals replaced deep-rooted vegetation, less water was transpired by the replacement crops than by the indigenous trees. This lower consumption of moisture resulted in an increase in the groundwater recharge. This concomitantly raised the saline water table and produced: (1) saline inflows into the area's streams that lowered the quality of stream flows; and (2) a rise in the saline groundwater, to within 2 m of the surface in many areas. This proximity to the surface allowed capillary action to bring the salts from the groundwater into the root zone. This explanation, first articulated in 1924, only partially explains the salinization problem in southwestern Australia (Conacher 1990: 7).

According to Conacher, a second factor contributing to the salinization problem was a change in the vegetation. By decreasing plant water demands and surface friction, a reduced plant cover increased overland flows and initiated widespread lateral subsurface water flows (throughflow). The result of this downslope increase in water movement, especially the throughflow, was that valley bottoms began to experience perched water tables of saline water (figure 3.10). As the throughflow moved down slope, it activated soluble salts that had been "locked" in the hillslope's substratum and transported them onto the valley floors. With the addition of these salts, the soils in the valley bottoms became too saline for wheat production. Thus, the change in vegetation from forest to grains and pasture initiated changes in the hydrologic cycle that have widely increased the supply of saline waters throughout a large area of southwestern Australia. The resulting secondary salinization has degraded the area to a sufficient magnitude as to lower the region's agricultural potential. Additionally, during dry periods, the degraded vegetation cover resulting from the saline conditions has initiated wind erosion as a significant problem in this area. The growing salinization problem in southwestern Australia represents an example of unintentional destructive change due to interactions between the human and physical domains (see chapter 6).

Interestingly, changes in vegetation cover often also initiate land instability problems in more humid settings. Since moisture is a critical variable affecting land stability in humid areas, during wet years or wet seasons erosional and mass movement processes potentially increase if vegetation changes alter local hydrologic systems so as to increase subsurface moisture. This specific condition is explored in the next section, which deals with the geomorphological domain and its relation to land degradation.

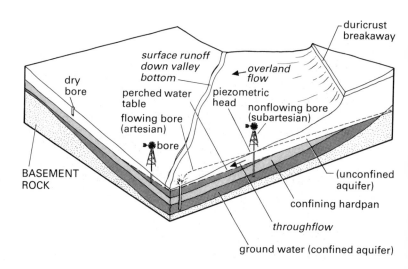

Figure 3.10
Subsurface water movement and salinization in Western Australia (after Guy Foster, Department of Geography, University of Western Australia)

Geomorphological Factors

Geomorphology is the study of the earth's landforms. The emphasis in the discipline has been both to describe landscapes and landforms as well as to investigate their genesis. Landforms are the result of two major processes. Tectonic processes, such as diastrophism and volcanism, originate from forces within the Earth's crust and mantle. Their net effect is to build up the Earth's surface. The second exogenic set of processes are derived from external forces resulting from surface and atmospheric phenomena. They include weathering, erosional, and mass movement processes. Their net effect is to remove earth materials and lower the Earth's surface toward sea level.

Until rather recently, geomorphic investigations largely ignored human activities as a significant geomorphic variable. The roles of wind, water, and ice, the natural exogenic agents, have been, and continue to be, stressed in geomorphic investigations. The importance of humans as a geomorphic agent is relegated to a minor role in most geomorphic investigations: as a typical example, Chorley et al. (1984) devote only 20 pages (in the appendix) out of 607 to the examination of human activities. Yet in many areas, humans are the dominant geomorphic agent. In the Netherlands, over 30 percent of the national territory is the direct result of the construction of public works. A significant percentage of its lands have been reclaimed from the sea while, through the building of extensive engineering works, the freshwater lake IJsselmeer has replaced its predecessor, the salty Zuider Zee. Along much of the northeastern seaboard in the United States, as much material is moved by construction machinery as by the area's rivers. For example, sanitary landfills have created hills where once depressions existed (plate 3.1), and expressway construction has cut through hills and filled

Plate 3.1
A landfill site at Northboro, Massachusetts

Plate 3.2
A road cut along Interstate 290

valley bottoms to alter terrain gradients totally in order to meet the needs of the automobile (plate 3.2).

In Chicago, the flow of the Chicago River has been reversed and is almost totally controlled by human manipulation. In Egypt, the damming of the Nile has resulted in significant erosion in the delta areas. And in Turkey, through a series of dams, some still under construction, the hydrology of the Euphrates River system is being altered. In our inquiry, the interaction of human activities with tectonic and natural exogenic processes will contribute to an understanding of the land degradation problem.

Degradation potential and tectonic processes

Plate tectonics and stability

With the acceptance of the validity of the theory of plate tectonics in the late 1960s, the general characteristics of the Earth's tectonic movements began to fit into a predictable pattern. The essence of plate tectonics theory is that the entire upper lithosphere (approximately from the surface to a depth of 100 km, which includes all of the crust and extreme upper mantle) is in motion. The motion of the lithosphere is manifested in seven major and nine minor rigid lithospheric units (table 3.2). Each unit, called a plate, moves as a single unit. Plates are dynamic not only in terms of motion, but also with regard to their size and shape. Since there are more than two plates moving independently of each other on the Earth's spheroidal surface, three major types of relative motion boundary conditions exist (figure 3.11):

- diverging or spreading edges, where the relative motions of the two juxtaposed plates are in the opposite direction

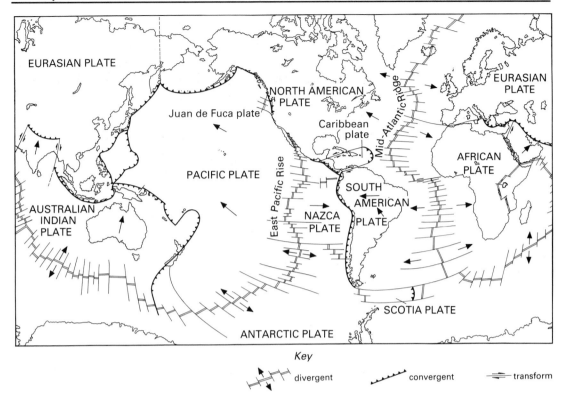

Key

✛ divergent   ⌒⌒⌒ convergent   ⇌ transform

- converging or consumption/subduction edges, where the relative motions of the two plates are toward each other
- slipping or transform edges, where plate margins slide past each other

Figure 3.11
The Earth's tectonic plate boundaries (after W. B. Hamilton, US Geographical Survey)

Each of these boundary conditions is associated with specific sets of tectonic activity that affect many characteristics of the surrounding lands. These tectonic differences are significant variables that planners of human activities need to acknowledge if land degradation is to be minimized. (It is only with respect to this aspect that we examine the theory of plate tectonics: a detailed explanation can

Table 3.2   The major and minor plates

Major	Minor
African	Anatolian
Antarctic	Arabian
Australian–Indian	Caribbean
Eurasian	Cocos
North American	Hellenic
South American	Iranian
Pacific	Juan de Fuca
	Nazca
	Philippine

be found in almost every contemporary introductory geology or physical geography text.)

Diverging or spreading edge boundaries

When plates move in opposite directions, they produce rifts in the Earth's crust. Volcanic materials rise to the surface in these rift zones to fill the space created by the diverging motion of the plates. Thus, zones of divergence are associated with volcanic eruptions. Predominantly basaltic lavas are always found in these zones. As all volcanic eruptions are heralded by earthquakes due to the rising magma fracturing the surrounding rock material, these regions are also zones that have a high frequency of earthquakes. Most zones along diverging plate boundaries are irrelevant with regard to the focus of this book, since they occur beneath the world's oceans and seas, and are hardly zones of land degradation! Yet where land masses exist in these zones, the dynamic nature of these areas present special conditions that need to be considered. Iceland and the island of Hawaii are two such areas.

Iceland, including its offshore islands, is about the only large land area situated along this type of plate boundary. However, the island of Hawaii, while located in the center of the Pacific Plate, experiences the same type of volcanic and seismic conditions as are associated with spreading edges. As the Pacific Plate moves NNW, the Hawaiian hot spot, a plume of rising oceanic magma, has remained stationary for over five million years. Upon reaching the ocean floor, the magma from this plume has deposited lava and created the Hawaiian island chain. Both Iceland and Hawaii result from volcanic eruptions dominated by basaltic lavas, and are in similar tectonic settings with regard to land degradation.

Compared to other lava types, basaltic lavas are relatively fluid (low viscosity). Because gases, which are part of magma, escape easily from low-viscosity lavas, minimal pressures build up during Icelandic and Hawaiian eruptions. The result is that these islands have volcanic eruptions characterized by their low explosive potential. With these basaltic lavas flowing at speeds measured in meters per hour, only immobile objects fail to avoid their pathways. For these two reasons, low explosiveness and low velocity, volcanic eruptions associated with basaltic materials are considered to pose relatively low risks, since loss of human life should rarely occur in these settings if governments are competent and citizens behave reasonably.

For example, on the island of Heimaey (Iceland) when the Helgafell volcano erupted on January 23, 1973, it was possible to evacuate the 5,000 inhabitants of the island without loss of life. Yet many farm fields, pastures, and urban lands were buried by the eruption's 5×10^6 m^3 (176,573,500 ft^3) of lava and pumice. These deposits removed these lands from production permanently. Farms were destroyed and a portion of Heimaey's economy was wrecked. Clearly, land was affected over a significant proportion of the island (Gunnarsson 1973); but from the perspective of this book, since the cause was volcanism rather than human activity, this was not land degradation.

Both in Iceland and Hawaii, attempts have been made to minimize the destructive effects of the lava flows by altering their actual and potential pathways. On Hawaii, earth embankments have been built and channels bulldozed in an attempt to divert future lava flows away from the town of Hilo and its surrounding housing subdivisions. These activities have met with partial success with regard to the volcanic eruptions that have occurred since the diversion structures were built. However, a major eruption would probably overwhelm these engineering works and engulf some useful land in the vicinity of Hilo. In contrast, on Heimaey during the 1973 eruption, the pumping of cold sea water onto the moving flow perhaps modified the flow's direction sufficiently so that it did not destroy the town's harbor (McPhee 1989). But this claim is conjectural, since the cooling efforts of spraying the cold ocean waters on the moving lava flow often had minimal effect on the direction of the flow during the five months in which this strategy was employed (Gunnarsson 1973). As far as changes in the flow's direction and preservation of the harbor due to spraying are concerned, McPhee's (1989: 179) comment with regard to "...the role of luck being unassessable, the effects of intervention being ultimately incalculable, and the assertion that people can stop a volcano being hubris enough to provoke a new eruption" is wryly pertinent.

Converging or subducting edges

When two plates move toward each other, their plate boundaries are zones in which earth materials collide. In these compression zones, rocks are deformed and become highly fractured. The collision zones are associated with active mountain building (e.g., the Andes, the Alps, the Caucasus, and the Elburz), earthquakes (e.g., Armenia, Iran, Alaska, and Japan), and often volcanic activity (e.g., Lassen Peak, Cotopaxi, and Tambora). Rhyolitic and andesitic materials, both more viscous than basalt, dominate the lavas produced in these settings. Because of their greater viscosity, the pressure resulting from the building up of gases within the magma can result in very explosive eruptions (e.g., Mont Pelée and Mt St Helens). While these eruptions are more dangerous than those associated with basaltic magmas, prior to the eruptions a series of earthquakes occur. Thus, if people take the precaution of evacuating the area, minimal dangers exist.

Since volcanic eruptions in these settings produce a greater amount of ash (pyroclastic) during their eruptions than those found in areas affected by plate divergence, often converging areas have thick unconsolidated deposits on the surface. If these deposits are disturbed, such as through road building, other construction activities, and the clearing of vegetation, their susceptibility to massive landslides is greatly increased during seismic events. It is the vibrations set off by the earthquake that initiate the slope failures.

Also, during periods of heavy rains, unconsolidated sediments can become saturated with water. Under these conditions, mass movements, such as mudflows, are possible. From the perspective of land degradation, these areas need to be utilized carefully in order to prevent environmental problems.

The June 1990 earthquake in northern Iran (Iranian and Eurasian Plates) illustrated some of the problems associated with human activities in areas of plate collision. For example, almost all of the mountain roads connecting northern Iran with the outside world were blocked by deposits resulting from mass movements triggered during the series of earthquakes that rocked this area. If the morphology of the slopes had not been altered by the road building activities, there is a high probability that the frequency of the mass movements would have been greatly reduced. The usefulness of portions of this region was decreased due to the interaction of human and tectonic processes. Land degradation clearly occurred during the period of earth tremors.

Slipping or transform plate boundaries
The third plate condition occurs where the dominant motion along the plate boundary is horizontal. These are the zones in which the plates are slipping and grinding past each other, areas in which massive fractures (faults) in the lithosphere exist. As the plates grind past each other, rocks are shattered and earthquakes occur. The southern portion of the Pacific coast of the United States is situated in this type of tectonic setting. San Francisco, built on the North American Plate, and Los Angeles, built on the Pacific Plate, are rapidly (from a geologic perspective) moving toward each other. One of the zones along which the plates are sliding is called the San Andreas Fault. This is a region that is highly susceptible to earthquakes and landslides under natural conditions. With the land and vegetation modifications resulting from human activities in this densely populated and rapidly growing area, this propensity toward instability has been exacerbated.

The San Gabriel Mountains, underlain with weak (e.g., shales) and highly fractured rocks, have a high potential for mass movements in their natural state. As Los Angeles has expanded onto the western slopes of these mountains, vegetation has been destroyed and slopes modified. The effects of any seismic event are magnified due to these changes. Additionally, with the destruction of vegetation, when heavy winter rains occur, mudflows become common. Areas affected by brush fires during the dry summer months are likely to experience mudflows during the wetter winter months. The connection between climatic and tectonic phenomena and land degradation in a setting such as Los Angeles is obvious. The question of how to prevent land degradation in such a tectonically active area while still meeting the demands of a rapidly growing urban area has yet to be resolved.

Degradation potential and exogenic (surface) processes

It is estimated that natural erosional processes result in 10 billion tons of sediment being deposited annually in the oceans (Judson 1981). Under existing practices of agriculture, grazing, and activities associated with urbanization, estimates of sediment deposited into our oceans range from 25 to a maximum of 50 billion tons per year (Brown and Wolf 1984). This increase in erosion indicates that human activities, as currently practiced, have escalated the erosional capabilities of the natural geomorphic agents of wind and water. The role of ice (glaciers), the other natural geomorphic agent, has not changed significantly due to human activities. Water is the dominant erosional agent in both humid and arid environments; wind is only effective under limited environmental conditions even on arid lands (Thomas 1989). In the following sections we explore some aspects of the conditions that lead to accelerated erosion and mass movements. A few common links between land alterations resulting from human actions and the increase in erosional and mass movement rates are presented. Hence some additional causes of the pattern of worldwide land degradation are introduced.

Surface water runoff and erosion
Fluvial erosion results from a complex set of relations among precipitation, gravity, vegetation cover, and the resistance (erodibility) of the soil to erosion (Wischmeier and Smith 1978; de Ploey 1990). Because these variables vary widely from place to place and the relations between these factors are complex, the absence of a database that includes these variables makes it difficult to develop strategies to minimize soil loss. This is a particular problem for the countries of the developing world. In the following four subsections we examine some of the critical erosional aspects of precipitation, gravity, groundcover, and erodibilty.

Precipitation   The erosivity property of rainfall, discussed in the first part of this chapter, is critical in affecting the soil erosion potential. As the amounts and intensity of precipitation increase, the potential for erosion grows. Because rainfall intensities, precipitation amounts, and their temporal patterns vary widely throughout the planet, humans must adapt different strategies for specific climates in order to minimize erosion. The humid tropical zones, in which both high rainfall totals and intensities are common, generally have high erosivity values and hence are characterized by high erosional potential. Tactics need to be put in place to neutralize this high erosional potential when lands are disturbed. A successful soil conservation strategy developed for the semiarid Great Plains, where a major amount of the soil moisture is from snowmelt, might not be effective along the semiarid southern coast of Jamaica, even though the net moisture conditions (semiarid) are similar. The lack of snowfall in Jamaica results in a significantly different precipitation pattern that must be addressed in any successful West Indian semiarid soil conservation strategy. This illustrates a climatic factor that contributes to the problem of successfully transferring land degradation strategies among "similar" (semiarid) but different (mid-latitude versus tropical) settings.

Gravity The importance of gravity in erosion is relatively obvious. Slope affects both the direction and the velocity of surface flows. On flat (0°) and very gentle slopes, water will be stagnant or sluggish and have no or little erosional potential. As slopes increase, the directions and zones of concentrated water flows are determined, and the potential for higher velocities and erosion increases. The erosional role of gravity is largely determined by an area's topographic characteristics. For example, steep highlands have a greater erosional potential than flat highlands. Therefore, once the steep lands are disturbed, they have a greater likelihood of experiencing accelerated soil loss than does a gentler area. It is the role of gravity that largely accounts for steep lands falling within the category of fragile land.

Vegetation (ground) cover Vegetation cover plays a critical role in affecting erosional rates. The alteration of groundcover by human activities is probably the greatest cause of increased erosional rates throughout the world. Vegetation has numerous properties that are beneficial to reducing erosion. First, the interception of rainfall – especially by leaves – reduces the total amount of water reaching the ground. For example, broad banana leaves can actually trap small amounts of rainfall, and a percentage of this trapped rainfall is directly evaporated without ever reaching the ground. Second, when rainfall strikes plants, they absorb a proportion of the falling drop's energy. This reduces the raindrop energy available to initiate soil movement. Third, when plant matter (including both living and dead matter) covers the ground completely, the soil is protected by this veneer and rainsplash is ineffective. Fourth, groundcover, which poses a resistance to surface water flows, increases surface friction and thus lowers surface water velocities. This reduction lowers the erosional potential. Fifth, plant roots both increase the resistance of the soil and remove a portion of the soil moisture. This latter property means that a greater amount of rainfall is needed to saturate soils, a condition under which they are less stable. Sixth, soil infiltration is often increased by vegetation. This reduces surface water flows. In summary, as vegetation cover increases, the erosional potential generally decreases.

This relation between groundcover and erosion is manifested in many temporal and spatial patterns of soil loss. In areas of annual crops, during the early growing season when much of the soil is bare, a moderate rainfall event can result in greater soil losses than a more intense storm during the summer when plant cover is high. In both urban and rural areas, construction areas are associated with high erosional rates. Typically, construction is associated with large areas of disturbed bare ground. Unless conservation strategies are implemented, construction areas are almost always major areas of sediment production.

Erodibility   Erodibility is the resistance of the soil to particle detachment and transport (Wischmeier and Smith 1978). A soil's erodibility properties are primarily determined by its texture (particle size), chemical properties, organic matter, and structure. These properties are dynamic and change both in response to natural and human phenomena. During a rain event, a soil's erodibility may increase due to surface seals or changes in particle orientation (Lal 1988). Plowing may destroy a soil's structure and make it more susceptible to erosion.

Soils vary in their susceptibility to erosion. Highly erodible soils have a potential that is about ten times greater than the most resistant soils. As the erodibility of a soil increases, greater care must be applied to prevent degradation. Some soils, such as those composed of fine sand and silt sized materials, are extremely vulnerable to erosion. The best strategy for areas with extremely highly erodible soils is to develop land activities that will not disturb the soil resource. Particular goals in such areas should be permanent complete groundcover and activities that reduce surface water runoff.

Wind erosion

Wind velocities associated with most meteorological phenomena greatly limit the wind's effectiveness as an erosional agent. Under extremely rare conditions, boulder sized materials may be moved by winds (Sharp and Carey 1976), but generally only sands (< 2.0 mm in diameter) or finer materials have the potential to be moved by wind. Usually it is only silts and clays (dust, < 0.0625 mm) that can become airborne. As an erosional agent, the effects of wind are restricted to areas where dry surfaces are void of vegetation. Under natural conditions, this largely restricts wind erosion to a limited number of settings within arid or coastal areas. However, when the vegetation cover is significantly reduced by agricultural, grazing, industrial, and urban activities, eolian erosion becomes possible. With regard to land degradation, it is especially in the semiarid zones and other climatic areas with pronounced dry seasons, such as in our southwestern Australia example, that wind erosion becomes a significant factor due to the removal of the vegetation cover by human activities.

Wind deposits of fine materials exist in all climatic settings. In humid areas, these deposits often occurred under drier climatic conditions than exist today. For example, in central Indiana in the humid American Midwest, forested hills of stabilized sand

dunes are found. These hills were formed during the drier periglacial conditions that existed in the region toward the end of the
Pleistocene. Today these sand deposits are fixed in location due
to the protection of the vegetation cover that has evolved under
humid conditions. Even when the vegetation cover is disturbed on
these hills, the dunes remain fixed due to sufficient ground moisture. This condition allows vegetation to be reestablished quickly
when it is partially removed. Wind erosion on these relic dunes is
minimal.

In contrast, in southern Mauritania – a considerably drier area –
because of overgrazing, previous fixed sand dunes have become
mobilized and are moving southward (Lewis and Berry 1988).
Because this region is relatively dry and, unlike Indiana, minimal
surplus soil moisture exists during most seasons, vegetation has
difficulty in reestablishing itself on the permeable sands once it
has been removed. As these recently mobilized sands move south,
they are progressively covering one of Mauritania's best agricultural areas, destroying the agricultural potential. Once stable conditions are upset, it is difficult and costly, although not impossible
(James 1992; Johnson 1992), to stabilize areas of fine surface
materials in dry settings. For this reason, in areas of marginal
moisture surplus such as Mauritania, it is extremely important to
maintain a good groundcover where fine surficial materials exist. In
these settings it is extremely difficult to stabilize an area once wind
erosion occurs, because of the resulting degrading quality of the soil
resource. First, the winds remove the finer, fertile components of the
soil, which reduces the soil moisture capabilities of the remaining
soils. Even in more humid settings, because wind selectively
removes the finer soil components, wind erosion must be combated
to prevent degradation. Second, the movement of the soil materials
prevents seedlings from becoming established without major human
interventions.

The importance of wind as an erosional agent is continuously
increasing both in magnitude and areal extent due to the misuse
of land. Modern agricultural methods on high-technology farms as
well as areas of low-resource farming are making wind an important agent in settings where it should be completely ineffective
(Changnon 1983). For example, occasional dust storms occur in
the American Midwest because of the exposure of vast tracts of
soil during periods of plowing and sowing. When farm units were
smaller, fields were smaller, and wind breaks along the boundaries
minimized the effectiveness of wind. Today, many of the wind-
breaks have been cut down to increase field size and to improve
the efficiency of the larger machinery now utilized. Under natural
conditions, bare ground would not exist in this area, and wind
would be ineffective as an erosional agent even during dry periods. The wind erosion that occurs in this portion of the American
farming belt has not resulted in major land degradation. To date,
most deleterious effects have been countered through fertilizer
applications and improvements in seed quality.

Mass movements

Erosion requires a transporting medium, such as water, air, or a dump truck, for the movement of earth materials. Mass movements are the detachment and downslope transport of earth materials directly under the influence of gravity. The landslide is probably the best known type of mass movement; but there are a multitude of other types of mass movements, ranging from rapidly moving mudflows to extremely slow creep. Regardless of type, all mass movements occur when stresses acting on a slope exceed the strength of the material. With regard to land degradation, it is usually the rapid types of mass movements that need to be minimized. When farm fields suddenly slide down a slope and the deposits bury other fields, not only is the degradation obvious, but it is dangerous too. Engineers usually evaluate the likelihood of rapid mass movements (slope failures) using the concept of the safety factor (S.F. = strength of shear resistance of material/magnitude of stress).

As the safety factor decreases, the probability of mass movements increases: when the value is less than one, mass movements become inevitable. Slopes are always being altered either through natural erosional/depositional processes, such as a stream eroding the base of a hillslope, or by human actions such as terracing and road cuts. Thus, the shear stresses acting on slope materials are always changing, sometimes slowly and sometimes rapidly. Likewise, the shear resistance of earth materials is dynamic. In general, regolith materials usually decrease in magnitude toward the surface. Chemical weathering processes account for this change. However, in addition, earth tremors, water saturation after rainy periods, heavy traffic, and the removal of vegetation and the decay of their root systems result in reductions in the shear resistance of earth materials.

Periglacial areas in which permafrost exists are a particularly fragile setting with regard to mass movements. Any disturbance in the tundra vegetation that ruptures the vegetation mat almost always lowers the local albedo. During warm periods, local thawing of the upper permafrost zone in these disturbed areas often results. These areas become saturated by the meltwater and slow earth movements are one of many geomorphic processes that are activated. Vehicles crossing the Alaskan and Canadian tundra during the brief warm periods have created permanent linear depressions that are filled with water. These "petrified" tracks result from thawing and local mass movements.

When extensive engineering works are built in permafrost areas, such as around Nome, Alaska, a hummocky landscape, known as thermokarst, can result from mass movements triggered by changes in the local heat budget (Ferrians, Kachadoorian, and Greene 1969). These features upset the local ecosystems and perhaps permanently alter this environment. As the tundra areas of Canada, Russia, and the United States experience greater human incursions, mass movements have the potential of making major changes in the topography and environmental quality of this zone.

Regardless of location, in today's world it is almost ubiquitous that the Earth's surface is continuously being modified to facilitate human activities. Steep lands are terraced to facilitate cultivation. Slopes are cut and filled to create a relatively limited range of gradients and curvatures on modern highways. Reservoirs also drastically modify local topography. All of these modifications set into motion complex changes that potentially could impact on the geomorphological stability of the area. For example, making a slope gentler could increase infiltration of water into the substrata. This could reduce runoff, and thus lower fluvial erosion. But the additional groundwater could increase mass movements by reducing the subsurface material's shear resistance. This likelihood of increased mass movements resulting from morphological change would increase even more if this alteration occurred in an active seismic zone or near an area of heavy traffic, since vibrations, due to earthquakes and traffic, are a common triggering mechanism of slope failures. Even widely accepted conservation strategies have the potential to initiate unintentional destructive change in the complex interactions that exist among crucial variables as humans attempt to reshape lands to better meet their demands.

Terracing and accelerated mass movements in the highlands of East Africa   If steep lands must be farmed, terracing has become the almost universal strategy to prevent land degradation. In many cases, with proper installation and continuous maintenance, terraces are a valid conservation tactic (Shaxson et al. 1989). Yet, like any terrain modification, because of complex feedbacks between processes and responses, the effectiveness of terracing to control environmental damage on steep lands often falls far short of expected results. This section illustrates how standard conservation practices, when not tailored to specific local conditions, can contribute to land degradation.

A primary intent of terracing is to minimize erosion on steep lands by controlling surface water runoff. Terrace construction reduces the slope angle of the areas where crops are planted. This slope reduction not only lowers the velocity of surface water flows but, by increasing infiltration, reduces the amount of water flows. Both of these changes reduces the erosional potential. In many highland areas of Kenya, agricultural expansion of steeper slopes has been marked by forest clearing and terrace construction. In most cases, erosional losses have been within acceptable levels (Lewis 1985). However, in some areas in which the hillslopes are underlain with impermeable and weak materials, the change in vegetation cover from forest to cropland and the alteration in the slope morphology to a slope completely terraced has encouraged massive landsliding. In these areas, the increase in water infiltration due to the reduced slope along the bench terrace, together with the lower evapotranspiration of the crops, results in the substratum becoming saturated.

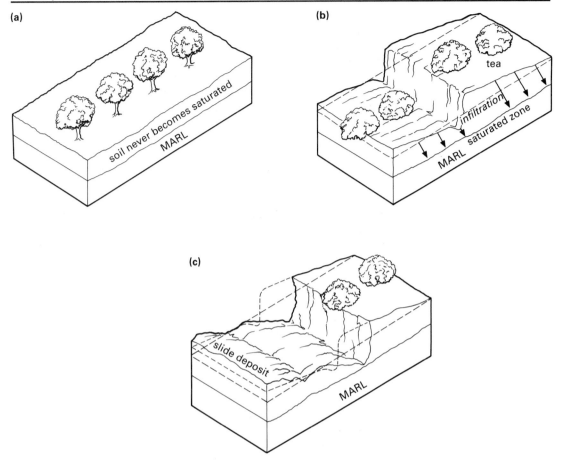

When the groundwater reaches the interface of the weak materials, in many cases a marl, a landslide occurs (figure 3.12). This type of slope failure could have been prevented if there had been proper drainage of the terraces to prevent excessive infiltration. However, because subsurface conditions vary widely in this area of highland Kenya, it is difficult to develop site-specific conservation strategies within existing financial and time constraints.

In northwestern Rwanda another relation between terracing and land degradation occurs (Lewis 1992). In this area, terracing has resulted in infertile soils being brought into the root zone along the back portions of the terrace (figure 3.13). This resulted in poor crop yields in these zones. To counter the poor soils in these areas, during the planting season the farmers remove a portion of the terrace riser immediately up slope and spread these soils over the back portion of the terrace. These removed soils act as a fertilizer and do improve crop yields. However, this human-driven soil movement process alone accounts for a downslope annual soil movement of 62 t/hectare (25 t/acre). This value is of sufficient magnitude to indicate that land degradation is occurring in this area of terraced fields. Thus, while terracing has curtailed fluvial

Figure 3.12
The interaction between land clearance, terracing, and landslides (after Lewis and Berry 1988)
(a) Under tree vegetation
(b) Cleared of trees, tea substituted, and land terraced
(c) Landslide due to greater infiltration of precipitation

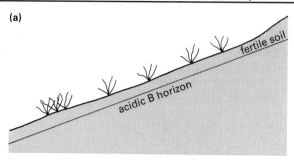

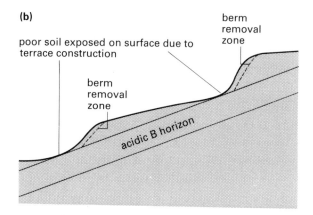

Figure 3.13
Terrace construction and degradation in Rwanda (from Lewis 1992)

erosion on these fields, the accelerated soil loss due to human activities has, to a very large degree, offset the benefits of the terraces. In this case, the constraints of the local farming system were overlooked when terracing was recommended as a strategy of land degradation control.

Summary

The risk of land degradation from any specific activity varies greatly from place to place. In this chapter we examined one set of critical factors affecting this risk, the physical setting of the area. An area's stability and resilience properties are determined by complex relations among its geomorphic, climatological, and vegetational attributes. Depending on the local characteristics of these variables, there will be a specific interaction between the natural and human systems. If critical thresholds are reached, land degradation will occur.

While generalizations about the risk of land degradation in a given type setting for specific activities are possible, they should never be considered more then a first approximation. For example, with regard to agriculture, high-energy settings, such as areas of steep slopes, would appear to have a higher erosional risk than low-energy areas such as plains. However, because land degradation results from the interaction of numerous physical associations, emphasis on any single property of the environment provides only a cursory insight into the relative physical settings susceptible to this problem. For example, many low-energy areas also have high erosional potentials. Semiarid plains, which are low-energy zones both with regard to gravity (low slope) and climate (low precipitation), are often as fragile for agriculture as are the high-energy environments of steep lands.

With any change in the land situation, a set of complex interactions usually occurs among biotic, climatic, and geomorphic variables. Some of these changes may be rapid and immediately obvious. The formation of a gully due to the removal of the groundcover on a slope of highly erodible soils illustrates this aspect. Other changes may occur slower over a longer period of time, but they are often equally destructive. A rise in the local groundwater level due to the construction of a dam and its reservoir eventually can contribute to an increase in mass movements by increasing hydrostatic pressures on slopes (magnitude of stress) while lowering the internal friction of the slopes' regolith (decreasing the shear resistance of the slope materials). The 1963 Vaiont landslide in northern Italy exemplified this situation: over 240×10^6 m^3 of material resulted in major destruction in the upper Piave River Valley three years after the reservoir was filled (Kiersch 1965).

The wide range of climatic, geomorphic, and vegetational properties found on this planet and the interactions among them indicate that a wide diversity of environmental conditions exist solely from the physical perspective with regard to land degradation potential. When human interactions are superimposed over the physical interactions, the environmental responses to change become ever more complex. While all parcels of land are susceptible to land degradation processes, because of the physical diversity, some areas are more prone than others to this problem.

While clearly not inclusive, the properties of precipitation and temperature that affect erosivity and the moisture balance play a crucial role in many of the degradation processes. In the vegetation domain, the protection that plants provide to the soil against erosion is very important; while in the geomorphic realm, areas situated in active tectonic zones are underlain with highly erodible materials or are situated in high-energy environments are settings where geomorphic processes are likely to be intense. The material presented in this chapter provides us with the base necessary for our examination of land degradation. In the remaining chapters of this book, the various dimensions of the land degradation problem, especially interactions between the physical and human spheres, are explored.

FOUR

The Physical Domain: Catalysts of Change

Introduction

To minimize or prevent land degradation, the laws that govern both the chemical and physical processes operating in natural systems need to be obeyed. Most often, land degradation is initiated when people intentionally or accidentally disturb existing balances among chemical and physical processes in the affected land system. Alterations in existing land systems, to meet specific human needs, often set into motion a complex interaction among the components of the system. In many cases, a degrading status in the overall production potential of the land system occurs even when the immediate goals of the change are met. To prevent degradation or to generate *creative destruction*, it is imperative to develop land-use practices that sustain or restore balances in critical components of the land system; "Not to do so is to court waste, damage, sometimes even injury and death" (Robinson and Speiker 1978: 90). In order to work in harmony with the cause/effect relations as they exist in nature, these relations must be both identified and understood. It is toward this goal that this chapter is written.

In chapter 3 primary characteristics and relations of climatological, geomorphological, and vegetational attributes important in understanding the land degradation problem were introduced. In this chapter other aspects of the physical domain are explored. This is partially accomplished by examining how low- and high-intensity geomorphic and climatological events act on the stability/resilience property of some critical components of land systems.

From the land degradation perspective, one of the most important components of the land resource is soil. This resource is crucial not only for humans, but for almost all terrestrial plant and animal species. When misused, it can even have severe ramifications in distant marine environments. One such example linking land degradation with negative coastal impacts is the decline of coral reefs off the coast of Kenya in the Indian Ocean (figure 4.1). Widespread soil erosion, which occurs throughout the Kenyan Highlands due to conversion of forest land into farm fields and to expansion of road building, has resulted in increased sediment load in many streams. Ultimately this enhanced sediment load reaches the rivers that flow into the Indian Ocean. A percentage of these sediments is then transported by offshore currents and waves into the zone of active coral growth. The excessive sediment results in an unfavorable environment for coral growth. The greater turbidity of the water that results from the greater delivery of sediment from the rivers is contributing to the destruction of the coral reef. With the degrading coral reefs, coastlines lose the protective shield that absorbed a large percentage of incoming storm wave energy. Two likely results of the degrading coral reef are that the less protected shorelines will undergo increased erosion in the future and that the overall coral reef ecosystem will continue to decline.

Soil is both a dynamic and renewable resource under almost all natural conditions. Not only does it contain biological activity but,

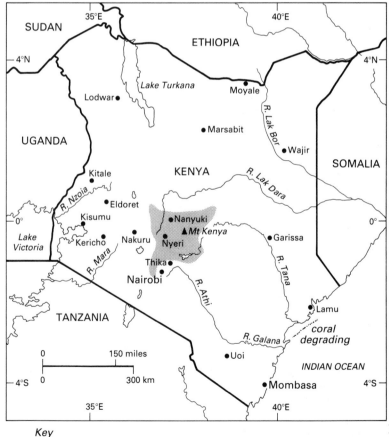

Key

▨ highlands – source of Tana soil erosion zone

Figure 4.1
The Central Highlands and coastal corals of Kenya

along with water, soil is crucial for most terrestrial animal and plant life. Both agriculture, pasturage, and forestry require the soil resource to remain in a good state for their very survival. Being a renewable resource, soil should maintain its fertility and hence productivity indefinitely when utilized properly. However, because of widespread exploitative and/or poor management practices, the soil resource, generally, is deteriorating worldwide. Even with massive conservation practices and technological interventions, destruction of this life-sustaining resource is increasing. Its decline is of such a magnitude that it represents a serious global problem. According to one reliable estimate, each year crop productivity on about 20 million hectares becomes either uneconomical or drops to zero due to land degradation related to poor management (UNEP 1980). The major cause of this land degradation is accelerated soil erosion. In some places climatic changes and geomorphic activity, such as increasing aridity or volcanic activity, are the cause of declining soil productivity and hence are beyond the control of humankind. However, the overwhelming cause of most soil degradation is an array of human activities. In chapter 6 specific linkages between many of these activities and degradation will be explored. As a prerequisite to that inquiry, crucial attributes of the soil resource are examined in this chapter.

The Soil Component

In most natural ecological systems, the loss of soil fertility due to both leaching (removal of soil nutrients by solution as water percolates through the soil) and surface (natural) erosion is offset by:

- the continuous recycling of nutrients through the decay of vegetal and animal matter
- the addition of new nutrients into the soil system due to the natural breakdown of soil parent material by weathering processes

However, in most agricultural and modern forestry systems, the recycling of nutrients is significantly reduced. Since the harvest is transported from the immediate area, significant quantities of soil nutrients are lost with each harvest. These nutrients are a component of the exported biomass. These decreases in the soil's nutrients require the replacement of the soluble minerals at fairly regular intervals through the addition of fertilizers. When supplemental fertilizers are not added, eventually fertility decreases and the soil resource begins to degrade.

Another contributing factor leading to the decline of the soil resource is the decrease or complete absence of an area's groundcover for a period of time as natural vegetation is replaced by agricultural and forestry systems. In most agricultural systems there is a complete absence of groundcover at least once a year during the field preparation and planting stages. The general ramification of this decrease in groundcover is accelerated soil erosion and the further removal of nutrients and other favorable attributes of the soil during the early portion of the growing season.

The nutrient or fertility property of soils, while clearly important, is only one of a set of critical factors affecting the status of the soil resource. Land degradation can occur even if no deficiencies in the nutrient status exist. The capacity of soils to hold moisture (porosity), along with their infiltration (permeability), aeration, and pH (acid, neutral, or alkaline) properties are other critical attributes affecting biotic activity. Without proper management conserving all of these critical factors, potential biological productivity declines. Under conditions of extreme abuse, soils are destroyed and previously productive areas become almost sterile (plate 4.1).

Within a national framework, the Caribbean country of Haiti is illustrative of widespread destruction of the soil resource. This has occurred to such a degree that land degradation is one of the factors that threaten the long-term viability of this island state. The widespread destruction of the soil resource, and the resulting declines in agricultural yields, is one cause that has contributed to strong out-migration from the Haitian countryside. The following examination of the soil system permits an understanding of the many relations between land use and land degradation.

Plate 4.1
Soil degradation in Haiti's highlands

The soil system

The term "soil" conveys different meanings for different disciplines. In engineering, often it implies no more then unconsolidated surface materials. In agriculture, it is the material in which plants grow. Thus, soil must be able to support plant growth. For the purposes of this book, the ability to support plant growth is a critical component. However, because soil is a dynamic, complex mixture of solids, liquids, and gases, any environmental changes in an area will likely have impacts on the nature of the local soil mixture. Sometimes these impacts on the soil are so severe that plant growth potential can be greatly reduced and in extreme cases may not be at all possible.

The composition of soils is complex. They are comprised of organic and inorganic matter, living and dead organisms, as well as gases, liquids, and solid materials. Our inquiry into soils intends to facilitate an understanding of how soils have the potential to change in response to the dynamic interaction between the five soil forming factors of parent material, time, climate, biota, and topography (Jenny 1941).

The parent material factor
Soils have the ability to produce and store plant nutrients through the diverse interactions of plants, bacteria, animals, air, temperature, moisture, and earth materials. The earth materials from which a soil evolves are known as the soil's parent material. While parent material and bedrock are often used interchangeably, this is incorrect since many soils are derived from unconsolidated materials, such as loess, sand dunes, river alluvium, colluvium, or glacial deposits. Some parent materials in their unaltered state can directly support plant life without undergoing any modification. Loess is one such material. However, technically loess is not a soil; it is a geomorphic deposit. Only when the upper portions of the loess deposit undergo modifications due to the interactions among the five soil-forming factors does it become a soil.

Earth materials are made up of minerals. Each mineral reacts differently to weathering processes. As weathering breaks down the parent material by physical, biotic, and chemical processes, a detritus is left that differs in both chemical and physical composition from the original parent material. The weathering processes are particularly important in soil formation when the parent material is rock. It is these processes that initially break down the rocks into unconsolidated fragments that facilitate the other soil-forming processes.

The properties of the parent material dominate a soil's characteristics during the **early** stages of soil formation. A limestone will produce soils having initially a high concentration of calcium; a sandstone will result in a silica-rich soil during the early stage of development. Textural properties of soils will also strongly reflect the attributes of the parent material during the early stages of soil formation. A coarse parent material, such as a sand dune made up of large sands, usually will weather initially into a coarse soil; a fine parent material, such as shales, even in the early stages will produce clay sized materials. However, with increasing age, as the weathering processes continue and the other soil-forming factors begin to interact with the youthful soil, the characteristics of the soil become ever more altered. With increasing time, the interactions among the other soil-forming factors decrease the early dominant role of the soil's parent material. In fact, with the passing of sufficient time, the influences of climate, topography, and biotic activity on the soil may completely efface all of the original properties of the parent material. Yet, in some environmental settings, such as on steep slopes and in arid areas, where the effects of weathering and biotic activity are curtailed, the characteristics of the parent material may continue to dominate the soil properties through time.

The time factor
The rate of breakdown and transformation of parent material into soil varies greatly depending on the properties of the four other soil-forming factors. Soil formation will be relatively fast if the parent material in the area is nonresistant, and if favorable climatic (high temperatures, humid) and topographic (well drained sites) conditions that are conducive to rapid weathering and high biotic activity exist in the area. A classic example of extremely rapid soil formation, occurring on a volcanic deposit in the warm humid setting of Krakatoa, is 350 mm of soil formation in 45 years (Mohr, van Baren, and van Schuylenborgh 1972). At the opposite extreme, when most parent materials are resistant and are situated in dry and cool climates (where the weathering rates and biotic activity are low), soil formation at rates of millimeters or less per 100 years would be the norm. In hyper-arid areas, regardless of the parent material, weathering rates are so low as to negate any soil formation.

Clearly, there is no simple relation between soil formation and time. However, in settings in which soil formation is slow, excessive soil erosion will likely have the potential for increasing the difficulty in reversing any degradation. Further complicating the temporal relation is that as soil development proceeds, its formation rate varies with time.

Nevertheless, the general rule is that most soil formation occurs slowly. From a geological perspective, where a time frame of thousands of years is the rule, the most severely degraded soil will redevelop except in extremely arid or cold regions. Even the most extremely degraded surface, such as the bare rocks exposed by massive soil erosion in parts of Haiti, will slowly break down and evolve into new soils over the millennia if human activities do not interfere with the natural processes. Yet, from the perspective of land degradation, because rates of soil formation in most environmental settings are imperceptible within the human life cycle, once a soil becomes severely eroded, it becomes a nonrenewable resource in the overwhelming number of cases. Even if agricultural activities in the Haitian countryside would completely cease and the land would be completely abandoned, as has already happened on many lands in the northwestern portion of the country, generations would be needed before the soil resource would regenerate itself.

The major type of soil degradation that so far can be generally offset economically by exogenous inputs is the depletion of nutrients through fertilizer applications. The maintenance of high agricultural yields in many parts of the United States, Canada, southern South America, and Western Europe is very heavily dependent on the application of fertilizer. However, the huge quantities of fertilizers added to soils in certain areas of northern Europe and the United States, as but two examples, are resulting in significant land and water pollution (James 1993). Hence, the prevention of the degradation of the soil is thus contributing to other aspects of land degradation.

The climatic factor

The primary climatic variables affecting soil characteristics are precipitation and temperature. With regard to soils, water is a powerful solvent. Since precipitation is almost pure distilled water, which is a mild acid, when it reaches the Earth's surface it is in an undersaturated state. As rain and snowmelt infiltrate into the ground, they slowly dissolve soluble minerals present in the soil. The higher the temperature of an area, the faster and more complete are the chemical and biotic processes that break down the existing minerals. Likewise, as precipitation increases, there is a greater quantity of water passing through the ground materials and hence a greater potential to leach these materials. Given sufficient time, in humid settings, most old soils will become relatively infertile due to the leaching of a significant proportion of their nutrients.

As water infiltrates into the ground, its dissolved load increases and its potential to remove minerals by solution decreases. If only leaching processes were acting in soils, it would be expected that through time, the upper portions of the soil profile would become largely devoid of soluble minerals and hence would be relatively infertile in humid climates. In the warmer humid climatic areas, where chemical processes are more intense, the zone of leaching would be both greater in magnitude and more complete than in the cooler humid areas. In dry areas, where minimal amounts of water pass through the soil, soluble minerals should be expected to remain in the soil mass. Clearly, this is a rudimentary explanation of the role of chemical reactions affecting soil formation. Yet, in a broad climatic context soils do reflect these roles of temperature and infiltration in their overall attributes. Soils in lowland humid tropical areas, with both high precipitation and temperature, generally have thick weathered profiles that are of low fertility. The net removal of soluble minerals from soils through the downward percolation of groundwater results in humid areas generally having acidic soils (low pH). Conversely, soils in arid areas, where infiltration to the groundwater system is minimal, and hence the removal of soluble minerals from the soil is slight, usually are rich in the soluble minerals. The net effect of this is that generally the soils of dry areas have a high pH, indicating their degree of alkalinity or salinity.

The biotic factor

Temperature and moisture also play critical roles in determining an area's biological activity and hence the biochemical and biophysical activity that occurs in a soil. As both temperature and moisture increase, plant and animal life increase. Hyper-arid areas, where minimal microorganisms, vegetation, and animal life exist, are locales in which soils are largely independent of this factor. In humid areas, the accumulation of organic matter reflects the balance between biomass production and decomposition. A warm tropical rainforest produces anything from three to four times as much vegetal matter per unit area as a cooler mid-latitude or highland tropical forest. However, microorganisms in the warmer and moist settings decompose this matter at roughly the same rate as their production (Lewis and Berry 1988). The net result is that minimal organic material accumulates in the soils of the moist, warm tropics. Greater amounts of organic matter are found in cooler forests where the lower production is more then offset by the lower decomposition rates. Nevertheless, even in warm humid areas where minimal organic matter accumulates in the soil, the production of plant biomass plays a critical role in countering the effects of leaching. The transportation of soil nutrients from the soil into living biomass through root systems, and the subsequent deposition of organic matter in the upper portions of the soil such as through leaf fall, is a critical recycling process that prevents leaching from eventually resulting in the complete absence of soluble nutrients in the upper portions of the soil profile in humid areas.

Besides the migration (recycling) of nutrients within the soil mass, vegetation plays another critical role in minimizing soil degradation from erosional processes. First, the interception of rainfall by plant canopies absorbs a large percentage of the raindrop energy. This lowers the rain energy available to erode the soil. Second, the veneer of decaying vegetal materials covering the ground protects the soil from the direct, erosional impacts of raindrops. Third, root systems increase the resistance of soils to erosion by increasing the internal friction of the soil. Finally, vegetation, by shading the soil from a proportion of direct sunlight, lowers the ground temperature and reduces the decay rates of organic matter. Because of the multi-dimensional effects of vegetation on the soil, changes in an area's vegetation cover almost always have direct impacts on the status of the soil resource.

The importance of organisms in affecting soil properties is another crucial component of the biotic factor. Organisms affect both the physical and chemical properties of soils in a multitude of ways. For example, the high amount and intensity of biotic activity occurring within many soils, such as termite and worm populations, are very beneficial to the potential productivity of soils. As soil passes through these organisms' digestive systems, minerals are altered and redistributed within the soil profile. The net result of the bio-chemical processes occurring in these organisms is to increase the fertility of the soil. Furthermore, worm casings and subterranean tunnels resulting from these organisms increase the infiltration and porosity of these soils, both of which increase soil moisture available for plant utilization. The net effect of these phenomena is to increase the water storage and fertility potential of soils, two very favorable soil attributes for agricultural activities.

The topographic factor

Topography has a strong modifying influence on soil characteristics. First, the slope (angle) of the terrain is a significant factor affecting a location's erosional, depositional, and drainage properties. Second, the spatial location of an area within a landscape influences the moisture conditions found at a site. The general pattern of ground moisture is to increase from hilltops toward valley bottoms. However, poorly drained (flat) uplands are an exception to this trend. Because of these two factors, topography plays a critical role in affecting the soil-forming processes, since erosion, deposition, drainage, and ground moisture are basic factors that affect soil processes.

Slope angle   As slopes increase, because of gravity the erosional potential acting on any soil mass increases. This relation, by itself, largely explains why soils on steep slopes are generally thin; on some very precipitous slopes, where erosion and mass movements remove soil as fast as it forms, the hillslope surface is bare rock. Conversely, as land becomes gentler, the likelihood of deposition or minimal removal increases. Thus, thicker soil profiles are the general rule on bottomlands and on many flat lands situated at the foot of steep slopes. The development of thicker soils in these settings results from a combination of two different processes:

- the deposition of transported soil materials originating on the steeper upslope areas where erosional and mass movement potentials are greater
- the formation of soil *in situ* – soils found in gentle settings on flat uplands also are often relatively thick, but these soils largely result from the development of the soil *in situ*, with minimal transported material accounting for the soil profile properties

Because minimal transportation is involved with upland soils developing in gentle areas, the fertility characteristics of these soils are largely dependent on the soil-forming processes and parent material interactions occurring in the local area. If their fertility decreases through misuse, natural processes will usually only slowly restore their fertility. However, lowland soils, where they are recipients of upslope-eroded soil materials, have the potential to maintain their fertility by the continuing deposition of the upslope-eroded materials, provided that the transported materials are fertile.

Slope steepness not only affects the thickness properties of soils through erosional and depositional processes, but also influences their soil moisture content. Precipitation striking steep slopes with thin soils will run off quickly. The effect of this rapid runoff is that only a minimal proportion of the rainfall can potentially infiltrate. The result is that, except in favored locations on the steep slopes such as in small niches, there is insufficient moisture available for vigorous plant growth. It is the interactions resulting from low soil moisture potential, poor plant cover, and high erosional potential that make it difficult for good soils to develop on steep lands. This factor alone makes most steep lands fall within a fragile environmental setting.

On gentle slopes in upland and bottomland settings, the percentage of precipitation that infiltrates into the soil is largely determined by the soils' infiltration properties and the intensity of the rainfall. Excessive runoff, resulting from the slope geometry, is not a critical determinant under these conditions. One result of the interactions resulting from slope steepness on soil properties is that low-slope areas are generally more resilient than steeper areas.

Spatial location   Water that infiltrates into the soil and is not either directly evaporated from the soil or captured by plant roots, slowly moves downward in the ground until it reaches a zone where all of the pores in the ground materials are completely saturated with water. This saturation zone is known as groundwater. The upper boundary of the groundwater is called the water table. Groundwater exists everywhere except in zones of extreme aridity or in lands that are underlain by impermeable rocks.

Normally, the pattern and location of the water table is a function of two major variables, climate and topography. In general, as climate becomes drier, the water table becomes deeper. In some extremely arid areas, water tables do not exist. Likewise, in a given area, during wet seasons or rainy periods, the water table rises. During these conditions, in depressions and other favorable locations such as in the immediate area of stream channels, the water table may reach the surface and create saturated conditions. Conversely, during dry seasons or rainless periods, its depth beneath the surface increases. Additionally, the overall pattern of the water table is strongly controlled by the area's topography; the water table approximates the terrain. Usually it is a subdued replica of the area's topography. In general, it reaches its highest elevation beneath hills and decreases toward valley bottoms. However, its depth beneath the surface is usually greatest under hills and least under lowlands (figure 4.2).

The water table's position is important with regard to soils in that the surface and subsurface moisture characteristics of an area play a

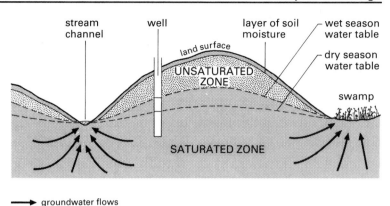

Figure 4.2
The water table and saturation zone of a hypothetical topographic profile (from Skinner and Porter 1987)

critical role in determining a soil's attributes. Moisture is an important determinate of the weathering processes affecting soil formation. Different soil properties evolve under various moisture conditions. Soils differ greatly between different moisture settings. When the water table intersects the soil profile, this represents a poor or impeded drainage condition. Under this condition, standing water is found at or very near the surface. Swamps, oases, lakes, and lands in close proximity to permanent streams are all situations in which the water table is at the surface. Soils in these areas are often waterlogged. Under these poorly drained conditions, free oxygen is absent within the soils, and the reduced anaerobic bacterial action that results strongly affects soil properties.

When the water table is below the surface, never comes in contact with any portion of the soil profile, and ground materials are permeable, then good drainage conditions prevail and oxidation processes proceed unimpeded. Under these conditions, the most important factor affecting the chemical weathering of the soils is the amount of water leaching the soil. Where moisture is not limited (humid locations), leaching will ultimately remove in solution most soluble components and result in soils having both a low fertility and pH. However, even in well drained topographic settings in dry areas, leaching is retarded due to insufficient precipitation. The result is that most of the chemically mobile soil constituents will be found in soils in dry areas which, under similar topographic settings, are deficient in them in humid areas. With regard to land degradation, this property of a large reservoir of soluble minerals stored in arid soils becomes a critical factor once irrigation is introduced. In particular, problems of salinization are related to this factor.

The term "catena" was coined for the repetitive regional sequence of soils correlated with topography. Because of the systematic interactions regionally between drainage, ground moisture, erosion and deposition, and vegetation with topography, there is a general relationship between the geomorphic surface (topography) and soils (Lepsch, Buol, and Daniels 1977; and see figure 4.3). The spatial

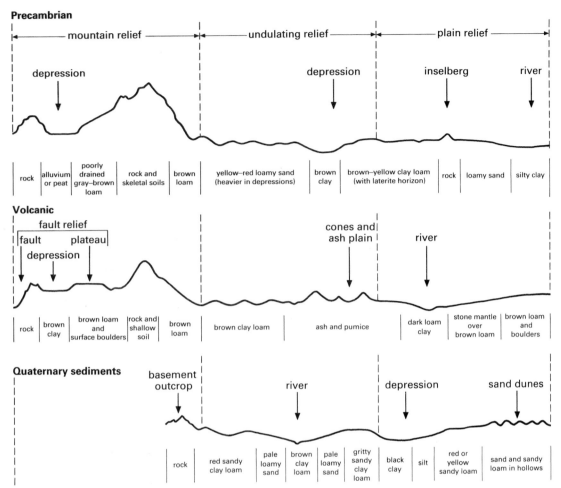

Figure 4.3
A typical soil catena in East Africa: range landscapes and their usual soil associations (from Pratt and Gwynne 1977). Sand dunes are shown here as part of a plain landscape, whereas well developed duneland would be classed as undulating relief

pattern of soil properties (catena) results from both site-specific attributes (static), such as slope angle and parent material, as well as dynamic causes such as downslope movement of soil materials (Young 1972). Because each catena is a dynamic phenomenon, it should respond similarly to specific human activities. Thus, recognition and application of the catena concept should provide useful insights into some of the causes of land degradation within a regional context. Clearly, each topographic setting will have a specific potential to degrade for a given activity depending on how it affects the relations among the static and dynamic causes of catenary differentiation.

Soils and some responses to low- and high-intensity events

The status of a soil at any given moment is the result of the continuous interactions between the five soil-forming factors. Two of the factors (topography and parent material) are relatively static. As a result, their roles in determining the condition of the soil resource can be considered to be constant through time. For example, topographic areas of steep slopes will remain steep and thus consistently are potentially a high-erosion zone. The biotic factor, under natural conditions, is also usually relatively constant. For instance, major vegetation types, such as a forested area, rarely shift abruptly under natural conditions to another vegetation type. An exception to this could be when a major fire occurs and a number of years are required before the forest cover can reestablish itself. Today, with the increasing importance of the human being as the dominant agent of vegetation change, the biotic factor is becoming increasingly variable.

Furthermore, the attributes of an area's biotic status are increasingly dynamic because humans are moving into ever more fragile zones; and hence people encounter less resilient settings. This is definitely one major reason for the soil resource becoming less stable. For example, as the tropical forest is cut down for farming and grazing purposes in a low-resource area, a whole host of interactions result. The net result is that many of the attributes of the existing soil, especially with regard to fertility, are altered in a negative direction (Woodwell 1990).

In the Amazon Basin, one critical change resulting from this transformation in the vegetation cover is the substitution of shallow-rooted grasses or crops in place of the deeper-rooted forest. This single change in a vegetation attribute inhibits the natural recycling of soil nutrients between the soil and living biomass. Previously, the deep and dense network of roots associated with the forest ecosystem captured a very large precentage of the soluble nutrients as they slowly percolated through the soil. Now, a larger proportion of the nutrients are not absorbed by the shallower roots of cultivated plants and are permanently lost from the area via groundwater and surface flows. Gradually, the soils become impoverished as soil nutrients are lost from the area by crop harvests and water outflows. In particular, available phosphorus is depleted. Because of a trend toward lower fertility and other soil changes negatively affecting growth, such as compaction and alterations in the soil temperatures, desired pasture grasses and crops become choked by second-growth weeds that require less fertile conditions, and the land is usually abandoned (Hecht 1989).

Is this process of land degradation resulting from a change in the biotic factor inevitable? Clear evidence exists that removal of the rainforest need not result in land degradation (Ruttan 1989). The main nutrient-loss pathway in the rainforest, as in most other agricultural systems, is the crop harvest. The major difference is that in the majority of lowland tropical rainforests there is a very small reserve of nutrients in the soils. Thus, from the onset of agricultural activities under these conditions, in order to counter the nutrient loss resulting from leaching and crop harvests, fertilizer application must be an integral and initial part of the new farming system if soil depletion is to be avoided. Rarely is this management practice initiated in conjunction with the alteration in the biotic factor.

In contrast to the other soil-forming factors, the climatic factor is inherently variable through time. Some years are wet; some are dry. Some years are cool; some are warm. Yet more important to the status of the soil resource than these types of annual oscillation is the fact that climates reflect the totality of an area's meteorological phenomena over a long period. Since average climatic values of a specific phenomenon rarely occur, the average parameters of a location's various weather elements, such as precipitation and temperature, are not particularly relevant in understanding land degradation. Average values mask the expected ranges (standard deviations) of meteorological phenomena that are likely to occur in any short time frame. Often it is when a weather element is in the high or low part of its expected range that the land resource is susceptible to change. For example, every climatic type experiences short-term precipitation variability, both in terms of intensity and amount. In fact, it is expected that precipitation variability will increase as climates become drier (Nicholson, Kim, and Hoopingarner 1988). When an area encounters a drier or wetter period, the resulting conditions (e.g., poor vegetation cover or saturated ground) increase the potential for dramatic changes in a specific area's soil even under natural conditions. When human activities have altered the natural setting, often the likelihood for change increases under these drier or wetter (but still within the range of normal) climatic conditions. This aspect of climate, an area's expected frequency and magnitude of precipitation phenomena, is now explored because of its relevance to land degradation.

Low-intensity (high-frequency) versus high-intensity (low-frequency) events

In any location, a wide range of moisture conditions are to be expected from year to year, month to month, and day to day. For example, the climatic record indicates that Lagos, Nigeria, has experienced annual rainfall values ranging from less than 1,000 to over 3,000 mm; and July monthly rainfall totals vary from a low of zero to a maximum of 912 mm (Nicholson, Kim, and Hoopingarner 1988). Analyses of rainfall data indicate that as rain events increase in magnitude and intensity they decrease in frequency. In the eastern United States this principle is reflected by frequent showers of low and moderate intensities, and only occasionally (usually less than once a year) intense hurricanes.

With regard to the potential for land degradation resulting from fluvial processes, it might be inferred from this relation that the infrequent but large rain event (e.g., hurricane) would be more significant than the everyday "average" events. Clearly, larger storms are associated with greater runoff and hence the potential for erosion is greater. When hurricanes do occur, landslides and massive erosion often do take place. While not minimizing the

importance of the infrequent event in contributing to land degradation, it is likely that the relatively frequent, moderate-intensity rain
events are just as important – if not more so – in most cases of land
degradation associated with fluvial erosion. Geomorphic research
indicates that, in the long run, moderate magnitude and frequency
events do most geomorphic (erosion/deposition) work (Coates and
Vitek 1980).

This finding, relevant to land degradation, is further substan-
tiated by results from agricultural research. Almost every year,
the greatest proportion of soil loss on American farmlands occurs
during spring rains. During this period, when the cultivated lands
are being prepared for planting, a large percentage of the soil is bare
and hence exposed to the direct impact of raindrops. Furthermore,
in the American farming belt, it is in spring that rainfall is most
frequent and the majority of the most erosive rains occur (Renard et
al. 1991). Typically, soil losses for any single year are not spectacular, but as these conditions of bare soil and erosive rains occur
almost every year, the long-term totals of soil loss are extremely
significant.

The majority of land problems resulting from wind erosion are
also associated with moderate-frequency wind velocities (Skidmore
and Woodruff 1968). Because the soil erosion resulting from these
normal conditions is not dramatic, remedial actions are often
delayed. Land does not abruptly cease to be unproductive. In the
short run, this type of chronic, areally pervasive soil erosion is often
countered through improvements in crop and animal breeding or
heavier applications of fertilizers. Yet eventually thresholds are
reached that often require uneconomical investments of capital or
labor to surmount. It was the continuing degrading condition of the
soil resource on American farmlands in response to normal rains
and winds that resulted in the formulation and adoption of the
federal Food Security Act of 1985. This act requires the identification of the agricultural lands with the greatest erosional problems.
For the majority of these lands, their degradation has developed
over a period of years in response to normal and expected climatic
conditions (low- and moderate-intensity/high-frequency events).

The status of the world's irrigated lands is another illustration in
the agricultural sector of poor human land and water management
practices, that interact with low-intensity geomorphic and climatic
processes and result in significant land degradation. Of the world's
235×10^6 hectares (580×10^6 acres) of irrigated land, about 10
percent are already severely degraded due to salinity. The long-
term viability of another 25 percent is questionable, as salinity
problems are becoming evident on these lands (*Economist* 1991).
While there are a number of causes of salinization, a major salinization process results from the continuous deposition of dissolved
compounds in the irrigated waters onto the soil (see chapter 3).
Eventually, the accumulation of excessive soluble salts in the root
zone results in a toxic environment for plant growth. In arid and
semiarid climates, daily evapotranspiration rates are sufficient to

concentrate the soluble components of the irrigated waters in the upper portions of the soil. Unless management practices are instituted to counter this phenomenon, such as preventing the accumulation of salts through drainage canals, it is inevitable that this type of degradation will occur. The continuous evaporation of irrigated waters and the slow accumulation of soluble minerals in the soil represents another low-intensity/high-frequency occurrence that slowly but surely results in the decline of the potential of irrigated agricultural lands in many of the world's dry climates.

In contrast to the chronic decline of resources in response to high-frequency events, which are often ignored or go undetected until the problem reaches a severe magnitude, abrupt land degradation in response to low-frequency/high-magnitude (intense) events quickly becomes evident and often is the focus of wide concern. Avalanches, gully formations, severely eroded stream channels, and a multitude of other phenomena having negative impacts on the land and water resources due to one hurricane became the focus of numerous investigations within days of the storm (Williams and Guy 1973). Catastrophic events, by their very acute nature, become the foci of attention. The human response to these incidents is often immediate. The same is true for human-initiated events that are of a high magnitude and hopefully remain of a low frequency. The Chernobyl disaster and its contamination of land and water resources have been intensely investigated almost from the very moment they occurred. The risks of environmental problems associated with nuclear accidents resulted in remedial actions in some countries even prior to any land degradation problems occurring in their national territory. For example, Denmark has adopted a policy not to build any nuclear power plants, and Sweden is gradually to phase out their existing plants. Low-frequency/high-magnitude phenomena quickly attract attention.

From the land degradation perspective, the far more ubiquitous, low-magnitude climatic and human-induced phenomena produce the majority of land degradation problems. Salinization and the decline in the overall fertility and structure of agricultural soils are two major examples of widespread land degradation resulting from normal, low-intensity actions. The Haitian countryside, with its generally severely degraded soils, did not result from any single event or factor. It is the culmination of a long and still continuing abuse of the land resources to meet the short-term needs of the population (Lewis and Coffey 1985). While not impressive on a case by case level, because of the vastness of the Earth's land areas affected by low to moderate magnitude events, these phenomena contribute to a very significant percentage of the world's land degradation problems.

The Climatic Component

Many aspects of the climatic component have already been introduced within the framework of the soil system. In particular, the roles of the variability, magnitude, and frequency of precipitation have been examined. With regard to land degradation, it is important to remember that the climate of any area is always changing. It is generally accepted that 30-year periods of continuous meteorological data are used to designate an area's "normal" climatic parameters. Since weather variables vary from year to year, a specific area's climatic "normal" depends on the specific preceding weather values of the immediate 30-year period: that is, what is meant by normal is not constant, but is dependent on the average values of the immediate 30-year period.

While the 30-year interval is arbitrary, direct and indirect measurements of climatic variables indicate that the climatic state varies on all time and space scales (Jäger and Barry 1990). Over the past century, a multitude of studies indicate that between 1900 and 1940 the global mean temperature increased; between 1940 through the early 1960s temperature decreased; and since the late 1960s a general warming trend has existed. Overall, this century has experienced approximately an 0.5 degree Celsius increase. According to Kondratyev (1988), the aberration of cooler climates during the middle of the twentieth century was due to the injection of NO_2 (nitrogen dioxide) into the atmosphere from atomic explosions. Other periods of recent worldwide cooling are often correlated with large volcanic eruptions such as Hekla (1783–4) and Tambora (1815). One conclusion that is feasibly inferred from climatic data is that we are in a period of climatic warming. Depending on one's own perspective, it is the result of either natural or anthropocentric activities. However, accumulating evidence indicates that human activities are playing an ever increasing role in climatic change (Woodwell 1990). As climates change, clear impacts on the land resource occur.

In northwest India and Pakistan, significant shifts of the moist/arid boundary have occurred over the past 3,800 years. In this zone, with the climatic shift toward aridity, desiccation of the landscape has taken place (Hare 1977). This has resulted in a general degradation of the land resource that is reflected in lower vegetation cover and overall lower biomass production. In many parts of Africa, the phenomenon of increasing aridity is almost 5,000 years old, even though within this time period there have been short interruptions of moister intervals. The net affect of this climatic trend, especially in the Nile Valley, Sahelian Africa, and other semiarid areas, has been progressive loss of natural savanna grasslands and increasing scarcity of water across a significant proportion of the African continent (Hare 1977).

In California, a short-term increase in aridity began in the late 1980s and continued at least until the winter of 1992–3. This drought had, and will continue to have, an effect on the land resource in a very densely populated portion of the United States. In 1990, California's rainfall was 60 percent of its average, and snowfall was 40 percent of normal. The resulting shortfall in water supply, along with ever increasing water demands due to population increase and economic development, has required strict and expensive water conservation. One aspect of the conservation need is a conflict between urban and agricultural water use. While shortages in water within the urban milieu, to date, only affect lifestyle, water scarcity in the rural sector due to severe irrigation restrictions results in crop losses and land being taken out of production. At least in the short term, biomass production has been reduced due to water rationing.

Some isolated cases of land degradation have been noted due to responses of the land systems to the policy of water rationing. In the vicinity of Palm Desert, lower rainfall along with water conservation have restricted sprinkling on the landscaping vegetation established to meet the aesthetic needs of the inhabitants of the area. Most of the exotic vegetation used in landscaping in the Palm Desert area usually has greater water demands than the native vegetation that it replaced. The curtailment of irrigating this vegetation has caused many of the shrubs to die. This, along with human activities that trample vegetation, has resulted in some sand dune activation in the vicinity of the city. The result is that in some places sand dunes have become active and have moved to the edge of heavily utilized urban lands. As the dunes migrate onto these lands they threaten their utilization (*Boston Globe* 1990). This clearly represents one type of land degradation resulting from an interaction between climatic fluctuations and land development (human versus natural land use).

The response of society to water shortages that occur whenever precipitation amounts are within the lower range expected in this region's climate will determine the future status of many of the utilized lands in the western portion of the United States. Will urban areas increasingly obtain a larger proportion of the limited water in this area? If so, how will farmers respond? Will farm abandonment occur and will the lands revert back to their natural status? Will the water shortages in the Southwest increase the probability of salinization on the irrigated lands in the region, as farmers try to keep their cropland in production with less water?

The stress being placed on the environment in this region by increased utilization of a restricted commodity, water, is a common phenomenon worldwide. For example, it is similar to the expansion of agriculture in the virgin lands of the former USSR, or of livestock and agriculture in the Sahel. As long as the climatic conditions are favorable (precipitation amounts are above average), the systems associated with the expansion of agricultural and pastoral activities work reasonably well. However, there is always the potential for the systems to be stressed when the moisture conditions deteriorate to below average conditions, but are still in the normal expected range of the area's climatic variability. It remains to be seen whether modern economic systems will develop strategies that incorporate the "random" but expected ranges of climatic variability to insure that land degradation is minimized.

The Biological Component

In the early phases of domestication as humans developed agricultural and livestock systems, plants and animals were largely selected to meet the specific human needs as well as the environmental situation in which the farmers and herders resided. During this era, only the agricultural systems that could adapt to the variations in an area's environmental ranges (conditions) survived. The gene pools of the plants and animals in these domesticated systems had a variety of traits that permitted continuity under both good and adverse conditions: that is, both stability and resiliency characteristics were strongly reflected in the gene pools. Production (yield) attributes were only one of many characteristics that were equally critical in crop/animal selection decisions throughout most of history. The diversity in the gene pools of the adopted plants and animals minimized risk at the expense of yield/production. This strategy attempts to insure that a sufficient harvest occurs over the wide range of local environmental conditions. Minimal external inputs were required in these traditional food systems. This strategy of minimizing risk at the expense of yield largely continued through the millennia with only incremental changes, until the end of World War II.

Even today, this approach is dominant in areas of low-resource agriculture (Third World nonexport-oriented), where infrastructures are weak and economic constraints are severe. Yet, it is clear that a worldwide strategy toward specialization had already begun during the nineteenth century as agricultural and livestock systems began to become increasingly commercial. For example, livestock and plants were introduced in Africa by colonial governments to replace local varieties. While the introduced types in many cases were potentially more productive, they usually required a greater degree of human intervention for their survival under African conditions than did the indigenous ones (Chasin and Franke 1980).

During the past 50 years, in contrast to stressing a sufficient harvest over a wide range of environmental conditions, the aim of selective breeding generally has been oriented toward maximizing yield and quality traits for a narrow range of optimal conditions. This trend is likely to accelerate even more in the future with the development of bioengineering, unless a major shift in research emphasis occurs. The vigor traits, which insured sufficient food production for survival in traditional systems under less than favorable conditions, have been largely reduced or removed, and traits leading toward high production in plants and animals under a narrow range of environmental situations have been the focus of most breeding strategies. The success of this approach is based on the assumption that – through technology – humans are able to counter and control environmental variability: that dry periods are offset by irrigation and supplemental water supplies such as through well drilling; pests and fungi are contained through chemical applications; and drainage, other engineering strategies, and fertilizers can "homogenize" terrain conditions to meet the critical environmental demands of these high-yield varieties.

The stress on production and environmental control has resulted in breeding strategies producing cultivars (plants) and stock (animals) that possess only a few of the traits found in the original species. These new strains, which result in highly specialized varieties, have a minimal genetic variation. Under the specific conditions for which the cultivars and stock are bred, high production results in the traits favored by the breeding strategies. While very little increase in the overall yield of dry matter has resulted from the introduction of new varieties of grains, the new varieties do produce a greater proportion of dry matter in the form most useful to our needs. For example, some modern wheat varieties have a large percentage of their dry matter made up of the head or seed component of the plant. Concomitantly, the plant's stock or straw segment has been significantly reduced, offsetting this increase. Thus, while no real increase in total plant yield has resulted, food production has been increased by the breeding strategy.

The green revolution largely was based on this general strategy of increasing the harvestable component of specific grains: that is, the work of geneticists focused on the redistribution of plant dry matter aimed at increasing the grain:straw ratio. In some areas, the introduction of these "miracle" grains has resulted in increased harvests but shortages of straw. One ramification of these shortages is that in some locales there are now constraints on the amount of vegetal matter available for mulches used in soil erosion control, increasing soil moisture, or fodder for livestock. Furthermore, in many cases the new varieties require fertilizers in order to attain their high yields, and they produce optimal yields only under a limited range of environmental conditions. Thus, when climatic conditions vary from the "ideal" conditions, or economic constraints limit environmental control (e.g., fertilizers and spraying are eliminated), production from these cultivars and stock decline drastically. The removal of the diversity in the plants' and animals' gene pools have greatly reduced their ability to adjust to less than optimal conditions. Stability has been sacrificed for yield. Modern agriculture based on high energy inputs of fertilizers, mechanization, chemical additives, engineering structures, and supplemental water clearly is not the cure-all that everyone hoped for during the early days of the green revolution. In fact, in places, the strategies required for success in utilizing this conceptual framework contribute to land degradation.

In the United States, to meet the increased investment needed for success in this modern form of agriculture, farms have gradually increased in size. The increasing size of the farming unit has taken place with a simultaneous increase in the size of farming machinery. While this efficient machinery permits an individual to increase the amount of land that can be farmed, it often results in less environmentally sound farming practices. Being less maneuverable than the smaller, economically obsolete equipment, one result of the shift to larger machinery has been that fields are cultivated relatively independent of the spatial variability of physical conditions found within individual fields. This is one factor contributing to the continuing soil erosion problem in the United States.

Another factor contributing to environmental problems is that crop rotations have largely been replaced by increased use of fertilizers and herbicides. Rarely do farmers implement strategies that vary the application of these materials to the site-specific environmental conditions found in individual fields, such as soil variability. To insure maximum yields, in most cases, these materials are overapplied. The runoff from the fields usually includes a higher than normal concentration of soluble components from these substances. Rural farm runoff is a major cause of nonpoint pollution that affects water quality in many farming areas of the United States.

Most natural land ecosystems are efficient in recycling nutrients. Small quantities of soluble minerals entering the water supply from any specific area are the norm. However, modern farming practices of monocropping and heavy applications of soluble materials often result in both increased soil erosion and excessive quantities of nutrients entering surface water systems. The erosion results from the soil not being protected by plant cover after the harvest; the increase in nutrient delivery results from the excess amounts of fertilizer application. The net effect of these two phenomena is that water bodies in many locales are experiencing increases in their sedimentation rates as well as nutrient enrichment.

These conditions facilitate the growth of phytoplankton and make the water more turbid. Not only do these changes decrease the amount of sunlight reaching the deeper aquatic plants, but the rapid growth of phytoplankton also results in accelerated amounts of detritus. The net effect of these changes is a reduction in the amount of dissolved oxygen in the affected water bodies. While total productivity within the bodies of water is not necessarily lowered by these phenomena, the productivity of plants and animals that humans desire, as well as the overall water quality, usually decreases (Nebel 1990). Eventually, the excessive delivery of sediment and concentrated nutrient supply can result in eutrophic conditions. As water is a critical component in most land systems, the result of the decline in the water resource represents a form of land degradation. When the eutrophic conditions affect reservoirs, especially in areas that need the water for irrigation, the reduction in water supply represents a direct link to land degradation. Biotic impoverishment as a result of diverse human activities is increasing globally (Woodwell 1990); land degradation is one result of this trend.

The Geologic/Tectonic Component

Tectonic forces are those that originate in response to internal forces within the Earth. The two major types of tectonics involve: (1) solid Earth movements (diastrophism), such as uplift and subsidence of the Earth's crustal materials; and (2) liquid Earth movements, such as volcanic activity. Catastrophism, the sudden and violent events resulting from internal forces within the Earth, has been a component of Earth history since almost its formation (Huggett 1990).

Recent manifestations of catastrophic tectonic activity that have implications for land degradation are the volcanic eruptions of Mt Unzen and Mt Pinatubo. While both mountains are in highly active tectonic zones, because the volcanoes had been dormant for a significant period, the inhabitants had developed intensive use of the lands surrounding them. The ash deposits resulting from the Unzen and Pinatubo eruptions have covered a significant amount of land in the immediate area of the volcanoes. On the farmlands surrounding Mt Pinatubo, the ash has buried the rice fields in this very fertile area of Luzon. This has destroyed the harvests for the year. Furthermore, it has necessitated, at least in the short term, the abandonment of a large proportion of this bread basket region's lowland farms. On Pinatubo, large amounts of pyroclastic materials, volcanic materials thrown into the air during eruptions, were deposited over hundreds of square kilometers in a 30 km (20 mile) radius of the mountain. On the steeper slopes this created unstable conditions. These ash deposits, interacting with precipitation and the steep slopes, triggered landslides that have degraded the affected hillslopes and altered the morphology of many lowland areas. In both Japan and the Philippines, the restoration of the affected areas to their former productivity will require time for the soil-forming processes to modify the parent material, as well as for the flora and fauna to reestablish themselves. From the human perspective, major investments in both capital and labor are needed before the existing land-use systems will attain their former levels of productivity. Clearly, in the short-term, and perhaps longer in many cases, agricultural activities will decline in the areas that received heavy deposits of volcanic ash from these eruptions.

These recent volcanic eruptions illustrate an impact of ordinary tectonic processes on the status of the land resource. The magnitudes of these eruptions have occurred frequently in the past and will continue to take place in the future. Because of the interactions among an area's climate, topography, and the nature of the materials involved, the relation of volcanic activity with land degradation is highly variable. With the articulation of plate tectonics theory, both the areas that will likely experience significant tectonic activity and the nature of this activity can be predicted in general terms (Cox and Hart 1986). The zones that are most susceptible to violent activity and hence degradation are the areas in which plate collision and continental-type materials are juxtaposed. In these settings, the eruptions are likely to be explosive and to be associated with large quantities of ash. This increases the probability of critical components of the land systems being destroyed with an eruption. Pompeii is such an example from the ancient world; the eruptions of Unzen and Pinatubo are contemporary examples. One result of these cataclysmic alterations is a degrading environmental condition from the perspective of the areas' inhabitants.

In contrast to volcanism in areas of plate convergence, eruptions associated with plate divergence are far less explosive, since a greater proportion of the materials being released during eruptions under these conditions occurs as lava flows, with a smaller proportion of pyroclastics. Nevertheless, destruction of the land resource is still possible in these settings. With the 1973 eruption of Helgafell in the Vestmannaeyjar Islands of Iceland, the lava flows from this eruption covered the surrounding farms and pastures under meters of lava (Gunnarsson 1973). The formerly productive pastures that provided grazing for livestock were replaced by freshly deposited basaltic materials. Because of the prevailing cool maritime climate of this area, the weathering of the lavas is slow. This has resulted in bare rock replacing soil as the surface material on the former pasture and farmland. Thus, for the foreseeable future, pasture and farmland have been destroyed. Depending on your perspective, the identical volcanic event can be viewed as either constructive or destructive. From the farming perspective, land degradation has occurred as former farms became totally destroyed. From the perspective of the area's fishing industry, the eruption improved the local environment because some of the lava deposits improved the harbor's capability to protect the fishing fleet from storms. From the geologic viewpoint, almost all volcanic activity is constructive, since it increases the amount of earth materials above sea level; and this counters the erosional processes. However, from the vantage point of this text, most volcanism likely results in at least a short-term decline in the utility of an area for most human activities.

Not all tectonic activity having land degradation ramifications is associated with catastrophic events. The continuing slow submergence of many coastlines results in increasing coastal flooding. Eventually these lands are lost to the marine environment. Such is the case in many zones along the East Coast of the United States, where downwarping of the coasts at rates between 1 and 5 mm/year is widespread (Brown and Oliver 1976). While engineering works can delay this process, eventually they must fail and land will be lost to the oceans.

Slow tectonic movements are not restricted to coastal environments. In central Africa, especially between the eastern and western Rift Valleys, slow crustal warping has resulted in alteration of the regional drainage patterns. In the Lake Victoria watershed, formally well drained areas have become wetlands as a result of local tectonics (Lewis and Berry 1988; Beadle 1974). However, because of the slow nature of these changes, which occur over a time period far greater than a human lifespan, changes in the productivity of these lands do not alter established land use and the activities of the region's inhabitants.

Summary

Natural land systems exist under a wide variety of geomorphic, climatological, and biotic conditions. Despite the controls and the variability and magnitude of phenomena that each of these three domains place on any given land system, most lands are resilient to stresses resulting from natural phenomena. Lands rarely degrade under natural conditions except when catastrophic events occur.

In more cases than not, when humans alter a natural land system they reduce its biotic diversity. The numbers of species found in human-disturbed areas are almost always less than under their natural state (Cairns and Pratt 1990). Similarly, the strategy in most modern agricultural systems has been to reduce biotic diversity and to have minimal genetic variation in the crops and stock emphasized. Crop rotations are replaced by artificial supplements and specialization in crops and livestock replaces diversity. Often, the trend toward homogeneity by human actions decreases the ability of the land system to respond to the variability of natural phenomena to which it is exposed. Prior to the establishment of farms in the American farming belt, a good groundcover existed in the spring of the year. Thus when intense spring rains occurred, the soil was protected and soil erosion occurred within tolerable rates. The likelihood of land degradation was minimal under these conditions even when a rare severe event occurred.

In areas of agricultural expansion in which minimal economic investment is available to offset the deleterious changes resulting from vegetation alterations, the probability of excessive impoverishment in the overall biotic system increases. Impoverishment triggers fundamental changes in the soil and water systems, and often results in a degradation of the land.

It is not inherent that, as humans alter land systems for their own needs, the land must degrade. Soils are a renewable resource and with suitable management strategies they can remain resilient to the expected ranges of an area's climatological and geological phenomena. The deforestation in Amazonia need not result in the degradation that all too often has occurred, and is still occurring (Hecht 1989; Fearnside 1990). With proper strategies that are related to the specific environments found in the rainforest, many areas could be developed in a sustained manner (Ruttan 1989).

This chapter has illustrated that a range in variability and magnitude of both climatological and tectonic phenomena are expected in many areas. To prevent land degradation from taking place, it is essential that strategies be developed that can absorb the normal expected range of these phenomena. Furthermore, it is crucial to identify the spatial variability of the crucial physical attributes in any area. For example, soils are not constant even in small areas;

but their variability is not random. Systematic patterns do exist with topography. The characteristics of an area's physical domain need to be understood, and the constraints that they place on the long-term use of an area need to be incorporated into any strategy if land degradation is to be prevented.

FIVE

Unintentional Destructive Change in Interactive Systems

Introduction

In almost any climatic and topographic setting in which there is human occupancy, regardless of the prevailing political and demographic situation, some land degradation is likely to occur. This widespread occurrence of land degradation, which not only threatens the quality of life but in some cases actually threatens the viability of the affected areas, cannot be explained by solely examining the physical, chemical, and biological processes and their interactions that result in land degradation. These natural processes are both reasonably understood and articulated. Furthermore, the technological interventions required to prevent or reverse the processes of land degradation already exist for most cases. In the agricultural, soil science, hydrological, engineering, and biological literatures, numerous strategies and techniques are well documented that are successful in arresting land degradation in the land, water, and vegetation domains from the technological perspective (OTA 1988). Yet, worldwide, it is clear that destructive processes are occurring at a higher magnitude and rate than the restorative and conserving interventions (Brown 1992). This fact alone emphasizes the fact that land degradation is as much a human behavioral problem as a physical/technical one. Human factors that are contributing to the degradation of our land, water, and plant resources under a wide range of diverse environmental settings are the foci of this and the following chapter.

Human Attempts to Stabilize Natural Dynamic Systems

Variability

The geologic, historical, and contemporary records all indicate that variability is a natural phenomenon in most Earth systems. From the planning and management perspective, variability creates numerous problems. A result of this is that humans are almost always attempting to minimize this variability in most of the Earth systems that they utilize in order to improve specific outcomes in areas as diverse as food production, energy generation, transportation, and water supply.

Depending on the environmental, economic, and technological situation, to increase food production humans have invoked numerous strategies to cope with dryness, wetness, steep lands, and variability in soil fertility. Improvements in transportation almost always result in the need to alter geomorphological phenomena in order to meet specific needs of the various modes of transport. Rivers, harbors, hillslopes, and valleys all have been, and continue to be, modified to meet the specific needs of transport systems. Any urban area requires a host of inputs and outputs in order to exist. These include basics such as food, water, and energy. Successful urban areas must minimize variability in the supply of these basics. Furthermore, urban areas inherently modify the natural environment by their very existence. As but one example, pavement and other construction processes alter infiltration and hence change critical aspects of the hydrologic cycle. Urban areas, regardless of their location, alter the vegetational, geomorphological, hydrologic, and atmospheric realms both directly and indirectly.

Most human systems, regardless of the environmental, economic, and cultural setting, attempt to minimize the variability of natural phenomena as a strategy by which to improve their livelihood. Because of the feedback that exists among the components of natural and human systems, a change in any component to meet specific direct goals often results in triggering a number of adjustments. Unfortunately, many of these adjustments result in land degradation. All too often, strategies emphasize altering only the components of greatest interest. In most strategies it is either assumed that the other variables will remain relatively constant or the components of the system that are not of direct concern are just ignored. In other cases, humans just neglect the ramifications of their environmental "control" because the changes that they cause are inflicted on peoples and places outside their immediate sphere of interest.

For example, much of California experiences a highly seasonal pattern of precipitation. The summer months are generally dry, as precipitation is concentrated in the winter months. One strategy initiated to meet urban and rural water needs throughout the year was to build a number of dams on the upland headwaters. The water captured behind these barrages and stored in the reservoirs during the periods of precipitation is then available for use during the drier months. In the process of capturing the rivers' waters, much of the river sediment that formerly reached the Pacific remains permanently trapped in reservoir sediments. Prior to dam construction, some of the sandy materials transported in the rivers reached the California coasts. This sediment was not only available for potential deposition, but actually comprised a significant supply of the materials deposited on the coastal beaches. With the deposition of these materials now in the reservoirs and not along the coast, beach deposition has been curtailed, but the erosional agents, such as waves, have remained constant. Without the deposition to offset the erosion, many of California's coastal beaches have degraded. The previous relations between coastal deposition, which were partially dependent on the California river systems for material supply, and coastal erosion due to waves and currents, were upset due to water demands of the population and the human response to satisfying these needs. Thus, the success of meeting a specific societal goal, in this case water demand, resulted in the unintentional degradation of some California beaches quite some distance removed from where the alteration in the fluvial system took place.

Environmental deficiencies

In order for living organisms to exist, minimal energy, nutrient, and material requirements must be available in the environment in which they are situated. Different organisms have different requirements. Likewise, any environment possesses only a limited range of energy, nutrient, and material attributes. Humans have developed a multitude of strategies to overcome the natural constraints of any given area in order to meet human demands.

Dating back to prehistory, one of the earliest modifications that continues to the present is that forests are cleared and replaced with crops. This alteration in the vegetation cover has resulted in converting the food energy of the forest, which is largely indigestible, into increased production of digestible foods such as fruits and seeds. Another widely utilized strategy having ancient roots is the transfer of water onto drylands to ameliorate the moisture constraints of these areas. Originally these transfers were local and utilized local groundwater or exotic rivers, such as the Nile, flowing through an arid land. Today, agricultural demands often result in water being diverted from river basins hundreds of kilometers from the irrigated areas. These water transfers onto drylands, supplementing local water supplies, permit both crop substitution and establishment of highly productive pastures. Increased yields in both agriculture and livestock have resulted from this tactic. This has resulted in significant food production increases for human consumption from many of the world's semiarid and arid lands.

In the area of food production, the success of the various vegetation change and water transfer strategies that have evolved for the wide ranges of environmental settings is reflected in the growth of human population. In the 10,000 years since the beginning of agriculture, the world's population has grown from a few million inhabitants to the billions of the contemporary world.

Worldwide today, there is not very much prime additional land that can be utilized to meet increasing food demands. The remaining nonagricultural lands, including former agricultural lands that have been abandoned, clearly are not lands with naturally high agricultural potential. Most remaining forested lands are:

- situated in fragile environments that have limited potential for crop production
- are needed to meet our needs for wood production, recreation, or to satisfy our desire to maintain wildlife habitats
- required to protect water supply for other urban and rural demands

In moisture-constrained habitats, the overwhelming majority of the remaining nonirrigated drylands have either poor soil conditions that negate irrigation or are too distant from an adequate water supply.

An additional factor contributing to limitations for areal expansion onto new lands to meet increased food demands is the reality that there is a continuing significant loss of land from rural activities due to urban expansion. Often, the lands being consumed for urban uses are potentially highly productive agricultural areas rather than marginal lands. Because of greater economic return in the urban sector, prime agricultural lands continue to be lost for nonagrarian uses. With existing land and environmental constraints, it is clear that future increases in agricultural and livestock production needed to meet increasing food, energy, and other material needs will largely have to be met through intensification of production on existing lands rather than areal expansion.

Intensification strategies often only utilize a limited number of the components of a natural system. Today, as humans utilize a segment of a natural system for their own purposes, on most lands they need to add supplements to the existing land system to meet their production requirements. For example, cultivating the same crop on the same land for more than one year in succession, a practice called monoculture, has become widespread in the farming systems of the United States. This is a practice that has been favored by mechanization, better crop varieties, and the development of chemicals that protect the crop from pests and weeds (Power and Follett 1987). Unfortunately, the successful monoculture of corn (maize) places high nitrogen demands on the soil. These demands are far in excess of what most soils can sustain, even in the short term. Thus, one common by-product of corn intensification is the need to add fertilizer supplements continuously to the soils to counter the limiting constraint of nitrogen shortfalls as well as of other crucial minerals.

Seed, plant, and animal varieties used in any locality must cope with the area's environmental constraints: that is, to achieve needed yields by utilizing an intensification strategy, alteration of or supplements to the existing environment are almost a requirement to counter an area's limiting factor(s). Two widespread intensification strategies to offset an area's limiting factors are to alter the environment and/or through breeding and biotechnology to develop hybrids. A variety of practices have evolved to alter the physical environment to make it better suited for the plant varieties and livestock that are desired for production. Examples of this approach include application of fertilizers, herbicides, insecticides, land leveling, drainage, and irrigation. While having beneficial effects in increasing food production, agricultural supplements are one of the major causes of water quality problems in the United States and elsewhere (i.e., northern Europe; James 1993).

The second major strategy is, through both plant and animal breeding or biotechnology, to develop varieties that can better adapt to the environmental constraints found in an area. Examples of this include the development of virus- and drought-resistant plant strains and the improvement of specific genetic aspects of animals, such as developing a larger goat. In the process of attempting to overcome specific environmental deficiencies affecting production, the strategies adopted often result in feedbacks into the environment that alter the *status quo*. In many places, the feedbacks resulting from the "improvements" that were intended to counter a specific locations' limiting factors have resulted in land degradation. In the remainder of this chapter, the unintentional destructive changes resulting from alterations of the Earth's lands to offset both environmental deficiencies and variability while meeting agricultural/livestock demands are examined in the context of land degradation. In chapter 6, human actions related to transportation, mining, and urbanization to counter both environmental variability and deficiencies, and some resulting environmental responses that eventuate in unintentional land degradation, are examined.

Food Production

Human efforts at food production strive to generate the maximum amount of sustenance for the maintenance of *Homo sapiens*. The success of these efforts is reflected in the dramatic increase in the size of the human population during our species' 3.5 million or more years of identifiable existence on this planet. Despite a general record of success, local failures in food production strategies and the social, economic, and political systems that are intimately intertwined with them have threatened the sustainability of the ecotechnologies that support existing populations. This threat of catastrophic collapse often originates in the very effort to manage and increase food production resources. Many examples of the unintentional destructive consequences of food production intensification exist. In this section we examine three instances of inadvertent *destructive creation*. Irrigation, animal husbandry, and rainfed agriculture constitute prime examples of intensified resource use gone awry in many areas. In these examples, solutions to the problems exist, and the final chapters in each tale remain to be written. Thus the cases cited represent stories in progress, with an uncertain outcome, rather than conclusive events.

Irrigation

Irrigation agriculture is an unfortunate example of the difficulties that arise when efforts at intensification of food production in dry environments result in unexpected negative consequences. We examined an historical example of this process in ancient Mesopotamia in chapter 2. In this section, two contemporary examples – the problems of which are generic and appear in different configurations in many parts of the globe – are examined. These are the efforts in Egypt to exploit more fully via intensive perennial irrigation the water and land resources of the River Nile, and the checkered contemporary experience of groundwater development for center-pivot irrigation in the North American High Plains.

Irrigation in the Nile Valley
A glance at the map (figure 5.1) indicates the importance of the River Nile for the sustenance of Egypt's large and still growing population. As has been the case since Biblical times, nearly all of Egypt's population is concentrated in approximately 3 percent of the national territory. Although the contemporary cultivated area has increased to ca.35,000 km^2 (ca.13,500 sq. miles) (Baines and Malek 1980: 16; Fisher 1993: 360), a 30 percent increase over that available to the ancient Egyptians, the total dependence of Egypt on the waters of the Nile remains the same today as it was 5,000 years ago. An elongated corridor of cultivated green in the delta and entrenched valley of the Nile stands in verdant contrast to the desolate desert, the expanse of which is only occasionally broken by flecks of oasis cultivation. The 27,000 km^2 (10,400 sq. miles) area of the delta, valley, and Faiyum depression constituted the maximum extent of cultivation in 1882 (Butzer 1976: 83). Approximately the same area made up the Greco-Roman ecumene when the ancient population reached its maximum size. This land area probably reflects the maximum extent of agricultural land use that could be achieved with a nonfossil fuel economy.

The Nile regime before regulation   The River Nile's hydrologic regime is extremely regular. Summer rainfall in the highlands of East Africa is the source of the Nile's streamflow. The White Nile, originating in the Lake Victoria basin, is the source for the Nile's base flow, while the Blue Nile, with headwaters in Ethiopia, provides the bulk of the Nile's annual flood. At Khartoum

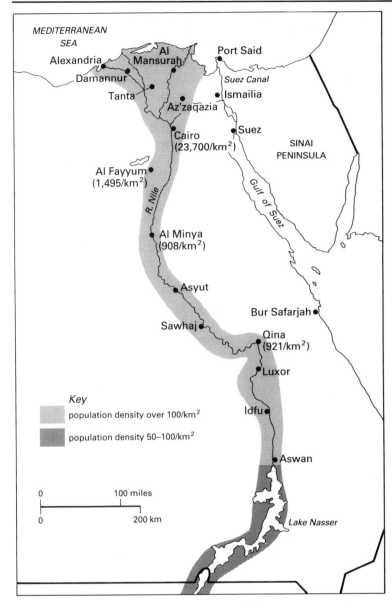

Figure 5.1
The population density in the Egyptian Nile Valley (after Munro 1988)

the two Nile branches join, and for its course across the Sahara to the Mediterranean the Nile rarely receives any appreciable additions of ground or surface waters. Usually, the flood reaches Cairo in August, peaks in September, and declines precipitously in volume in October and November. From February to June, the Nile is at low ebb, a mere trickle of its floodstage self. Ancient Egypt's political authority and social stability were directly tied to the regular and predictable occurrence of the Nile flood.

Until the nineteenth century, Egyptian life and livelihood was directly dependent on the annual, unregulated Nile flood. Farmers utilized the floodwaters by means of a basin irrigation system (Willcocks 1889). In this system, the Nile functioned as both main water delivery canal and as principal drainage ditch. During each flood episode, the floodwaters breached or overflowed the river's natural levees and inundated the adjoining floodplain. Because the Nile floodplain is extremely narrow throughout most of the river's course in Egypt, in most years the majority of the plain was inundated by the flood. At the height of the flood, only higher levee tops, village sites which were always situated on high ground to avoid floodwaters, and distant portions of the floodplain remained above water. The meander scars of abandoned stream-courses retained water for longer periods once the flood began to decrease. Farmers learned to augment this floodwater retention capability by erecting low, linear mounds that carved the flood-plain up into a series of smaller basins. The increased water retained in the basins enhanced the amount of infiltration that occurred and increased the soil moisture store available for agriculture. Annual floods also distributed substantial amounts of silt on the fields. These silt deposits were important for agricultural sustainability by both maintaining soil texture and supporting soil fertility.

The entire agrarian system was simplicity personified. Little physical infrastructure was needed to bring water to or to drain water from the fields. When the crest of the Nile flood passed, the river channel served as the main drain by which the receding waters were removed from the fields. The natural rhythm of riverine ebb and flow established the parameters of the region's single cropping season. For once the floodwaters receded, bringing supplemental moisture to fields was a monumental task, beyond the technological and organizational abilities of the Nile's early inhabitants. During the low-flow periods, most fields were too high above the level of the river and too distant to make raising the water possible.

Yet the desire to intensify crop production and to improve on the productive capabilities of the natural regime were evident at an early date. Attention initially focused on developing double cropping along the levees adjacent to the river. The *shaduf* (a hand-operated lever action water-lifter), first introduced sometime between 1800 BC and 1250 BC, enabled about 10 percent of the floodplain to be double cropped (Butzer 1976: 46). Adoption of the *sagiya* (an animal-driven water wheel) in Ptolemaic Egypt enabled larger quantities of water to be lifted higher, and permitted a further 10 percent of the floodplain to be cropped in the low-flow period. Beyond these limits, areal expansion and intensification was unable to proceed until the nineteenth century. While Nile silt increments were sufficient to maintain soil fertility in the areas inundated only with the Nile's annual flood, especially if grains and legumes were rotated, additional use of night soil, pigeon droppings, and fallow was essential to maintain the fertility of the double-cropped areas close to the Nile (Butzer 1976: 89ff). Additional gains in productivity depended on the manipulation of farming techniques and the introduction of new crops (Watson 1983) within the constraints of the time-honored basin irrigation system.

The Nile after regulation   Completion of the High Dam at Aswan in 1967 was the latest stage in a century and a half of effort to escape the constraints imposed on agricultural productivity by the natural river regime and the basin irrigation system. Population growth, particularly in the twentieth century, placed pressure on Egypt's limited agricultural lands, and made it easy to regard the floodwaters that escaped unutilized into the eastern Mediterranean as "wasted." Moreover, although the Nile river's flood regime was conspicuous in its regularity, low-flow episodes have occurred that resulted in food shortfalls (Bell 1975; Riehl and Meitin 1979). For periods as long as a century, lower Nile flows had devastating impacts on the integrity of Egyptian society and population (Butzer 1976).

Traditional Egyptian basin irrigation was vulnerable to the impact of low River Nile flow for two reasons. First, the high degree of reliability of normal-magnitude floods discouraged investment in the elaborate infrastructure of canals, diversion structures, and drainage ditches that would be needed in order to extract irrigation water by gravity flow from the Nile. Second, existing water-lifting technology was too limited in scope to irrigate a sufficient area to compensate for low-flood episodes. Storage of food reserves to offset lean years was the strategy adopted to cope with production shortfalls in low-flow years.

Efforts to increase the productivity of Egyptian agricultural space began in the 1840s with the construction of a barrage at the head of the delta. Other dams and barrages followed elsewhere in the delta and floodplain. The primary aim of these structures was to divert river water to the fields during the low-flow months. In accomplishing this, Egypt was able to increase agricultural production by planting more than one crop each year on the same parcel of land. The basin irrigation system began to disappear and was replaced by a perennial irrigation regime.

The Aswan High Dam was the latest stage in a century and a half long process. The High Dam at Aswan went well beyond previous efforts to regulate the Nile regime and increase its agricultural productivity. Its aim was to store the floodwaters that previously had passed "unutilized" and "wasted" to the sea. By releasing this water throughout the year, the intention was to increase the multi-cropping of Egyptian farmland and to extend the perennial system to all existing cropland. In addition, new land, particularly to the west of the delta, was targeted for development with the newly saved Nile water. Hydroelectric power generated at the dam was expected to dramatically increase the country's available power, and to serve as a magnet for power-hungry fertilizer, chemical, and processing industries.

Although largely foreseen, the environmental impacts of the Aswan High Dam are considerable. Examination of those impacts has dominated discourse, and the positive contributions of the dam to Egypt's industrial development have received less attention. Ideology, emotion, and the sheer magnitude and complexity of the problems involved undoubtedly contribute to this imbalance. Here we contribute to the one-sidedness of the debate by discussing only those land degradational impacts that result from the High Dam's construction.

The land-degrading impacts of the High Dam are of two types: (1) those that result from the establishment of *sacrifice zones* as a consequence of the dam's construction; and (2) problems that have emerged to threaten both short- and medium-term sustainable agriculture in Egypt's *critical agricultural zones*.

Several *sacrifice zones* are associated with the High Dam at Aswan. The most immediate of these is the portion of the Nile valley now inundated by Lake Nasser (Lake Nubia in the Sudan). Rich riverine agricultural lands were flooded and over 100,000 Nubians were relocated (White 1988: 7). While health conditions in the new agricultural settlement schemes may have constituted an improvement, it is not clear that the productivity of the newly developed lands matched that of the drowned fields. Certainly, the Nubians' attachment and commitment to their new areas was considerably less than to their traditional homeland (Fernea 1973). Substantial change took place in the plankton and fish populations of the Nile once the dam was closed, with many species disappearing. In the short term, at least, the fish catch, largely *Tilapia* and *Alestes*, has exceeded expectations, but this has been at the cost of the diversity of the fish population, and may also be placing serious strains on the replacement capacity of the species most commonly exploited (White 1988: 9). In contrast, the seasonally fluctuating edge of Lake Nasser has created a new habitat for wildlife, and the same water and vegetation resources have attracted nomadic herders. In sum, lands sacrificed to the construction of the dam and its storage reservoir have had significant social and economic opportunity costs (particularly in moving or protecting the region's most significant archaeological monuments), but have been at least partially offset by gains elsewhere. This is not surprising, since the places and populations affected are relatively small in scale and planning for compensatory development was integral to the process of erecting the High Dam. As one moves further from the site of the dam, it becomes easier to sacrifice more distant populations and environments for the expedient interests of the beneficiaries of the project.

Still reasonably close to home, but more distant from the dam site and therefore more difficult to address, are the coastal erosion problems that have emerged since the dam's construction. The threat of coastal erosion was first articulated to a wide audience by Mohammed Kassas (1972), who contrasted the millennia of delta growth through deposition with the post-Aswan environment.

It was the loss of sediments now trapped behind the High Dam that starved the delta of essential building material. Without an annual supplement of silt and sediment to maintain land levels and to provide material for the construction of coastal bars and beaches, the sea began to advance (Hefny 1982; National Geographical Society 1992; Stanley and Warne 1993). A rising sea level due to global warming also appears to be implicated in the delta's problems (Milliman, Broadus, and Gable 1989). White (1988: 36) indicates that coastal erosion appears to be restricted to a few districts only. However, when these menaced areas are the narrow barrier bars and spits that separate the delta's brackish lakes and lagoons from the open sea, not only is the loss of an important inland fishery threatened but also the breeding function of the coastal lakes for the marine fishery is lost. Moreover, conversion of the lakes and lagoons into open arms of the sea would place at risk the extremely low-lying reclaimed agricultural land at the northern fringe of the delta (Kassas 1972: 187). Today the northeastern portion of the delta is sinking at an annual rate of 0.5 cm. This subsidence has existed for over 5,000 years, but in the past it was offset by the deposition of Nile sediments. With dam building, especially the Aswan High Dam, the sediment supply was curtailed. If current rates of deposition, coastal erosion, and subsidence continue, it is estimated that by the year 2100 the delta's coastal margins will retreat between 20 and 30 km (12.5–19 miles) inland from west of Alexandria to east of Port Said. Up to 26 percent of the most productive rural and urban areas within Egypt could be lost if present trends were to continue for the next century (Milliman, Broadus, and Gable 1989: 343). The "emergence" of this *sacrifice zone* due to policies aimed at increasing agricultural production in the Nile Valley needs to be addressed (figure 5.2).

An example of more distant degradational impacts is that of the fishery losses that have occurred in the eastern Mediterranean.

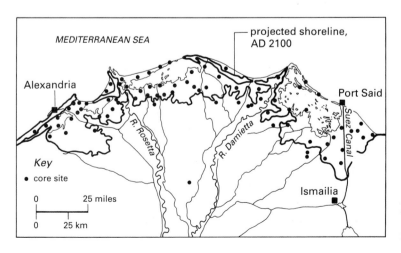

Figure 5.2
The projected submergence of the Nile Delta

Annual fish yields that in the mid-1960s were 31,000 tons had declined to 11,200 tons in 1982, while the sardine fishery experienced a five-fold decrease from 15,000 to 3,000 tons per year (White 1988: 34). These changes were already well advanced within a few years after the completion of the Aswan High Dam (George 1972). White also reported a 30 percent decrease in the pre-dam shrimp catch. All of these declines are attributed to the elimination of Nile floodwaters reaching the Mediterranean. The absence of mineral and organic nutrients flushed by the Nile flood into the Mediterranean starves marine food chains of fundamental resources at their base. Decreases in yield are an inevitable consequence. The increased salinity of the Mediterranean in the vicinity of the delta, undiluted by the Nile's fresh water, changes the habitat for many desirable commercial species and contributes to their decline, just as it encourages the invasion of more salt-tolerant Red Sea species, for whom there is much less market demand. Because Egyptian fishermen were only a minor component of the Mediterranean fishery, most losses experienced in the Mediterranean were absorbed by non-Egyptians. Because Mediterranean fish harvests were offset by an increase in the fish catch from Lake Nasser, which the Egyptians largely controlled, it was easy to treat the eastern Mediterranean as a *sacrifice zone* of only minimal importance.

The same situation applies to Egypt's efforts to enhance the amount of water available for consumptive use. In excess of 10 percent of the water stored by the High Dam is lost each year to evaporation and to seepage into adjacent groundwater aquifers (White 1988: 8). This loss of water to nonproductive uses constitutes a sacrifice imposed by the decision to build the dam. The loss is especially significant because both Egypt and the Sudan are approaching full use of the River Nile waters allocated to each by the existing international agreements (Waterbury 1979; Haynes and Whittington 1981; Whittington and Haynes 1985), although the imminence of the crisis has been questioned (Adams 1983). Impending limitations in water supply have spurred plans to develop and control the water resources of the entire basin. Many of these plans envisage over-year storage of water in the lakes of highland eastern Africa (Tana, Victoria, and Mobutu) and the reduction of water losses *en route* to Egypt (Whittington and Guariso 1983: 185–98).

The first large-scale effort was directed at the Sudd, a region of vast swamps in southern Sudan covering more than 5,000 km^2 (1,930 sq. miles) (figure 5.3). Sustenance for the marshes of the region, and of the agropastoral livelihood systems of the region's inhabitants, requires 14 billion m^3 (18 billion cu. yd) of White Nile water annually (Waterbury 1979: 89ff). From the standpoint of Egyptian water managers, this is a wasteful use of water needed to meet Egypt's future water requirements. Moreover, plans to store water in highland lakes, far from the high evaporation rates of the Egyptian Sahara, make little sense if the water gained thereby is

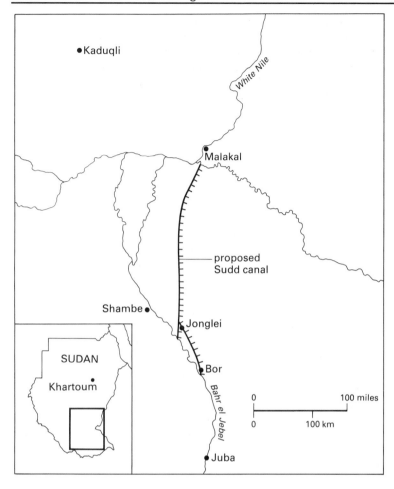

Figure 5.3
The proposed Jonglei Canal

filtered through a giant swamp, which acts like a huge sponge, soaking up water on its way to Egypt. The solution proposed was the Jonglei Canal, a bypass structure intended to divert water around the Sudd with minimal losses. The expectation was that the project, when completed and allowing for evaporation and in-stream losses along the way, would deliver 1.9 km^3 (2,500 cu. yd) of new water at Aswan (Whittington and Guariso 1983: 49). To date, the Jonglei project has not been completed. Opposition by local groups and the continuing civil war in the southern Sudan have created sufficiently unstable conditions to halt the project, at least temporarily. For the time being, the profound transformation that the Jonglei Canal would have wrought has been prevented from converting the Sudd into a *sacrifice zone*. Additional plans currently under discussion would extend the Jonglei project, regulate the Machar Marshes on the Ethiopia/Sudan border, and initiate similar control practices along the Bahr el-Ghazal (Whittington and Guariso 1983: 50ff), all of which would require international agreements between the respective countries, and all of which would extend the *sacrifice zone* concept to new areas distant from Egypt.

The *critical zones* that nourish Egypt's large and growing population have also experienced degradational pressure since the construction of the Aswan High Dam. In the effort to make the Nile's agricultural soils more productive, processes have begun that destabilize and degrade these environments. Because Lake Nasser acts as a huge sediment trap, Nile water now contains a markedly lower level of dissolved solids and has reduced turbidity (Whittington and Guariso 1983: 87). Sediment declines from 1600 ppm to 50 ppm are reported (White 1988: 11). Not only does this change in stream characteristics have a severe impact on the number and types of aquatic organisms that can survive in the new conditions, but also the river possesses much greater potential to erode. Thus, both channel degradation and bank collapse have occurred, erosive processes that threaten the functions of dams, barrages, and irrigation canal intakes. Costly corrective measures to protect essential structures are required to offset the ramifications of this channel instability.

Within the floodplain irrigation districts themselves (Grove 1982), and the desert reclamation areas to which Nile water is transferred (UNDP/FAO 1978), a litany of complications such as salinization, waterlogging, silting, and alkalinization threaten these *critical zones*. Kishk (1986: 228) estimates that from one-third to one-half of all irrigated land experiences salinization problems. Compounding these difficulties, the availability of water throughout the year leads inevitably to drainage complications. Efforts to extract multiple crops from the same field each year provide neither rest for the land nor opportunities for the land to dry out. In many instances not even the most careful management can prevent a rise in groundwater level, and some areas in Upper Egypt have experienced water table rises of almost 2 m (2 yd) (UNEP/GEMS 1991: 14). The delta has had less substantial water table rises, but in Lower Egypt the water table is closer to the surface, so that even a modest rise is a potential crisis. Waterlogging and an elevated water table often are closely linked to salinization, since as the subsurface soil becomes increasingly saturated the farmer has less ability to wash crop-threatening salts from the upper levels of the soil horizon.

An additional threat to Egypt's *critical zones* comes from the linked changes that accompany the shift from seasonal basin to perennial irrigation. Silt entrapment behind the Aswan Dam removes the annual silt increment formerly deposited on floodplain soils. While the fertility enhancement represented by these deposits was overrated, silt was essential to improving the tilth of riverine soils (White 1988: 36), to diluting the windblown sandy material transported into the Nile Valley each year from the adjacent desert (Kishk 1986: 230), and to providing needed trace elements such as iron, magnesium, and zinc (Whittington and Guariso 1983: 88). Although the magnitude and spatial extent of fertility loss are uncertain (White 1988: 36), the impression that such degradational changes are occurring is widespread. Preventing fertility declines caused by the new multicropping techniques, the loss of Nile silts, and reduced organic manure inputs (a consequence of diminished fallow land on which to maintain animals) requires increased use of inorganic fertilizers. The use of nitrogen and phosphate fertilizers has increased from a 1951–2 level of 0.88×10^6 tons to 6.19×10^6 tons in 1987–8 (UNEP/GEMS 1991: 14). The widespread use of pesticides and insecticides to protect the crop environment contributes an additional residue element that enters widely into both surface and groundwater.

The result has been a serious decline in the quality of water available throughout the Nile Valley for agricultural, urban, and industrial purposes, with particularly serious implications for water purification (White 1988: 36). No area has experienced greater water quality degradation than has the delta. Because water used for irrigation in the delta is often reused water from agricultural districts further to the south, contaminant concentrations are higher. High levels of organic matter and nutrients in the water, stagnant water with limited velocity, and limited dilution capacity create ideal conditions for eutrophication and the growth of macrophytes and algae (UNEP/GEMS 1991: 14). In effect, the entire delta is in danger of becoming a *sacrifice zone* as increasingly poor quality water is distributed in the district for agricultural purposes. The delta is experiencing problems analogous to a clogged toilet into which waste continues to be deposited. While treating the eastern Mediterranean as a septic tank into which contaminated water can be safely discharged is hardly a long-term solution, the absence of the cleansing action of the annual Nile flood is beginning to become increasingly apparent in perhaps the most sensitive environmental zone in Egyptian agriculture.

Two other land-degrading pressures are threatening Egypt's agricultural lands. These forces are at best only indirectly linked to the High Dam at Aswan. The increased presence of water hyacinth (*Eichhornia crassipes*) is an important threat to land productivity (Hefny 1982: 25; White 1988: 36; UNEP/GEMS 1991: 14). A South American aquatic weed introduced as an ornamental and accidently released into the wild, the water hyacinth's spreading distribution is difficult to stop. It flourishes year round in the nutrient-rich waters of irrigation canals, basins, and reservoirs made possible by the High Dam, and consumes large amounts of water intended for irrigated crops. In an increasingly water-limited agricultural regime, the presence of water hyacinth is a serious threat to future productivity.

Urban expansion generates pressures that directly threaten agriculture's *critical zones*. In Egypt this takes place in two ways. One is direct physical expansion of urban settlement onto adjacent farmland. Cairo is the major, but by no means the only, urban center to alienate farmland for nonfood-production uses. Kishk (1986: 229) estimates that an average of 120 km^2 (46 sq. miles) of land are lost to urban expansion each year. From the standpoint of economic rent, such losses may represent a more productive use of the land. But from the perspective of sustainable agriculture, particularly in an extraordinarily constrained environment such as Egypt's, the constant loss of highly productive critical resources and the need to bring increasingly less productive land into production as a replacement is nothing short of tragic. Another urban pressure on critical agricultural land resources is brick manufacture. Although theoretically illegal since 1985, mining of agricultural soils to a depth of 1 m for brick-making alienates 120 km^2 (46 sq. miles) annually (Kishk 1986: 229). There is every reason to assume that continued brick manufacturing losses still occur, albeit at a smaller scale.

Nile Valley irrigation: summary The Aswan High Dam has had a profound impact on Egyptian economy and society. The power resources and industrial opportunities created by the High Dam have transformed the Egyptian economy. Over-year storage of Nile floodwaters has altered the natural rhythm of the river and enabled year-round use of limited agricultural soils. Conversion to perennial irrigation has increased the effective use of Egypt's limited agricultural land by as much as three times through multi-cropping. This has made it possible to at least keep pace with population growth.

However, there is a dark side to this story, one that threatens to create a profound act of *destructive creation*. Lands flooded by Lake Nasser and eastern Mediterranean fisheries were consciously *sacrificed* to create the dam. Coastal environments along the delta's Mediterranean shore, as well as the agricultural zones of the delta itself, are at risk to erosion caused by the absence of flood-borne sediments. Water quality declines and groundwater table rises from excess accumulations of polluted water threaten the delta. Areas within both regions are on the verge of becoming *sacrifice zones*. Other *sacrifice zones* may well be created in Sudan, Ethiopia, Uganda, and Zaire in the future as a consequence of efforts to increase the supply of water available for Egyptian agriculture and industry. Yet more alarming are the changes within the core of Egyptian space that threaten *critical zones* – the Nile's vital agricultural soils – with salinization, waterlogging, nutrient and chemical pollution, and alienation for urban and industrial uses. The magnitude of the difficulties generated for sustainable agriculture and the scope of land-degrading forces unleashed by the High Dam over 25 years ago are beginning to become ever more apparent. These threats are the malevolent aspects of the abundant opportunities created by the High Dam, and justify regarding the creation of the Aswan Dam as an act of both actual and potential *destructive creation*.

Irrigation in the Great Plains
Irrigation in the Great Plains of the United States takes place in a completely different physical and social environment compared to

irrigation in the Nile. The Great Plains are a vast tract of subhumid and semiarid grassland, extending from Texas in the south to Montana and the Dakotas in the north (Mather 1972; and see figure 5.4). Drier along the western edge, where the region is strongly influenced by the rainshadow effect of the Rockies, the Great Plains become moister along the eastern margin where they blend imperceptibly into the lowlands of the humid Corn Belt. Typical of dry areas, low annual rainfall totals are linked to great interannual variability throughout the Great Plains. Gaines County in western Texas, to cite but one example of a pattern characteristic of the area, experienced a low precipitation total of 168 mm (6.6 in) in 1956 and a high of 960 mm (37.8 in) in 1941 (Sheridan 1981: 90). This general regime of high variability has remained constant throughout the period of historical record, as witnessed by data for Antelope County in northeast Nebraska. In this county, 292 mm (11.48 in) are recorded for 1894 and 950 mm (37.04 in) in 1903 (Center for Rural Affairs 1988: 19). Permanent streams are infrequent in the region, and where they do occur they experience

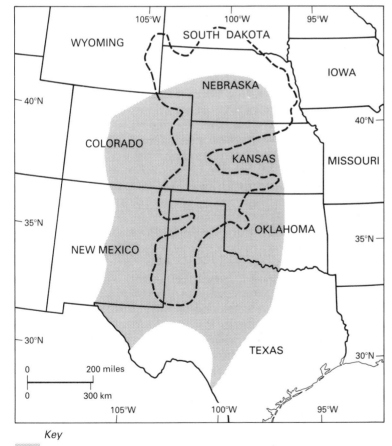

Key

▓ area in which significant groundwater overdraft is occurring

⌐ ⌐ boundary of Ogallala Aquifer

Figure 5.4
Declining groundwater in the Ogallala groundwater system (after US Geological Survey 1982)

marked seasonal variations in stream flow. High stream flows occur primarily in the spring, the result largely of excess moisture due to the melting of snows accumulated during the winter months. During the summer growing season little if any surplus precipitation occurs, and flows in the perennial rivers largely reflect the inflow of groundwater into stream channels.

In contrast to limited stream flows, underground water resources were abundant in a substantial portion of the region prior to over-pumping for agricultural activities (Kromm and White 1992). These groundwater resources are associated with the Ogallala Aquifer, a complex formation of sedimentary deposits accumulated over the past 17 million years from eroded materials originating in the Rocky Mountains. Prior to pumping, these formations stored vast amounts of water, which in scale exceed those of Lake Huron (Lewis 1990: 42), and underlay much of the central and southern plains. The High Plains aquifer system may store vast amounts of water, but the bounty is unevenly distributed. While the average thickness of water-bearing layers is 61 m (200 ft), this ranges from less than 15 m (50 ft) in Texas to nearly 366 m (1200 ft) in Nebraska (Powers 1987: 2). Although some of the water contained in the Ogallala Aquifer is fossil water, much of the water resource is renewable. The recharge rates are dependent on a small and variable rainfall as well as on snowmelt to produce sufficient moisture to infiltrate into the groundwater store. These rates are highly sensitive to soil texture and the character of substrate formations. In all cases, the recharge rates are very slow when compared to the continuing withdrawal rates for the aquifer. Recharge acts on a time scale of centuries. The maximum recharge rates of the Ogallala occur in the highly permeable Sand Hills of Nebraska. Here 15 cm (6 in) of recharge occurs on an annual basis. In the southern subregions of the Great Plains, the aquifer recharge is less than 1 cm (1/2 inch) per year (Powers 1987: 2); yet, due to the higher temperatures of the southern plains, this is the subregion with the highest water demands to meet agricultural needs.

The first Americans of European heritage to encounter the Great Plains were unaware of the rich groundwater resource beneath their feet. Judging by presumed surface conditions, vegetation, and previous education and experience in a wetter habitat east of the Mississippi, the eastern-educated elite regarded the Great Plains as a desert (Bowden 1969). It is not clear that this elitist view, despite some support from early exploratory and mapping parties (Alford 1969), ever influenced the geographical ideas of the first farmers to enter the region. These people brought with them crops and farming systems adapted to the moister prairies of the Middle West and strove to make known practices work in the drier plains environment. The myth of an American Desert east of the Rockies is more the product of the booster literature of railroad companies and land developers (Bowden 1975), who used the pioneer heroics of the first wave of "desert"-conquering settlers to encourage potential new settlers to perceive the potential of the plains environment in more optimistic terms. After several false starts in which farming systems inappropriate to long-term climate conditions were practiced, the region evolved a mixed system that combined animal husbandry with grain farming. Wheat was grown in the drier west, corn in the moister eastern fringe, and beef cattle were raised everywhere. In the drier western and southern districts of the plains, and especially after the Dust Bowl experience of the 1930s, extensive mixed farming systems that left large areas of native grasses in place became the rule.

This system was technologically limited by the ability of farmers to extract groundwater (Glantz 1990: 16). The windmill had made permanent settlement possible on the plains, but the amount of water brought to the surface was limited. Only enough water for settlers and their animals could be extracted from near-surface aquifers. The deeper and more abundant Ogallala resources were not exploitable by wind-powered technology, and most farmers were unable to gain access to enough water to practice irrigation. Drought risk encouraged a flexible, diversified form of dryland agriculture, one that was sustainable and resisted excessive soil erosion as long as the conservation techniques and increased scale of farm operation mandated by the Dust Bowl experience remained in place.

Changes in technology, agricultural economics, and government tax policy encouraged massive exploitation of the Ogallala Aquifer after 1950. The introduction of more powerful diesel and gasoline fuel-driven turbine pumps made it possible to extract water from deeper underground. In 1953 a new irrigation technology was developed to take advantage of the newly accessible groundwater resource. This was the center-pivot irrigation system (Splinter 1976), so called because it linked an overhead sprinkler boom to a central water source (plate 5.1). The overhead boom was mounted upon a series of wheeled towers and was anchored at one end to the water source. The entire structure rotated around the central pivot and was designed to irrigate approximately one quarter section (160 acres/65 hectares). Because the center-pivot irrigation system inscribed a perfect circle, in practice the corners of the quarter section could not be reached by the boom. This left approximately 11 hectares (27 acres) in each quarter section without the benefits of the system's artificial rain. The American penchant for giantism has led to the construction of center-pivot systems capable of irrigating 214 hectares (530 acres), as well as trailing arms capable of swinging out to irrigate the normally ignored corners (Splinter 1976: 93), although these variants are not the general rule. In the space of two decades, the rectilinear landscape of many parts of the American West had been transformed into a spectacular array of circles (Sutton 1977).

The major advantage of the center-pivot system was its great structural flexibility, which made it possible to irrigate tracts of land that otherwise were topographically unsuited for gravity irrigation farming (Center for Rural Affairs 1988:2). No massive land

Plate 5.1
Center-pivot irrigation in the
Kufra Oasis, Libya

leveling was needed to make center-pivot irrigation possible. This
made it feasible to convert marginal range- and cropland, without
incurring large land preparation costs, into more intensive uses in
order to produce a higher-value product. Lewis (1990: 43) suggests
part of the economic incentive for the conversion when he claims
that irrigation can produce yields 600–800 times greater than dry
farming technology. In Nebraska and the moister eastern plains
land was converted to corn, while in the southern and drier
plains cotton was the crop of choice. Another advantage of the
system is its low labor requirement. Once established, it operates
automatically with the aid of a computer to synchronize water
application along the boom and to initiate movement to a new
position on the circle. Expensive, time-consuming construction
and maintenance of canals and ditches, or the physical movement
of static pipe systems, are eliminated. Because the water is applied
automatically and does not depend on rainfall, the farming system
becomes drought-proof, guaranteeing good yields regardless of
precipitation conditions. Moreover, fertilizers, herbicides, and pesticides can be applied through the irrigation system, thus further
reducing labor costs.

The emergence of a sophisticated technology to exploit groundwater for irrigation purposes occurred at a favorable time.
Although the capital costs of installing center-pivot irrigation systems were considerable, ranging from $35,000 to 45,000 per quar-
ter section (65 hectares) in 1975 (Center for Rural Affairs 1988: 5),
prices for crops were also high. Farmers, bankers, and agricultural
extension agents all anticipated large profits for the foreseeable
future. Government tax policy created a favorable investment
climate by providing investment tax credits, favorable equipment depreciation schedules, tax concessions on land purchases,

depletion allowances for groundwater withdrawals, and a generous capital gains policy. Crop price support programs, crop insurance, disaster relief payments, and low-interest mortgage loans all favored farming over ranching. Widespread speculation in land became common, because simply plowing up land and planting it with wheat doubled the land's value (Huszar and Young 1984: 233). Banks jumped on the bandwagon by lending large amounts of money to farmers who wanted to intensify their production. In addition, the attitude toward regulation of groundwater withdrawals was sympathetic to rapid development. Common law treated groundwater use as a right belonging to the owner of the land under whose property the water was located. Any reasonable use of this water was permitted provided that the similar rights of neighbors were not infringed. Changes in this legal status for groundwater were slow to develop (Emel and Brooks 1988; Emel, Roberts, and Sauri 1992), were fiercely resisted in some areas, and only came about as depletion reached alarming proportions and conflicts over water use increased.

The boom times atmosphere that developed prompted rapid expansion in irrigated acreage as rangeland and marginal cropland were converted. In the space of three decades irrigated acreage increased fivefold, reaching over 6×10^6 hectares (15×10^6 acres) in 1980. However, the stability of the irrigated acreage created is threatened by a number of serious degradation impacts. By far the most serious is the dramatic reduction that is occurring in groundwater reserves, and the corresponding decline in the water table that accompanies aquifer depletion. In effect, farmers are mining the groundwater in the same way that coal miners mine coal. This is because the rate at which nature replaces withdrawn groundwater is far slower than the rate of depletion. The rate of withdrawal is more threatening in the southern plains where the amount of groundwater is more limited. Predictions that by the year 2020 the southern Ogallala will be reduced to one-third its former volume (Glantz 1990: 19) are reflected in decreased water pressure in existing wells and a decline in the water table by as much as 30 m (100 ft). Less drastic declines are reported for the northern High Plains (Evans 1972: 52). Were current rates of withdrawal and recharge to continue, Press and Siever (1974: 238) estimate that it would take several millennia before pre-center-pivot irrigation conditions could be reestablished.

In this instance, a *sacrifice zone* is being created that will impact future generations more severely than present users because, while present users will be protected by social assistance programs from the worst effects of the impending impacts of *destructive creation*, subsequent inhabitants of the region will find their range of resource use options drastically reduced. Dire predictions of complete collapse of the Ogallala are undoubtedly excessively apocalyptic, because as groundwater availability decreases, the cost of extracting it increases. This encourages a search for new water supplies, all of which are currently much too costly to contemplate

(Evans 1972; Powers 1987: 5) even if present users of distant surface water were willing to contemplate their release for transfer to the plains.

Increasingly constricted supply also encourages greater efforts at conservation and water-use efficiency. Solar-powered surge valves, runoff pits to collect and recycle irrigation and rain-water, and water applications that inject water directly into the soil rather than into the air are all reported (Lewis 1990: 43). Water losses from sprinkler irrigation have declined from 40 percent to 5 percent, while the estimated net depletion of the Ogallala Aquifer in at least one conservation district in Texas has declined from 278×10^7 m^3 (2.25×10^6 acre-feet) in 1963 to about 370×10^6 m^3 (300,000 acre-feet) in 1985 (Wyatt 1988: 2). These measures promise to reduce the rate of Ogallala Aquifer degradation, but they do not reverse the damage done to its ability to sustain irrigated agriculture and other livelihood options in the long term.

Groundwater is connected to surface-water bodies such as rivers, lakes, and reservoirs. As water table levels decline, the amount of water available to nourish streams and lakes decreases. Particularly in sandy soil areas, more water infiltrates and less is available for nearby water bodies. Streams carry less water and reservoirs store smaller amounts of water for surface irrigators and municipal water systems (Glantz 1990: 19). The needs of these populations, as well as the habitat of aquatic wildlife, are sacrificed to the demands of the groundwater exploiters located elsewhere in the region.

Linked to the more intensive center-pivot irrigation systems and the cash crops that they generate is the use of a package of production factors that are intended to produce higher yields. Pesticides, herbicides, and fertilizers are often lavishly applied, usually in conjunction with the irrigation water. The use of chemicals is part of a general tendency to push highly capitalized farms as close to their production limit as possible in order to generate income flows that will contribute to debt reduction. Crop monoculture and the reduction or elimination of crop rotations mandate the use of chemicals to compensate for the loss of natural regenerative processes. Nitrate fertilizers are a particularly prominent source of groundwater contaminants (Center for Rural Affairs 1988: 13, 17; Lewis 1990: 44). Degradation of groundwater quality seems to be a particularly dangerous example of biting the hand that feeds you.

Degradation of the groundwater resources of the Ogallala Aquifer is accompanied by increases in soil erosion as a result of changes in land management. Many of the soils brought into intensive production by the center-pivot technology are of low inherent fertility, or are light or sandy soils that are readily susceptible to wind erosion. Often these soils were used as rangeland and were protected from excessive wind erosion by a permanent vegetation cover. Conversion to irrigation increases the erosion risk. The development of center-pivot systems capable of operating on 30 percent slopes meant that many areas that were topographically unstable and prone to erosion were brought into production. The large wheels necessary to carry irrigation pipe supporting towers across the landscape often created field scars capable of evolving into gullies (Center for Rural Affairs 1988: 16). Piper (1989: 70) estimates that 1.9×10^6 hectares (4.8×10^6 acres) of cropland in the Great Plains have suffered damage since 1955. Although yield reductions due to soil erosion have been modest (table 5.1) due to the use of inorganic fertilizers, off-farm damage to recreation facilities, navigation, reservoir siltation, and residential areas is nearly 20 times larger (Colacicco, Osborn, and Alt 1989; Piper 1989: 74). Changes linked to the conversion process also increase this erosion threat. Planting of shelterbelts along field edges was a major strategy for reducing wind erosion vulnerability after the Dust Bowl (figure 5.5). The desire to build larger center-pivot systems made many of these shelterbelts obstacles to pivot operation and they were frequently removed (Center for Rural Affairs 1988: 16). If global warming increases the aridity and drought vulnerability of the plains, the potential for accelerated wind erosion of dry, exposed soils should increase (Glantz 1990: 20).

Table 5.1 Yield reductions in the Great Plains due to soil erosion (percent)

Crop	Northern Plains	Southern Plains
Corn	1.9	0.8
Soybeans	2.4	—
Wheat	1.1	1.6
Legumes/hay	0.2	0.8
Cotton	—	6.4

Source: Colacicco, Osborn, and Alt (1989: 36).

Figure 5.5
Change in shelterbelt frequency in Antelope County, Nebraska, 1963–86

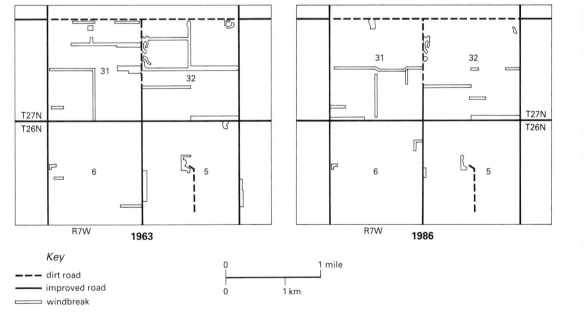

Key

- - - dirt road
——— improved road
⎓⎓⎓ windbreak

0   1 mile

0   1 km

Regardless of whether or not this proves to be a realistic concern, there is little doubt that groundwater-based irrigation has exposed a *critical zone*, the region's agricultural soils, to both actual degradation and potential threat. Combined with the pressure placed upon groundwater resources, the expansion of irrigated farming has been an act of *destructive creation* with limited prospects of developing a sustainable agricultural system. The most likely prospect for the region is a gradual reconversion of the agricultural habitat back to a much less intensive mixture of wheat cultivation and livestock rearing, that will be handicapped by the diminished resource base of a mined groundwater and varying degrees of an eroded and less fertile soil.

Animal husbandry

Animal husbandry is an important component of agricultural systems. Not only do domesticated animals provide important food and protein resources, but also they supply fiber, leather, tools, fertilizer, transportation, and traction power to the human communities that utilize them. Animals are kept in either of two modes: as a component of the crop production system of largely sedentary groups; or as the primary focus of activity of more mobile herding communities. Traditionally, both farmers and herders raised their animals using extensive techniques. In extensive systems, it is the animals who move to the fodder resources, thus expending energy to reach and convert plant material into products that are valued by and useful to humans. This is as true for the farmer's chickens who roam the barnyard scrounging for grain and grubs as it is for the nomad's camels or sheep. While efforts to raise animals more intensively are ancient, the costs and transportation problems created by attempts to bring food to stationary animals made large-scale intensification exceedingly difficult to achieve. Only in the past two centuries has significant intensification proven possible, and in many instances these efforts to achieve concentrated intensive production have generated significant environmental problems.

Both *creative destruction* and *destructive creation* processes have operated in animal husbandry. Creative destruction has been the least common occurrence in recent times, but it has taken place in several settings in which traditional pastoralists have altered environmental conditions in their favor over long periods of time. Pastoralists have accomplished this by promoting the growth of grass and particularly valuable fodder trees and shrubs at the expense of less desirable woody vegetation. Similarly, sedentary farmers have created sustainable agricultural systems based on creative destruction principles in which animal manure has been a vital component of the ability to maintain long-term productivity. We will explore several examples of this creative process in this section. Unfortunately, destructive creation is the process frequently encountered in animal husbandry. This land-degrading tendency is often associated with futile efforts to intensify production. These intensification efforts may arise spontaneously from within the animal-keeping community, but more commonly they flow from land-degradation pressures set in motion by the misguided efforts of distant groups to enhance production (Grainger 1990: 76). In many instances, these intensification pressures convert *critical zones* for the animal husbandman, usually the dry season pasture resources, into *sacrifice zones* that are extremely difficult to rehabilitate. We examine several examples of destructive creation in both highly industrialized and modestly industrialized countries in the second part of this section.

Creative destruction in animal husbandry
Pastoral nomads practice an extensive mode of animal husbandry. Animals and herders move together between seasonal pastures in order to find adequate grass and water. Most often the lands that pastoral nomads use are too risky for settled communities to exploit. Either water is too scarce, rainfall is too scanty and variable, growing seasons are too short for crops, slopes are too steep, or soils are too infertile to make farming a viable enterprise. Under such restrictive conditions, specialized herders of animals are often able to extract a living from these types of agriculturally marginal environments by moving their herds frequently from one pasture to another (Johnson 1969).

By engaging in frequent seasonal locational shifts, herders are able to avoid prolonged impact on vital *critical zones*. By far the most important zones for nomadic pastoralists are the dry or winter season pastures that nourish their herds when natural fodder is limited. Equally important are the watering points in these dry season zones, and the grazing areas within reasonable distance of each well, spring, perennial stream, or cistern. For "vertical" nomads, who rely on movement up and down slope to escape seasonally dry conditions, the *critical zones* are the high mountain pastures where temperatures are cool, grass is relatively abundant, and water can be found. These pastures set an upper limit on the number of animals that the community can keep. Good quality dry season pastures insure that newborn animals will likely survive and will be available to exploit the wet season pastures in the lower elevations at other times of the year. Institutions have evolved to control the numbers of animals that can be kept in these critical pastures (Bencherifa and Johnson 1991), and pastures can be closed by tribal elders if grazing conditions are too poor to sustain the herd and still provide grazing in subsequent years. Often the animals that are moved from lowlands to highlands have developed into special breeds. Barth (1961: 6) reports that when the Basseri tribal migrations were halted in Iran, massive mortality in the tribe's sheep herds occurred. The sheep were incapable of withstanding either the winter cold of the uplands or the summer heat of the lowlands. Movement between the two areas was essential for the well-being of the herds. In effect, the animals and the herding system had coevolved into a structure that required movement for survival.

Activities that expanded the amount of grassland at the expense of forest in such uplands can be regarded as an act of creative destruction as long as the impact favored more grass for more sheep and goats, without setting in motion negative feedbacks such as soil erosion that destroyed the growth environment for the desired pasture grasses. Equally important were the institutions and values that promoted responsible use of the environment, protected useful fodder trees, and established guardians to control use of limited resources (Hobbs 1989).

In "horizontal" nomadic systems the critical resources are not alpine pastures but rather the grazing land around wells and the pastures in wetlands and on floodplains of permanent streams. Tribal groups return to these areas after the rainy season is over, floodwaters have retreated, farmers' crop cycles are complete, and fields are in fallow. Because farmers cannot keep enough animals to use all of the potential dry season fodder, and because only one crop can be grown on cropland each year due to climatic factors, a niche exists which the pastoral nomad traditionally took advantage of. Protecting access to that niche was of vital importance to the herder. In the traditional system, access was never threatened to these lands during the post-harvest period because the farmer needed the dairy and meat products and manure of the migratory herds to enhance human nutrition and maintain soil fertility. In the delta of the Niger River in Mali, these relationships were codified into a legal system that protected the rights of all parties and formalized the timing of their respective periods of use (Gallais 1972; Bremen and de Wit 1983). In eastern Libya, formalized understandings existed between noble and client lineages about who had to leave drought-stressed *critical zones* adjacent to permanent watering holes in order to insure that there always was a balance between human population, animal numbers, biomass carrying capacity, and water availability (Peters 1968). Even more elaborate stratified social structures among the Touareg had in part similar objectives (Bernus 1990: 152–5). The result of these devices was to protect the critical dry season pastures from severe pressure in stress periods, and reduce herd sizes to levels that would permit regeneration when a cycle of better conditions returned.

Good conditions during the rainy season were handled in just the opposite fashion. When pasture was rich in any one tribe's or clan's territory, other herding groups were allowed to enter the district and help consume the bounty. Access to high-quality pasture, particularly in the spring when lambs and calves are growing out, is essential to a nomadic pastoral operation. However, the sporadic spatial distribution of rainfall in drylands insures that some herders will lack such grazing in their territory in any given year. Thus, allowing other groups access to an abundant seasonal and temporal resource when and where it exists enables a group to claim access to the resources of its neighbors at a future time when its local resources are scanty. The result is the open grazing system described by Perevolotsky (1987) in the South Sinai, one that is replicated in most traditional grazing areas throughout the world.

In many traditional nomadic pastoral systems fire played a major role in environmental management. Burning the previous year's unconsumed dry growth of perennial grasses just before the next rainy season is a common practice in much of sub-Saharan Africa (Carr 1977: 17–19). Although poorly timed and too frequently repeated burning can destroy vegetation cover and promote soil erosion (Goldammer 1993), the wise use of fire has many advantages (Stocks and Trollope 1993: 320–1). It stimulates perennial grasses to produce new shoots, thus jump-starting growth. The ash provides a modest nutrient surge to support this growth. Fire also discourages the growth of shrubs and trees in a grassland, since most woody plants lack a bark that is thick enough to insulate the plant from the heat of the fire. By destroying trees and seedlings, fire reduces the moist, shady habitat that is most conducive to tsetse flies. It is in these tsetse-free areas, particularly in regions of higher rainfall, that domestic livestock are concentrated (Bourn 1978). When fire-influenced grasslands become less hospitable to tsetse flies, the vector for the transmission of trypanosomiasis (sleeping sickness), both humans and animals benefit (Lewis and Berry 1988: 76). Large tracks of African environment have been cleared of tsetse habitat by the use of fire. The result is an act of *creative destruction*, in which overall woody biomass is lowered by the destruction of trees and shrubs in order to promote the growth of grassy habitat that is more useful to animal herders. Over time, grassland plants, herd animals, and herders increase at the expense of other species. They coevolve through a frequently repeated series of destructive acts which, provided that they are neither too intense nor accompanied by too rapid an increase in herd size, produce a sustainable pastoral system. The artificial nature of this balance between grass, trees, herds, and humans is illustrated by the changes that occur whenever the pressure of fire is removed from the system. Lewis and Berry (1988: 373) report that the diversion of manpower from the herding economy to the mines in Zambia resulted in reduced frequency of fire. Without fire to sustain the grassland habitat, shrubs and trees regenerated rapidly, tsetse-borne sleeping sickness expanded, and domesticated livestock experienced a sharp decline in numbers. Efforts to use chemicals as an alternative means by which to destroy tsetse habitat have had limited success, and pose serious risks for the environment (Linear 1985).

Destructive creation in animal husbandry
Unfortunately, *destructive creation* is a more common process in dryland pastoral systems than is creative destruction. This trend has accelerated in the twentieth century as the balance of political and economic power has shifted away from pastoral groups in favor of settled farming and urban populations (Johnson 1993b). There are

many reasons why degradation has become the dominant trend in pastoral rangeland environments.

The most important factor is the loss of *critical zones* in the pastoral system to agricultural development. These zones do not have to be large in area to have a major impact on the viability of pastoral communities. If they are important areas of dry season pasture, the loss of a small *critical zone* can have a far-reaching impact. The conversion of river floodplains from a single annual crop system to a multicrop system is a case in point. This most often occurs when the desire to intensify agricultural production leads to the construction of a dam to store water for year-round irrigation. Not only is former dry season pastureland lost beneath the reservoir, but also conversion of farmland below the dam from basin or recessional agriculture into perennial irrigation makes it impossible for migratory herders to graze on seasonally fallow land or post-harvest crop residues. Any additional recessional grazing created along the banks of the reservoir is inadequate compensation for the lost seasonal floodplain pasture and is usually accompanied by greater exposure to water-borne disease (White 1988). Pastoralists are then thrown back into year-round dependence on rangeland resources that historically have only been asked to support them for part of the year. The inevitable degradation that results makes pastoralists and their habitat the *sacrifice zone* for the production gains that take place in the irrigated agricultural sector and/or the generation of hydroelectric power sent to distant locations to satisfy a multitude of urban/industrial demands or power generation for irrigation pumps.

The expansion of dryland farmers into pastoral zones is another example of this process. Because rain-dependent farmers do not have access to enough surface water for irrigation, they are limited to zones in which they can hope to receive enough rainfall to support a crop in a majority of years. These areas may be marginal from the standpoint of the reliability of cereal crop cultivation, but they are often the best, most *critical zones* for pastoral activities (Sollod 1990). Their loss has contributed in large measure to land degradation in northern Niger (Bernus 1990), the lowland fringes of the Kenyan Highlands (Campbell, 1981, 1986; Little 1987), and the High Plateau of Algeria (Bedrani 1991), and this story can be repeated in many of the Third World's semiarid rangeland provinces where farming and herding can both be practiced.

Changes in pastoral practice and technology that encourage excessive concentration of people and animals can also promote grassland degradation. The most common culprit in such instances is the introduction of borewells that are able to tap groundwater resources that were unreachable by traditional techniques. These borewells replace shallow traditional wells constructed by hand. Overgrazing always is a problem around any well, but the limited supply produced by traditional well technology discouraged the concentration of too many animals at any well (plate 5.2). Also, local herders maintained control over access to their wells. The

Plate 5.2
Land degradation around a borewell in Kenya's Central Province

introduction of government-funded wells was accompanied by the loss of control by local animal herders over who could bring their animals to the well, and over how many animals could be watered.

Permanent settlement near traditional dry season grazing and water also was very limited in the past. However, the development of secure, good-quality water resources attracted settlement by herders and the creation of infrastructure (schools, health clinics, police posts, vehicle repair facilities, etc.) to serve the increasingly permanent population. Few former nomadic pastoralists were willing to give up their animals when they settled if they could possibly avoid it. Therefore, grazing pressure around wells increased dramatically, producing a small pimple of desertification potential in the midst of an area. Frequently the herds circulating around these deep-well sites were owned by merchants who preferred to invest capital in traditional means of production such as animals. Often masked by apparently decent grazing during periods of adequate rainfall, these splotches of degradation burst into full view whenever drought eliminated annual grass production and caught large herds of water-demanding sheep and cattle in exposed locations (Bernus 1980). In the Sahelian drought of the early 1970s, the results proved disastrous for many herds, as well as very costly for the environment. There are reports from the central Sudan that, despite the occurrence of better rains in the late 1980s, some of the most desirable fodder grasses have yet to reappear around important well and village sites (Khogali 1991: 89). In contrast, in other parts of central Sudan a lightening of pressure has resulted in at least partial recovery of the vegetation (Olsson 1985).

The pressures that encourage excessive concentration of people and animals near spatially limited resources are not restricted to developing countries. They are found in the rangelands of industrialized states too. In such situations, the problem of degradation is increased by the commercial context in which the herder must operate. In the Gascoyne Basin of western Australia, sheep ranchers must generate income from sales of sheep and wool. Whenever prices drop for either wool or mutton, off-station sales are difficult to justify economically. When depressed prices coincide with drought, the station operator is encouraged to keep more animals on the station than can be sustained without degradation. Owners are reluctant to destock their operation because they hope to bring sufficient animals through the bad period in order to capitalize on what they anticipate will be a future rise in price. Thus, animals are retained for economic reasons rather than reduced for ecological reasons. These animals are not kept on the lowest productivity pastures, but rather are held on the land with the highest potential to produce fodder. The ironic result is that the concentration of animals on the best land produces the greatest degradation on the most productive land, while the least productive land remains relatively unscathed (Walls 1980), a cycle that is typical of much of pastoral Australia (Williams 1978).

A third major factor promoting *destructive creation* in rangeland is the loss of control over their range resources that most traditional pastoral communities have experienced. In most rangeland, tribal lands were nationalized in the period immediately following independence from colonial rule. The common property of a collectivity became the common property of the nation as a whole. Most governments have found it difficult to substitute a reasonable system of control for the former tribal system. For many pastoralists, especially in the drier rangelands, trees are an important source of fodder in dry seasons and in drought periods. Protecting trees from excessive cutting is essential to pastoral survival (Hobbs 1989: 104–6; Barrow 1990). Both because most modern pastoral development strategies have tended to ignore the importance of trees and because the loss of proprietal communal rights to rangeland removes traditional managerial constraints on the exploitation of rangeland, dryland arboreal resources have experienced considerable degradational pressure during the post-independence era, especially in Africa. This is particularly true in the case in which charcoal is a major source of domestic fuel, and drought offers no alternative for pastoralists but to consume their trees for non-pastoral purposes (Dahl 1991).

In many countries, as a consequence of new tenure arrangements following the assertion of national control over rangelands, blocks of land were set aside for group ranches or reassigned for special ranching schemes designed to produce meat for export to the markets of industrialized countries. These ranches seldom worked well for two reasons. First, they took important sections of range out of use by traditional producers, thus forcing them to concentrate on producing the same number of animals on less land. The resulting degradation in range quality outside the fence was usually attributed to nomadic pastoral mismanagement of a common property resource, when in fact in represented a *sacrifice zone* created by the favored ranching scheme and other governmental actions. Second, ranches were fixed entities with rigid boundaries in a highly variable and fluctuating environment. Whenever rain failed to fall within the fence, herds were trapped, unable to move legally to adjoining areas that had received better rainfall. The strategies developed for ranches generally ignored the *genius loci principle* by not incorporating the highly variable nature of precipitation in drylands. Degradation then often occurred within the fenced perimeter. In Syria, nationalization of pastoral areas set off a scramble for access to the land by private individuals without effective constraint on herd numbers. Only by reestablishing control through the mechanism of pastoral cooperatives based on the old tribal system and traditional concepts of environmental protection (*hima*), an acknowledgement of the *genius loci principle*, was further rangeland degradation arrested (Draz 1977, 1990; Shoup 1990).

Destructive creation in animal husbandry systems is the product of a failure to view rangeland development as part of a larger system. By treating development segmentally rather than holistically, changes in agricultural systems have cut traditional links to pastoral areas and turned them into *sacrifice zones*. When new technology, such as diesel pump driven borewells, has been introduced, it has often operated in a vacuum, divorced from meaningful association with local communities. The result has been the creation of *local sacrifice zones* adjacent to the new technology and the initiation of severe degradational pressures. Stopping and reversing this degradation has proven to be very difficult, and has only occurred in rare instances when developers and managers were willing to listen to the *genius loci principle* and base change upon time-honored precepts that are in tune with local society and environment.

Rainfed agriculture

Rainfed agriculture, which is the world's overwhelming type of farming, depends solely on natural precipitation to grow crops. In contrast to irrigated farming, in which management can control the timing, intensity, and volume of water reaching any field, rainfed agricultural systems must be able to cope successfully with the inconstant nature of precipitation that exists in most of the climates in which this form of farming is practiced. Rainfed agriculture exists through a spectrum of diverse geographic settings. Topographically, it is practiced on slopes ranging from extremely flat lands, such as along the Mississippi floodplain, to lands steeper than 30 degrees, such as in many parts of the highlands of East Africa, the Andes, and the Himalayan zone. Climatologically, it is practiced in areas ranging from locales that experience almost daily rainfall and/or large amounts of precipitation to zones that receive relatively infrequent and small amounts of annual precipitation. Farming in these latter zones, under conditions in which annual precipitation is 500 mm (20 in) or less, is classified as a particularly risky type of rainfed cultivation, namely dryland agriculture (OIA 1978). In the remaining portions of this section, rainfed agriculture and land degradation are examined in one dryland and two humid case studies.

Dryland agriculture

In practice, dryland agriculture exists in areas experiencing precipitation in excess of the previously mentioned 500 mm criterion. This is especially true in warm tropical areas that have high potential evapotranspiration rates. In India, the dry agricultural areas have precipitation values ranging from 350 mm (14 in) to 1,500 mm (59 in). Because of the seasonal distribution of rainfall in northwestern India, even the areas of higher precipitation (1,000–1,500 mm) experience 8–9 months of potential evapotranspiration rates exceeding precipitation (Jodha 1988). Thus, even in these zones, moisture conservation measures, a critical component in dryland agricultural systems, are crucial for successful agriculture. Because moisture is the critical limiting factor in areas in which dryland farming is practiced, when farmers are unable to manage their crops in synchronization with an area's precipitation variability, often crop failures and/or land degradation occur.

The environmental settings of dryland farming are usually risk-prone in the agricultural context. Because of limited moisture, without supplemental irrigation the likelihood of crop failure during periods of drought is high in these agricultural zones. As discussed in chapter 3, the expected variability of precipitation in dry climates is large. Thus, crop failures or poor harvests during the drier years are an expected phenomenon. Successful farming strategies in these areas require that during drought the soil resource must remain protected for use during the wetter years. When crops fail or their growth is retarded during dry periods, plant cover is diminished. One result of this decrease in crop cover is that the soil is less protected from both wind and water erosion during drought conditions. Plowing and the removal of the grassland cover for crops makes dryland agricultural lands susceptible to erosion during periods of drought. The result is that accelerated soil erosion is often associated with dryland agriculture during dry periods unless conservation practices remain implemented during these periods of moisture stress. When conservation is sacrificed for short-term demands, many lands in these areas are susceptible to land degradation. Accelerated soil erosion reduces the resilience of vegetation to recover after each drought period.

To prevent land degradation from occurring under dryland farming, fallow periods were often one accepted strategy that evolved in these dry but agriculturally feasible settings. In the contemporary world, because of numerous factors, periods of fallow are often reduced. This puts many traditional agricultural areas, where dryland farming was successfully practiced, at risk. Fallow and other traditional coping strategies in the farming system of the Luni district of India are examined in the next section. Prior to the twentieth century, these adaptive strategies had resulted in a sustainable dryland agriculture. Alterations in the traditional system in response to contemporary conditions have upset a stable dryland farming system, and severe land degradation has been the end product.

The Luni Block of Rajasthan   The Luni Block is a 1,989 km^2 (770 sq. miles) tract of land situated in a largely semiarid area of western Rajasthan. Agriculture is the dominant activity. Prior to the 1930s, successful rainfed agriculture was practiced throughout this area (figure 5.6) without significant deterioration in the resource base. Today it is an area that clearly has experienced land degradation. Formerly productive fields and pasture have become wastelands covered by dunes or rocky soils, or have experienced salinization.

The Luni Block is clearly situated in a marginal climatic area for farming, but successful agriculture existed in the area prior to the European incursion and up to the 1930s. Records show that a stable agricultural system existed in the Luni Block from at least

Figure 5.6
The Luni Development Block in Rajasthan, India (after Walls 1980)

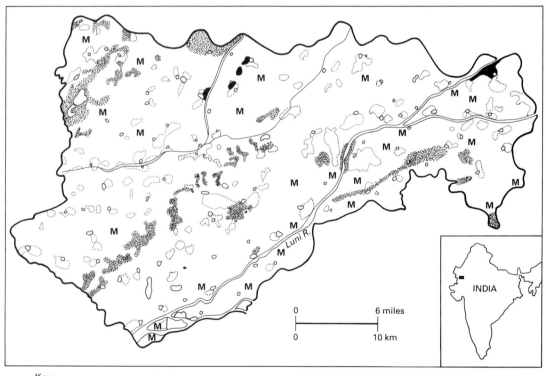

Key

⬭ settlements

⬚ permanent pasture

M monoculture

⬛ double cropped area

▨ waste

as early as the nineteenth century. Rainfall is almost totally concentrated in three months, June through August. This concentration is crucial for agriculture since it results in a short growing season, which is sufficient for dryland farming despite the low annual rainfall of 310–390 mm (12–15 in) (CAZRI 1976). With precipitation having a coefficient of variation between 51 and 56 percent, both drought and flood are common. Farmers had to develop strategies to cope with both the low precipitation as well as the area's highly variable nature of moisture. As a consequence "...the highly adaptive and evolved form of present day agricultural practices also speak that many centuries of human experience must have gone into its making" (CAZRI 1976: 13).

Climatic conditions have remained constant from the 1800s through the modern period. Especially pertinent to the land degradation/desertification problem that this area is now experiencing is the fact that there is no indication of any progressive aridity: that is, there is no evidence of climatic degeneration occurring throughout the area (Walls 1980: 181). Climatic change is not the cause of this area's deteriorating land resource.

Prior to the 1930s, the population was relatively stable. Increases in population due to high birth rates were countered by frequent disease epidemics. For example, between 1911 and 1921, the area's population decreased due to numerous outbreaks of disease (CAZRI 1976: 10). Population in the Luni Block, at least from the 1800s up to 1930, hovered at around 30,000 (15.1/km^2; 40/sq. mile). The introduction of immunization, and more recently antibiotics, into the area clearly illustrates the effectiveness of modern medicine in controlling diseases and curtailing premature deaths (Walls 1980: 181). However, no concurrent innovations were introduced to curtail the area's high birth rate, which was needed to insure a stable population prior to the introduction of modern medicine. One result was that by the 1970s, the area's population had more than tripled, to about 96,000 (48/km^2; 125/sq. mile). One repercussion of this introduction of some components of modern medicine into this hazard-prone and agriculturally marginal area was that the local resources had to support an ever increasing population.

The rapid increase in population during the twentieth century is likely the significant catalyst for the introduction of numerous modifications in the traditional farming system that, prior to these changes, had resulted in a long-term stable system. Four critical alterations were particularly important. First, migration took place onto more marginal, drier lands in the western portion of the Block. Because these lands were more arid, they had been avoided previously by farmers, as there was too high a probability of suffering crop failure from shortfalls in precipitation. Their usage had been limited to either grazing or shrub forest, which provided complementary and needed goods to the nearby farmers. Second, the period of fallow, crucial for allowing the soil to repair itself, was shortened or eliminated. Third, a change in the livestock mix occurred: buffalo and sheep declined while goats increased (CAZRI 1976: 9–10). Goats, being very efficient browsers, not only can thrive where other livestock cannot but, if not properly managed, their efficiency can cause further degradation by stripping the land of a large proportion of the remaining vegetation cover. Fourth, the increased gathering of firewood to meet the needs of the growing population furthered the decline of the area's vegetation cover (Walls 1980: 187; and see table 5.2).

Table 5.2 Exploitation of woody biomass (metric tons)

	1963	1973
Fuel wood requirements	32,703	42,732
Other wood demands (building, fences)	1,662	2,338
Remaining trees, shrubs or forest, fallow, and degraded land	153,695	81,214

Source: Central Arid Zone Research Institute (1976), p. 22.

Besides population pressure, other factors contributed to the expansion of agriculture into ever more marginal areas. Among these were the government's need for cash crops, short-term pressures of market economies, and the lure of quick money to be gained by bringing new areas under cultivation (Walls 1980: 183). These economic incentives remove traditional constraints that encourage rural folk to limit exploitation of resources to what is needed for basic food and income needs. Thus, even on land that remained in long-term fallow or common rangeland, farmers, particularly those with resources to invest, were encouraged to invest in animals that produced milk and meat products that were in high demand in urban areas. The result has been steadily increasing pressure through overstocking (Jodha 1985: 261–2).

Also important were the changes in the social conditions of arid India. Jodha (1985) implicates land reform as a prime factor in promoting land degradation in Rajasthan. Before land reform in the 1950s, feudal landlords controlled access to common property resources by imposing heavy taxes and rents on peasants who wanted to convert rangeland into cropland. The high cost of the conversion made it economically undesirable to open new cropland in marginal dry areas. Land reform removed this constraint on use, and the village councils that were supposed to function as substitutes for the old landlords proved unsuited to the task. Changing caste relationships also played a role (Jodha 1985: 258–9). Many Untouchables who had been restricted to lower-status craft occupations were liberated when the social control of the landlord class declined after 1950. Their ambition, frequently realized, was to become farmers – even if it meant accepting suboptimal farmland redistributed from formerly communal property. Members of the

Rajput (military) caste also lost work in the 1950s, as the armies of princely states were eliminated or consolidated into the national forces. These higher-status individuals affiliated to the landlord class were frequently compensated with land resources drawn from the drier rangeland. Technological innovations contributed to conversion into cropland, since tractors were introduced into the region in increasingly large numbers after 1950. No longer dependent on animal power to plow land during the short planting season, farmers were able to bring into production increasingly large tracts of land, forcing animal herders to concentrate on reduced amounts of the poorest quality land. Thus, social and technological change along with population growth played a role in both Luni Block's and Rajasthan's pattern of *destructive creation*.

Traditional coping strategies and change in the Luni Block   The lands comprising the Luni Block (figure 5.6), while harsh and fragile, had successfully been farmed in a sustainable manner through the development of a complex traditional agricultural system. Because the region is extremely flat, soil erosion by overland flows of water, while important, is not the major process that threatened the viability of farming in the Luni Block. In this dry setting, wind erosion had to be curtailed if long-term agriculture was to be successful. Furthermore, because of the variable nature of precipitation, sufficient food, energy, and other material needs for the individual farm units had to be insured, even if some fields experienced crop failure during the year due to climatic factors. A set of crucial strategies evolved within the traditional system that permitted the short-term food and material needs of the inhabitants to be met, while preserving the land resource over the long term.

Four major components of the traditional mixed cropping – livestock agricultural system were: (1) cultivated fields; (2) pasture and fallow lands; (3) xeromorphic forest; and (4) water resources. The rational management of the interactions between these four components was crucial to the maintenance of a sustainable agricultural system. The primary function of cultivated fields was to produce foodstuffs. Wheat, pearl millet, and sorghum, all crops that can survive dry intervals during the growing seasons, were the main crops. Each of these crops not only provided grains for human consumption but their residues also were utilized for livestock fodder. Fields in fallow and pasture not only provided grazing for livestock but also, under this less intensive land use, soil fertility was restored. Thus these lands, which in the future would be cultivated, formed an integral part of the farming system. Furthermore, provided that overgrazing did not occur, fallow and pasture lands, with their complete groundcover of grasses and shrubs, were an effective conservation practice that diminished wind erosion. The xeromorphic thorn forest lands provided both firewood and building materials for the inhabitants of the Luni Block. The trees also curtailed wind erosion on these lands, which were marginal for agriculture. Especially important is the role that their roots play in binding the soil. Both surface and groundwater were utilized as water supplies for the inhabitants in the Block. Surface water was harvested and stored in ponds. Each village had either one or two of these ponds; and in an average year, enough water was stored to last from 3 to 6 months (CAZRI 1976: 33). After the ponds dried up, the villagers dug shallow wells near the pond sites or relied on deep brackish wells until the advent of the next rainy season.

Each of these components, the cultivated fields, fallow and pasture lands, the xeromorphic forest, and the water resources, have experienced a drastic decline in their potential since the 1950s. CAZRI's conclusion is that their decline and the resulting land degradation is primarily the result of the area's rapidly increasing population. This growth required the inhabitants of the Luni Block to alter significantly the sustainable traditional farming system in order to insure their short-term survival (CAZRI 1976: 25–35).

To meet the ever increasing demand for foodstuffs, cultivation was altered in two significant ways. Both of these changes were clearly destructive (Walls 1980: 187). First, a "phenomenal increase" (CAZRI 1976: 13) of cultivated land occurred in the arid western districts of the Luni Block. Given that fertilizers were too expensive, in these areas, even if rainfall were adequate (which it generally is not), productivity would rapidly decline due to the low fertility of the local soils. In the process of converting these lands from grazing to farming, the grasses and shrubs that protected the soil were drastically reduced. Furthermore, the thin soils were broken into small clods during field preparation for planting. Both of these actions increased the probability of wind erosion. Second, farming on the traditional cropped lands was intensified and extended eastward. It is especially noteworthy that the fallow period was reduced. Between 1964 and 1975, the land under fallow was reduced by almost a factor of two (46,892 hectares versus 24,940 hectares; CAZRI 1976: Table 5) even though agricultural expansion to the west had occurred. Thus fields were increasingly brought back into cultivation before their fertility was restored. This decline in fertility resulted in a reduced vegetation cover provided by the crops. Again, this resulted in conditions that were more susceptible to wind erosion. The expansion of agriculture toward the east, without making modifications for the different drainage conditions found in this portion of the Block, altered the natural drainage in this zone. This caused the local water table of brackish water to rise. This change has resulted in 40.3 km^2 (15.6 sq. miles) of land being lost to future use due to salinization (Walls 1980: 189).

The Luni Block's pasturelands experienced direct consequences from agricultural expansion onto former pasture and intensification on the traditional farmlands. First, the amount of land available for livestock was reduced. Previously, a large proportion of farmland was in fallow and thus available for grazing. The conversion of the more marginal permanent pasture to cropland exacerbated this trend. The net result was a reduction in grazing lands without a corresponding reduction in livestock. Overgrazing occurred on lands that were very susceptible to wind erosion. To offset the decline in forage biomass due to the overgrazing, the inhabitants began to shift from sheep to goats in order to meet their immediate needs. However, while goats, unlike sheep, can thrive on a degraded pasture, care must be taken to prevent them from further degrading the pasture. With minimal slack in the system, and a continuing increase in the population, no significant controls on goat grazing appear to have evolved (Kumar and Bhandari 1992). Thus, while the expansion of the goat population met the immediate needs of the growing population for a short time period, overgrazing continued at a greater intensity. The resulting decrease in vegetation cover increased the risk of significant wind erosion and permanent land degradation, particularly on the lighter, sandier soils common in the region (Kumar and Bhandari 1993a).

Overexploitation of the area's woody biomass is a long-term process that population growth may have accelerated but did not initiate (Kumar and Bhandari 1993b). These problems are particularly concentrated around surface water collectors and borewells, and in villages in which people and domestic livestock are most densely concentrated (plate 5.3). Trees and shrubs have been cut indiscriminately from fields, governmental plots, and other unoccupied lands over a long period of time to meet needs for: (1) fuel; (2) top-feed for livestock; (3) fencing; (4) hut construction; and (5) employment of a large segment of the population. Exploitation of the woody biomass by the people is complete. Not only do they remove the above-ground component, but roots are dug up and

Plate 5.3
Concentrated human and animal use of a Rajasthan well site

extracted too! This latter practice loosens the soil. Both the decrease in vegetation cover and the removal of the roots result in an environment that is more prone to wind erosion.

The previously described changes in the Luni Block have resulted in negative impacts on the area's limited water resources. Even if all other components of the environment had remained constant, satisfying the demands of the increased population would have been difficult given the limited water resources found in this dry setting. Due to the flatness of the area, fluvial erosion is not a major problem with regard to agriculture. However, it does have an adverse impact on the water supply. Water harvesting to capture the maximum surplus water during the short growing season and its storage in ponds was a crucial strategy in the traditional system. These stored waters met local water needs during a significant proportion of the long dry season. Every village relied on one or two ponds for this function. With both overgrazing and a decrease in fallow, sedimentation rates in these ponds increased as the harvested waters carried a larger load of eroded soils (suspended sediment) than previously. In response to this change, villagers have had to initiate dredging activities in the ponds. Dredging has largely offset the accelerated sedimentation, but there has still been a drop in pond storage capacity in some of the villages. The decrease in the storage capacity of some ponds along with the increased population throughout the Luni Block has aggravated the region's water scarcity. One response to this shortfall was a greater reliance on brackish groundwater to satisfy needs during the dry months. Increased use of wells has resulted in an overall lowering (overexploitation) of their discharge potential (CAZRI 1976: 34), as well as heightened pressure on vegetation near the well. This has lowered their potential for future utilization.

Land degradation/desertification in the Luni Block From an agricultural perspective, the Luni Block is situated in a hazardous setting. Climatically, it is situated at the extreme minimal limit of precipitation for rainfed agriculture. Its predominant soils are light to medium in texture. When exposed to wind, most of the soils are highly susceptible to erosion and are prone to dune and hummock formation. In addition, a large percentage of this region's soils are underlain by a thick calcium carbonate caliche layer. Despite these risky conditions, a successful rainfed agricultural and livestock system evolved over the millennia.

By the mid-1950s, in response to rapidly changing conditions, the traditional agricultural system could not meet the demands of the inhabitants living in the Luni Block. The human responses to these demands, agricultural expansion and intensification, placed stresses on the environment that were beyond its capacity to absorb without change. In particular, decreases in soil fertility and overgrazing resulted in a dramatic decrease in the area's groundcover. Successful dry farming in this setting requires fields to be as clear of vegetation as possible during the dormant season so as to conserve the maximum soil moisture for use during the growing season. Traditionally, the effectiveness of wind erosion on these cleared agricultural fields was minimized by: (1) woody vegetation along field boundaries; (2) the scattered vegetation pattern resulting from a mixed land use of fallow, pasture, and cultivated fields that limited large contiguous areas of bare soil; (3) the avoidance of farming the most sensitive lands; and (4) controlling overgrazing. These components of the traditional system were all altered to meet the immediate survival needs of this area's growing population.

Feedbacks among these changes often increased the effectiveness of wind erosion, resulting in a degrading land resource. For example, since livestock numbers remained large, the expansion of farming onto the drier pasturelands in the west resulted in greater pressure on the remaining pasture. The wind erosion that occurred on these new agricultural lands (due to insufficient rainfall as well as poor crop cover owing to the soils' low fertility) resulted in these "new" farm lands being largely abandoned by the mid-1970s (CAZRI 1976). Unfortunately, because of their severe degradation (desertification), manifested by a dune and hummocky landscape, exposure of the subsurface calcium carbonate caliche layer, and sparse xeromorphic vegetation, the area did not revert back to serving as pastureland. A crucial component of the sustainable traditional dry farming and livestock system had been lost during an extremely brief period of agricultural expansion, when a minimal short-term gain of additional crops occurred.

The traditional farming areas of the Luni Block did not escape severe land degradation. The decrease in fallow resulted in poorer harvests and lower vegetation cover (table 5.2). Overgrazing throughout the area had the same net effect. By 1976, as a result of the increased effectiveness of wind erosion, 26.4 percent of the area had been converted from a potentially agriculturally useful flat plain to an extremely infertile hummocky and sandy plain. In the Luni Block, as agriculture expanded and became intensified, the whole area became a sacrifice zone for the immediate and ephemeral gains of agricultural output. The result was the replacement of a sustainable agriculture and livestock system by a highly degraded land resource, without the potential to meet the needs of the inhabitants in the foreseeable future. What was an extremely rational and successful sustainable agricultural system evolved into an unstable and degrading system, both because of the areal expansion of agriculture and changes introduced into the system in response to the demands of the increasing population. This growth overloaded an ingenious dryland and livestock agricultural system that had evolved, under relatively stable population conditions, over centuries. Stressed beyond its capacities to produce sufficient food and other requisite products, the Luni agropastoral system was no longer able to meet the present and future needs of its local population.

Rainfed agriculture in a humid mid-latitude area: the American Southeast

The destruction of farmland due to wind erosion, and the resulting Dust Bowl in the semiarid American southern plains during the 1930s, is ranked as one of history's three worst ecological blunders due to agricultural activities (Borgstrom 1973: 23). However, throughout the 1930s, the humid American Southeast, and not the southern plains, experienced the most extensive and severe soil erosion (Healy and Sojka 1985). Farming with soil misuse in many previously forested areas throughout the region's 11 states (figure 5.7) resulted in millions of hectares of former cropland no longer being suitable for farming. Up to 1938, a significant proportion of the nation's 20.25×10^6 hectares (50×10^6 acres) of farmland abandoned due to a degrading land resource was in the Southeast (Barnes 1938). Today, excessive soil loss remains a significant problem in many parts of the Southeast, especially the hilly Piedmont (USDA 1988).

Land degradation differs markedly in this area compared to the Luni Block. In contrast to most semiarid areas, where land degradation is often manifested by a desolate landscape with few options, in this humid area of the United States land degradation eliminated many land-use options. However, it did not usually result in a land resource devoid of any potential economic or agricultural productivity. As will become evident in comparison to the Luni Block, the Southeast has largely been a region that has suffered serious but

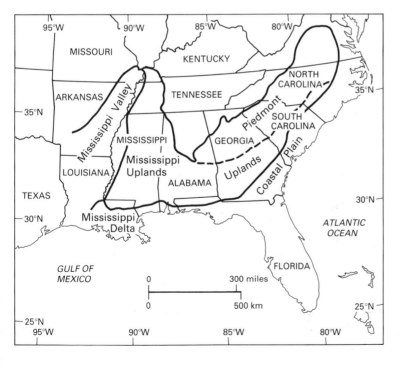

Figure 5.7
The physiographic regions of the southeastern United States (after Fenneman 1938)

relatively moderate land degradation compared with the absolute devastation in many parts of the semiarid Luni. This significant difference in the land degradation outcome is a result of both physical and human contrasts between a portion of semiarid India and a humid area within the United States.

The Southeast region can be divided into five major subdivisions from an agricultural/land resource perspective: (1) the Mississippi Delta; (2) the Southern Piedmont; (3) the Coastal Plain; (4) the Mississippi Uplands; and (5) the Mississippi Valley (figure 5.7). It is within the hilly Piedmont subdivision that particularly severe agricultural and other man-induced soil erosion occurred (Bennett 1939; Trimble 1974). It is this area that is the focus of our second inquiry into unintentional destructive changes resulting from human misuse of the land.

The southern Piedmont The southern portion of the Appalachian Piedmont begins in southern Virginia and stretches some 1,200 km (750 miles) in a southwesterly direction, ending in eastern Alabama. It is a rolling hilly region sandwiched primarily between the higher and steeper Blue Ridge Mountains in the west and the lower and flatter Atlantic Coastal Plain on the east. Local relief ranges from 15 m (50 ft) near the coastal plain, gradually increasing toward the west, to 60 m (200 ft) along its western border with the Blue Ridge (Powers 1966). Its underlying rocks are primarily igneous or metamorphic. These are covered with a deeply weathered veneer averaging between 8 and 16 m (26–52 ft), but in some areas reaching up to 30 m (98 ft) in thickness (Thornbury 1965). The Piedmont's climate is both humid and mild (average annual rainfall 1,250–1,500 mm (49–59 in); frost-free period 200–240 days).

Prior to European settlement the Piedmont was almost 100 percent forested. Only small isolated clearings likely used for farming by the scattered indigenous American Indian population broke this pattern. "Erosion appears to have been negligible on the Southern Piedmont of aboriginal times as indicated by the clear streams, the presence of dark, mature bottomland soils, and the fact that present erosion rates of undisturbed forested areas are minimal" (Trimble 1974: 34). Beginning around the 1700s and continuing up to the 1930s, exploitation of Piedmont soils for short-term profit became the *modus operandi* throughout most of this region.

From its inception, Piedmont agriculture developed into an extensive versus an intensive type of farming system. Unlike the Luni Block, an abundance of inexpensive land was available for the local inhabitants. With relatively high labor costs, farmers developed cropping practices that concentrated on yields per farmer and not production per area. Land was cleared of the forest, farmed, and abandoned once yields did not satisfy needs. Farms, when deserted, were left without vegetation cover to protect the remaining soil. Thus abandonment, in itself, was a highly deleterious farming practice that exacerbated soil loss. The exposed soils on the abandoned lands quickly eroded due to the abundant rainfall and the soils' relatively high erodability properties. This continuous misuse of land over a 150-year period was made possible by readily available fertile land immediately west of the Piedmont (Trimble 1974: 42). Once "exhausted" for agricultural needs, land was sold and new territory was cleared to the west (Craven 1926). This was the general cycle of tobacco, cotton, and mixed crop farming throughout the Piedmont. Immediate profit with minimal regard to the land resource and its sustainability is one legacy of the colonial and early American periods in this region. Soil erosion was a widespread problem during the early history of the United States (Hambidge 1938). Of all the regions, the Piedmont may have experienced the nation's most serious erosion. The widespread occurrence of clay soils, the deeply weathered subsoil, the small amount of level land, and ubiquitous poor farming practices resulted in widespread degradation, at times culminating in gully formation (Fenneman 1938). The lack of a national conservation ethos resulted in 95 percent of the Piedmont's uplands experiencing some degree of soil loss, 65 percent of the Piedmont losing its topsoil and more than 10 percent of the area also losing its sub-soil. About 1.1 million hectares (2.7 million acres) of land had active gullies before 1930 (Bennett 1929). Degraded landscapes with gully scars, poorly vegetated hillslopes and muddy rivers, due to the accelerated erosion occurring throughout this region, were the rule throughout the Piedmont.

One result of the exploitive nature of the farming system and the resulting land degradation was that between 1925 and 1960 harvested cropland decreased in every county, except one, in the southern Piedmont. Acreage in row crops declined by over 90 percent in many parts of the region (Trimble 1974: 97–8). Yet, unlike the Luni Block, where erosion and salinization result in a devastated land resource that leaves only reduced economic opportunities for future generations, today in the Piedmont only rare and isolated patches of bare soil can be found. With complete abandonment of land in this region, even though severely eroded, climatic conditions and the remaining weathered subsoil permitted natural processes to convert the area through a slow transition process from weeds to brush and finally back to forest. This natural land healing development was additionally reinforced, beginning in the late 1930s, through the intervention of major national conservation programs and by the introduction of a changing crop mix on the remaining cultivated lands. Gully and sheet erosion were curtailed through both engineering works and the planting of soil-conserving vegetation. In particular, kudzu, pasture, pine trees, soybeans, and wheat replaced cotton and corn (maize). These conservation interventions have resulted in a region where once again streams are generally clear, forest and pasture occupy a majority of the former highly eroded croplands, and tilled cropland reflects a "new" farming system that emphasizes a variety of conservation practices and new crops. No longer is short-term exploitation of the region's remaining agricultural lands the norm. Interestingly, a study undertaken in the 1980s found that a higher economic return occurs when pasture and cropland are converted to pine plantations in some parts of the Piedmont (USDA 1983). Thus the changes in land-use practices are rational both from economic and rehabilitation perspectives.

However, today's land use in the Piedmont, albeit productive, still strongly reflects its colonial and American heritage of land misuse and land degradation. First, existing soils on the uplands and hillslopes throughout the Piedmont have a lower potential than they possessed prior to the clearing of the area. Thus, potential options that are only viable with soils having a higher fertility are lost. Second, the sediment transport today, while comparable to the rates that occurred under the virgin forest, has initiated a new cycle of stream erosion in parts of the region. The excessive erosion prior to the 1940s resulted in thick alluvial deposits along valley bottoms throughout the area. Because of the effective conservation practices that have rehabilitated the degraded uplands and slope areas, the resulting clearer stream flows have resulted in renewed stream competency. "Streams (have) incised themselves into the modern alluvium lowering their beds as much as 3.7 m (12 ft). There has been intensive erosion of the friable stream banks (composed of post 1700 sediments) taking trees, fences, and good pasture or cropland" (Trimble 1974: 118). This contemporary instability along some of the bottomlands in the Piedmont, which is the result of the interaction between the prior land degradation in the uplands/slopes area and the concomitant deposition of sediment along the valley bottoms (deposition in the form of unintentional creative destruction), and the modern rehabilitation practices throughout the region, reduces the productivity of these floodplains. In particular, wetlands are being created. While less productive from the human perspective, these new wetlands open up niches for other plant and animal communities. This continuing process of wetland expansion illustrates some of the complex interactions that exist in nature. The contemporary conservation strategies directed to rehabilitate the eroded upland and slope zones have inadvertently initiated accelerated fluvial erosion on the productive bottomlands. From the human perspective, the resulting expansion of wetlands along the valley bottoms is resulting in a degrading land resource that had originally been "improved" at the expense of land degradation in the region's upland and slope zones.

The Piedmont, unlike the Luni Block, represents relative versus absolute land degradation. The humid climate found in the southeastern United States clearly results in a more resilient environmental condition than in semiarid northwestern India. It is more forgiving of human transgressions. Even though upland and hillslope soils were severely degraded and their moisture storage capabilities lowered, because moisture is not in short supply – a ramification of the humid climate – vegetation was able to reestablish itself in a relatively short time. Furthermore, because the southern Piedmont exists within the context of a strong national economy, resources derived from outside the area were made available to facilitate the rehabilitation process. Finally, while the Piedmont is again a highly productive area, it is less productive than it would have been if the degradation dating back to its early American history had not occurred. In this sense, the full range of natural resource based options was sacrificed for the enjoyment of present advantage. Instead of exporting environmental costs to a different spatial unit, these costs were transferred across time and inflicted upon future generations. This is a variation on the *sacrifice zone* pattern that is more characteristic of destructive creation. In this sense, the resource-use practices of the southern Piedmont reflect the intergenerational impacts of destructive creation that are more characteristic of the contemporary era.

Rainfed agriculture in a humid tropical area: the Amazon River Basin
Large-scale deforestation for agricultural development has been a widely recommended development strategy for the Amazon Basin since Alexander von Humboldt suggested it at the turn of the nineteenth century, during his exploration of this area. Yet attempts to clear large areas of rainforest in Amazonia, with the intent of replacing the sustainable, but low-intensity, slash and burn agriculture of the local population with pasture, or with plantation or large-scale agricultural development, have almost always failed. "Amazonia today is a very different place than it was before AD 1500 not because the climate or topography has altered appreciably but because the cultural increment has drastically changed. The degradation that has taken place in the habitat... during the past half-century provides a clear demonstration of the (human/physical) environment relationship in its most disharmonious form" (Meggers 1971: 4–5). Today alternative strategies, besides agricultural or pasture development, are being proposed for the utilization of tropical rainforests as a way of avoiding the environmental degradation that usually occurs after clearing the forest of its natural vegetation cover (Committee on Selected Biological Problems in the Humid Tropics 1982; Pollack 1992).

The following brief inquiry identifies some unintentional destructive changes resulting from the numerous attempts to develop the basin agriculturally. It illustrates the need to be aware of the complexity of environmental impacts associated with vegetation cover changes if long-term negative effects of human activities are to be avoided in Amazonia and many other tropical rainforest areas.

The general physical properties of Amazonia Amazonia is the lowland – 200 m (660 ft) above sea level or less – portion of the Amazon Basin. This excludes the Guyanan, Brazilian, and Andean highland portions of the basin (figure 5.8). The great plain

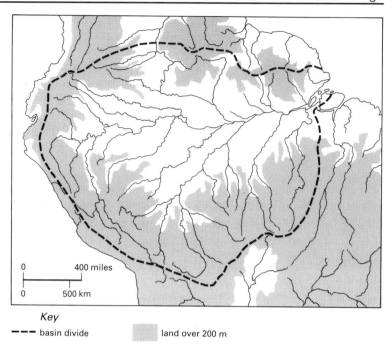

Figure 5.8
The Amazon Drainage Basin
(after Turner et al. 1990)

Key

- - - basin divide   land over 200 m

that comprises Amazonia is subdivided into three major topographic subdivisions:

- contemporary floodplains and swamps (*igapo*) that are slightly above the average river level
- river terraces (*caatinga* forest) formed during the Pleistocene
- the Amazon "Plateau" (*terra firme* forest), representing almost 98 percent of the area, at altitudes up to 200 m (Salati et al. 1990)

Unlike the other lowland areas, the *igapo* lands are flooded annually. The floodwaters bring sediments that largely originate as an erosional product in the Andean Highlands. Containing a variety of nutrients, these sediments bring to the *igapo* soils an annual flush of nutrients similar to the Nile floodplain prior to construction of the Aswan dams. However, unlike the Nile Valley, these lands are in a humid tropical setting with poor drainage throughout the year. This aspect complicates their use for agricultural activities, even though they are generally the best soils for agriculture in all of Amazonia.

The terraces and plateau lands are largely comprised of mature clayey and sandy soils. The early optimistic view of these lands was that they had an unlimited potential once cleared of their dense, lush vegetation. Today, it is a widespread perception that most agricultural ventures in the plateau have failed partly because these soils are extremely deficient in plant nutrients. This shortage in soil nutrients is the result of the continuous removal of soluble minerals (leaching) due to the heavy rainfall (annual average 2,000–3,000 mm, or 80–120 in) and high temperatures found throughout the area (Meggers 1971: 14). This perception of extreme nutrient deficiency and its direct link to the failure of agriculture is examined here immediately after the introduction of the physical properties in the section devoted to agriculture and land degradation. No matter what the causes of failure were, there is no doubt that most agricultural development schemes for the plateau lands have largely failed and that degradation has generally resulted (Hecht and Cockburn 1989).

With its equatorial location, low altitude, and prevailing high quantities of water vapor throughout the year (high cloud cover, and high absolute and relative humidities), average monthly temperatures vary less than 3°C over the annual period for the vast majority of Amazonia (Salati et al. 1990). The highest monthly temperatures (27–32°C, 81°–90°F) usually occur around the start of the rainy season; the lowest temperatures (24–30°C, 75°–86°F) reflecting increased cloud cover and higher evaporation rates, exist during the rainy season. Average rainfall varies between 2,000 and 3,000 mm/year (80–120 in/year), with the highest values occurring in the eastern (coastal) and western (Andean margins) portions of the region. Tropical rains are produced largely by convection throughout the region. This results in a regional rainfall that is generally intense and has a high erosional (erosivity) potential.

Approximately 50 percent of the solar energy reaching the rainforest is consumed by evapotranspiration processes. According to some studies (Salati et al. 1979), these high evapotranspiration rates are crucial in maintaining the precipitation patterns of the region. Their calculations indicate that only 50 percent of the water vapor resulting in precipitation in Amazonia originates directly from moisture evaporated from the Atlantic. The remaining precipitation (50 percent) is from moisture derived internally within the basin from the recycling of the Atlantic moisture by evapotranspiration. That is, the current humid climate found throughout the region is a result of a complex interaction between climatic (e.g., winds and solar energy) factors and aspects of the rainforest (e.g., transpiration, interception/evaporation, and infiltration). While the rainforest vegetation of Amazonia is primarily the reflection of the climate, the rainforest itself plays a critical role in maintaining a humid climate, especially in the central portion of the region, which is both distant from the moisture supply of the Atlantic and the orographic effects of the Andes (Salati et al. 1990).

The forest is situated in an environment in which there is always sufficient temperature and soil moisture to permit uninterrupted metabolism of the plant and animal domains. Soil moisture storage and the vegetation cover itself play a critical role in maintaining sufficient soil moisture during the short dry season that exists throughout most of Amazonia (Richards 1952; Tricart 1965). With no unfavorable season affecting its growth, the forest – but not some individual trees – is evergreen. Leaf fall from plants is not determined by a seasonal attribute such as temperature (middle latitudes) or moisture deficit (savannas); it is determined by specific attributes related to the growth pattern of the individual plants. With each plant having its own biological rhythm, there occurs a continuous supply of falling leaves to the forest floor. The massive seasonal defoliations so common throughout the middle latitudes do not exist in Amazonia. With the high temperatures and moist conditions found along the forest floor, plant litter supply and decomposition are in an approximate balance. The result is that litter is very thin throughout the area.

Species diversity is the rule throughout most areas of Amazonia. More than 60,000 species of plants have been estimated to grow in the area (Salati et al. 1990). This vegetation diversity partially reflects microenvironmental variability in topography, soil, and water table (groundwater) conditions. The result is that in most areas there is no strong dominance by any one species. This aspect alone creates a problem for the economic utilization of the forest for tree products. The low concentration of any single species under natural conditions increases the cost of bringing forest products to market. Another property of the rainforest is that because of the heights of the trees (over 40 m/130 ft for the tallest), the multistoried aspects of the forest canopy, and the very dense tree canopy, very little direct sunlight reaches the forest floor. This results in a generally sparse underbrush, except along wide rivers where the sunlight reaches the ground through the opening provided by the rivers. This interception of direct sunlight by the plant cover is crucial in preventing the soil surface from attaining high temperatures. This helps in the maintenance of a high soil moisture during the drier season by lowering direct evaporation losses from the upper soil.

Agriculture and land degradation in Amazonia The primary matrix of stability in the Amazonian ecosystem is a delicate set of reciprocal interactions among the climate, vegetation, soil, and associated organisms (Committee on Selected Biological Problems in the Humid Tropics 1982). Any unilateral change in one of these factors without compensating changes to maintain the existing interactions usually results in deleterious conditions. The natural vegetation plays a critical protective role with regard to the soil by protecting it from splash erosion, preventing the raindrops from compacting the soil, slowing runoff, and reducing the concentration of overland flow. Once the forest is removed and nontree crops are substituted, the soil climate is completely modified. Because sunlight can now directly reach the soil surface, with the removal of the complete forest canopy, the soil surface dries out after a few rainless days and soil temperatures rise. The cessation of leaf fall results in the supply of organic matter being interrupted and the decomposition of the remaining organic matter often accelerates. On many rainforest soils these changes often result in a hard thin crust developing on the surface. All of these changes encourage increased runoff once a rain event occurs; thus, the potential for soil erosion increases. Even in gently sloping areas, because the weathered regolith is often very thick and rainfall is intense, rapid degradation of the soil can occur. Gully formation and increased mass movements all too often result (Tricart 1965).

Clearly, large-scale agricultural development of the rainforest is a risky strategy, especially when annual crops are substituted because of the associated alterations in environmental conditions that accompany agriculture. In contrast, the traditional slash and burn agriculture practiced by the indigenous population and small traditional farmers minimized upsetting the existing relations among vegetation, soil, climate, and organisms over large areas. Their irregularly cleared small fields only permitted localized runoff. Furthermore, the surrounding forest provided enough shade to minimize temperature changes. Seed sources remained in close enough proximity to cleared sites to facilitate recolonization. Particularly critical was abandonment of the fields, usually taking place within a three year period of use. Abandonment prevented significant alterations in the soil's textural and fertility properties. Minimal change in soil and temperature conditions allowed the forest to regenerate itself (Jordan 1989).

With the exception of the soils situated on the floodplain, the soils in Amazonia receive minimal new nutrients to offset any losses occurring due to poor farming practices. The widespread belief is that the low nutrient supply in the rainforest soils, due to the heavy leaching associated with the ample moisture supply and the high annual temperatures, is the reason why rapid declines occur in crop productivity. This is often cited as the reason why slash and burn agriculture and agricultural attempts to develop the area have resulted in land abandonment after a few years (Richards 1952; Tricart 1965; Meggers 1971). Yet on soils rich in nutrients throughout the humid tropics, such as recent volcanic-derived soils in Central America, the same drop-off in productivity often occurs. Thus, the causes of field abandonment must occur in response to other factors too. Recent studies indicate that nutrient stocks in fields at the time cultivation is abandoned, even on relatively infertile soils, are higher than in the undisturbed surrounding forest soils (Jordan 1989). Others have found little evidence of nutrient stress at the sites that they investigated prior to field abandonment (Proctor 1983). According to Jordan (1989), nutrient stress is only one factor to consider in explaining crop productivity declines on rainforest soils. Soil erosion, changes in pH, pests, and the invasion of weeds and shrubs are some other critical factors contributing to crop productivity declines. After a few years of cultivation, weeds and other successional vegetation increase in density, and their success is at the expense of the planted crops. Thus, while crop productivity clearly decreases in the slash and burn agricultural system after a very brief period, this decline is offset by increased productivity of vegetation not coveted by the inhabitants. To counter this decline in crop yields requires more labor and capital than just clearing a new plot of land. Field abandonment allows the weeds and other nondesired plants to colonize the disturbed areas. This sets into motion a complex set of interactions that permit a rainforest to regenerate itself (Jordan 1989).

Human actions, such as contemporary major land clearing and the establishment of "permanent" pasture, alter the environmental setting significantly beyond the minor perturbations associated with slash and burn agriculture. The resulting changes in the nutrient supply, soil texture, microclimate, and other factors result in significant changes in the environment that prevent regeneration of the rainforest (Hecht 1981). In other words, land degradation results. Activities in some parts of Amazonia appear to have reached this level.

With large population growth and a shortage of arable land for the poor small farmer in the heavily populated areas along the periphery of Brazil, the rainforest was viewed by the national government in the 1960s as the logical area for both agricultural expansion and economic development. In a 1966 speech, President Castello Branco outlined a number of objectives and strategies to develop Amazonia agriculturally (Davis 1977). The goal was to establish an integrated development plan that would permit Amazonia to become an integral contributor to the Brazilian economy, meet the food needs of the growing population, and serve as a new settlement area. One of the first steps of the Branco government was to authorize the building of the Transamazon Highway to improve the accessibility of the area. For a multitude of reasons beyond the scope of this book, the "strategies" of the government failed to result in a sustainable land use in the vast majority of the area (Hecht and Cockburn 1989). In fact, just the opposite occurred in many areas, namely severe land degradation.

Degradation in the basin resulted both from development of a basic infrastructure that was built in order to facilitate accessibility into previously remote areas as well as the agricultural practices of the new settlers. Paralleling the Amazon River, construction of the Transamazon Highway linking Recife and Joao Pessoa (Brazilian northeast) to the Peruvian border and the Brasilia – Cuiabá – Porto Velho highway became roads leading to disaster (figure 5.9). In particular, the Brasilia – Porto Velho road resulted in major ecological problems, both direct and indirect, due to the opening up of vast new areas to migration and settlement. The infrastructure impacts were exacerbated due to poor road design and an inadequate database. Road construction directly resulted in significant degradation. In the process of bulldozing the land for the roads, the civil engineers learnt that the topography was not as flat as their planning strategies had assumed (Smith 1982). The result was that exposed hillsides suffered extensive erosion to such a degree as to prevent vegetation from reestablishing itself in many cases. Also,

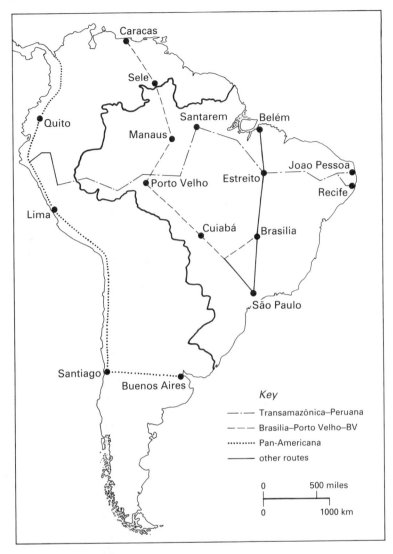

Figure 5.9
*The Transamazon Highway
and other regional routes (after
Turner et al. 1990)*

bridge construction resulted in the destabilization of many stream channels.

Government policies encouraged agricultural, cattle ranching, mining, and logging projects in the areas opened up by these roads. Implementation of these policies resulted in over 40,000 km² (15,000 sq. miles) being deforested during the decade from 1980 to 1990 in the single state of Rondnia (Salati et al. 1990). Over 30 percent of Amazonia has been estimated to be already damaged in varying degrees of magnitude as a result of these policies (Malingreau and Tucker 1987). A multitude of problems have been documented, ranging from mercury pollution associated with mining to land degradation resulting from ranching (Hecht and Cockburn 1989). Because of the magnitude of the area cleared for agriculture and ranching, these activities have been particularly harmful. According to Denevan (1981), the greatest threat to the rainforest is not agriculture but ranching. Pastures established after land clearing and planted in grasses after a few years begin to decrease in production. In this decline, pastureland parallels the pattern followed by cropland. However, unlike the traditional agricultural system in which the land was abandoned when yields declined and the rainforest was permitted to reestablish itself, ranchers adopted another strategy. They often burnt the old pasture to obtain a new flush of nutrients for the soil. This permitted new grasses to sprout with sufficient nutrients for livestock. However, each burn resulted in slightly less nutritious grasses. After a ten-year period, the nutrient level in many of the areas has decreased to such a low level that not only must the pastures be abandoned, but also it is highly questionable if the rainforest will ever be able to recover naturally (Jordan 1989). Clearly, land degradation has resulted from this short-term strategy in many parts of Amazonia. In many of the lands opened up for development in Amazonia, the region's environmental constraints were not taken into consideration. The result was the sacrifice of a viable ecosystem with potential for other opportunities – many not yet discovered – for a limited number of exceedingly short-term gains.

Summary

The irony of *destructive creation* in food production systems is that land degradation is most often derived from efforts to increase food production and security. Frequently, efforts are made to remove or ignore variability in both natural and human systems, and these only lead to vulnerability when drought or economic recession occur. Often, misguided attempts to improve the human condition initiate negative impacts on the environment because insufficient attention is directed to the spatial and temporal consequences of new developments. Equally serious is the impoverishment of one portion of the environment and one group of people without adequate compensation and without creating stable, sustainable systems to replace the components that are lost. Rather than creating *sacrifice zones* that are degraded in support of an improved *critical zone* in another location, critical resources are often sacrificed to the short-term needs of a particular community or class. In most instances, pursuit of immediate self-interest results in *opportunity reduction* for future generations. Both technology and values are implicated in these examples of land degradation. Changing norms usually reduce traditional constraints on resource use faster than new effective protective institutions can be put in place. Increased technological capabilities offer the lure of dramatic increases in agricultural output, and are put into use before their full implications for environment and society are assessed. These same degradational processes can be seen in nonagricultural systems, which are the focus of the following chapter.

SIX

Unintentional Destructive Change in Interactive Systems: Impacts of Water Demands, Energy, Transportation, and Urbanization

Introduction

In chapter 5 we addressed some situations in which unintentional land degradation resulted from human activities. In particular, direct impacts on the land resource associated with the agricultural revolution and the resulting expansion of agriculture were explored. One special category of resources not examined, but pertinent to rural land use in many areas of the world, is common property resources. In contrast to private ownership, these resources are held and managed in common by groups of users. Common property resources were widespread prior to and during the agricultural revolution. While they represent a very ancient system of land use that pre-dates the transition to the industrial age, common property resources still persist today; but their survival is often threatened by the forces unleashed by the agricultural, industrial, and technological revolutions. The strong emphasis on privatization of resources as a rational means of accelerating development is one force that places pressure upon common property resources under contemporary conditions. The relations between common properties and land degradation are the focus of the first part of this chapter.

Following the industrial revolution of the late eighteenth century, nonagricultural impacts affecting land degradation began to increase in importance. Today, with new technologies resulting in more powerful and sophisticated machinery, the ability of human actions to alter the Earth's surface significantly in order to satisfy modern needs is even less constrained. For example, in Japan an island has been created in order to permit the construction of a new international airport. The magnitude of the rock materials utilized in the landfill for creating this island necessitated the removal (levelling) of numerous hills on some other small Japanese islands. These hills became sacrifice zones for the new land mass created for the airport complex. In today's world, it is economic and quality-of-life considerations, not technical ability, that act as the major brakes upon transforming the Earth's surface to better meet specific human needs. In altering the Earth to facilitate satisfying the array of human desires, different sets of unintentional land degradation occur that were unknown or were of relatively minor magnitude prior to the industrial revolution.

Since the beginning of the industrial revolution, energy requirements and the demand for raw and finished materials have continuously increased. As the industrial revolution has evolved, in order to satisfy these energy and material demands, regions and countries have become less self-sufficient, until today phrases such as "the global market" have become common utterances among academics, business people, and governmental bureaucrats and politicians. The complex movement of raw materials and semifinished and finished goods from source and manufacturing areas to markets is ever increasing. Hence, improvements in and expansion of transportation systems, energy production, and mining have accelerated to meet the needs of modern societies. With their expansion and refinements, it was inevitable that direct and indirect impacts on the land resource would occur. The creation of one land mass and the degradation of another one to satisfy some of the air transport needs in Japan is one manifestation of how human actions have been modifying the Earth to satisfy the demands of the industrial age. Furthermore, with urbanization exploding in tandem with the improvements in the agricultural, industrial, and technological domains, another force has been added in altering the Earth's surface. After common property resources are examined, in the following sections of this chapter we explore inadvertent land degradation processes that are related to the expansion of urbanization, transportation, energy production, and mining – results of human responses to industrial and technological growth.

Common Property Resources

Common property resources constitute a special category of environments that are at risk to degradation. The common property resources examined in this section are contributors to food production. However, this involvement differs from the direct cultivation of crops: that is, it usually involves the capture of an animal and/or the exploitation of a resource that is fundamental to the production activity – but that is not the primary focus of attention here. For example, fishermen devote their attention to fish, and only infrequently to the water medium in which the fish live. Yet the water in which the fish swim is as much a common property as are the fish themselves. In this sense, the common property resource of the water is an indirect, enabling resource rather than a direct object of human interest in the fishing ecosystem. In this respect, common property resources more closely resemble primary extractive enterprises such as mining rather than an agricultural undertaking. For instance, fishing and the management of wild animal populations are characteristically capture ventures rather than the controlled agricultural activities in which domesticated livestock or plants are directly manipulated by humans. Similarly, the grass, marsh, forest, and water resources that are managed as communal property share with wild aquatic and terrestrial animals a common feature: they are in many places treated as a free good, such as air, open to use – and abuse – by all.

Characteristics of common property resources

Common property resources (CPRs) exhibit three important characteristics. First, CPRs are controlled by a group of people for the individual and collective benefit of that group. The size of the controlling group can vary from a small, informal, unrecognized gang to a larger, more formalized clan or tribe, or even to the nation as a whole. In the latter case, the state – as the representative voice of the people – claims the right to exercise authority over the "common" space and to manage the resource in what the state and its experts consider to be the best interests of the country. National parks, national forests, state-managed rangelands, and coastal waters (within 320 km, or 200 miles) are typical examples of nation-scale communal resources. The Maine lobster gangs described by Acheson (1988), who defend their economic interests by destroying the gear of unauthorized interlopers, and whose management system has proven effective despite the worries of professional managers, are representative of the informal but persistent patterns of CPR management typical of local-scale resource use.

Second, in most CPR systems it is difficult to restrict access to the resource. Local-scale social groups attempt to do this by carefully defining, usually on a genealogical basis, who is part of the community that has a right to exploit the group's resource. "Outsiders" who try to gain access to the resource commonly cannot do so unless – by marriage, long apprenticeship, or contract – they are able to establish a working relationship with members of the group controlling the resource. But for members of the fortunate group, access to the resource is literally a birthright. As long as the individual possesses the right genealogical qualifications, and has the requisite skills, technology, labor, and capitol, barriers to using resources are few. In theory, this makes it difficult to avoid overexploiting the resource base in question. In practice, social controls often operate, together with capital and labor limitations, to reduce access to and abuse of common property resources to sustainable proportions.

A third feature of common property resources is that extraction of the resource by the individual inevitably means that the resource base available to others is reduced. Thus, the grass consumed by the animals of one herder is no longer available to nourish the animals of another herder. As long as the intensity of use is low, this characteristic of "subtractability" (Berkes 1989) has no negative impact on either the environment or the group. When intensity of use increases, competition between individual resource users can produce both conflict and resource degradation. Usually, communities using CPRs have developed rules that promote reasonable equity in access to the resource while minimizing conflict and preventing degradation. Were groups not able to develop such institutions, it is unlikely that these communities would be able to survive. Since traditional CPR user groups have been able to maintain themselves for long periods, it follows that they have managed to create sustainable systems of common property management.

To a considerable extent, sustainable CPR systems strike a balance between communal responsibility and private advantage. Some scholars have stressed the difficulty of striking such a balance. Hardin (1968) argues that individuals will always be tempted to extract more from the common resource store than is their fair share, because to do so maximizes personal benefits while at the same time it shares costs among the entire community. Thus, when placed on common pasture, individually owned livestock will produce milk and other products for the owner but will consume grass that is potentially available to all. As more able and successful herders place more and more animals on the common pasture, increasing amounts of the resource are diverted to their gain. Any decrease in forage quantity and quality is a cost that is shared by all. According to Hardin, the result of this self-aggrandizing tendency is the inevitable deterioration of the resource base. Hardin called this triumph of greed and personal gain over communal good, environmental quality, and sustainable use the "tragedy of the commons."

He believed that rapid population growth was a major stimulus to adverse change in common property systems, and conspired with the structural problems indicated above to destroy prospects for sustainable resource use. Hardin's solution to the problem was either privatization of the common property resource, in order to make the user bear the real costs of excessive use, or the centralization of CPR management in the hands of governmental agencies with sufficient strength and wisdom to protect the public good. Whether through privatization or centralized management, the essential objective advocated by Hardin is to prevent open access to CPRs.

These three general features of common property resource systems, with respect to land degradation, are examined in the remainder of this section. Common property aspects of fishing, wild animal exploitation, wetlands, forests, grasslands, and water resources are briefly discussed. They are considered in the context of whether they are large-scale or small-scale CPRs.

Management of large-scale CPRs

Large-scale CPRs frequently come closest to meeting Hardin's concept of a "tragedy of the commons." This is a consequence of the difficulties encountered in large-scale systems in restricting access to the resource. Some implications of this problem with regard to land degradation are now considered.

Fish resources   The world's primary source of fish for food is the ocean, followed by freshwater lakes. In most of the Western world, the fish stocks that are located in these water bodies are treated as common property resources and, until the recent establishment of a 200 mile (320 km) economic exploitation zone controlled by riparian states, only limited and generally ineffective controls were established on the size of the catch from any particular fishery. The result of this strategy of generally unrestrained hunting of fish has been a drastic and steady decline in global fish stocks. The path of degradation in the size of the fish population has followed a standard pattern, with overexploitation concentrating on the most desired species until their numbers drastically decline. Only then, and often too late, are efforts made to limit the size of the catch. This is followed by a switch to a second set of somewhat less desirable fish species. Ultimately, fish such as shark and monk fish, which were once despised or limited to ethnic cuisine, become the gourmet choice of the elite. Eventually these "second choice" fish also begin to exhibit a dramatic decrease in size and availability, such as the contemporary drastic reduction in the shark population. These trends are paralleled by technological innovations such as larger and more seaworthy boats, improved sonar to locate fish schools, and factory ships to process and preserve fish catches far from the fisherman's base. Such developments enable fishermen to travel further and to catch fish that their artisinal ancestors would have ignored. In the process the fish resource is drastically degraded.

Management of fish resources is difficult because territories often are hard to delineate and control, and because the fish are often difficult to locate and often migrate from one area to another over their life cycle. Thus access to the resource is difficult to control and, once caught, the fish take benefits primarily the one who caught it. Yet examples of successful management of fish resources do exist. Ruddle (1989) demonstrates that Japanese coastal fishermen have a complex system of village tenure over and rights to nearby fisheries. This is accomplished in large part because the Japanese legal system makes no distinction between tenure on land and tenure rights at sea, and because modern national legislation controlling exploitation of the marine fisheries resources has been based on customary practices. These practices guarantee sharing of sea space on the part of different communities, today organized as cooperatives, on the basis of different techniques employed to catch fish and of the right to take only certain types of fish. Outsiders granted access to tenured sea space were generally restricted in the timing of their access, were limited in the technology that they could employ, and were required to pay a fee for the fishing privilege. Contemporary fishery cooperative associations are zealous in defense of their customary rights, a practice that makes it difficult to develop comprehensive management plans for coastal fisheries (Ruddle 1989: 183). However, like the lobster "gangs" along the Maine coast, such local-scale systems have proven capable of preserving the basic fish resource. Where local cooperative systems lose control of the CPR, as is the case in some but not all of the the coastal fisheries in Turkey (Berkes 1986), overexploitation by large-scale industrial fishing enterprises results in a drastic decline in the desired fish species.

Water resource degradation

The fish resource is only as viable as is the quality of the water medium within which the fish live. There are many examples in which degradation of water quality and quantity has had a negative impact on the aquatic environment. Except in some dry environments, where the principle of first use in time conveys a proprietal right in subsequent allocation of use, water is most commonly treated as an open access common property resource. Problems in overextraction of groundwater from the Ogallala Aquifer, discussed in chapter 5, are an example of how overzealous use of a common property resource can create difficulties for long-term sustainable use. Historically, rivers and lakes have been used as basic waste disposal systems by the communities that have grown up along their shores.

Such is the case with the small-scale tanning industries in the North Arcot district of Tamil Nadu, India, the effluent from which has had a serious polluting effect on the region's water resources (Bowonder and Ramana 1986). This impact is particularly severe in Tamil Nadu because the region is a low-rainfall area. Some 300 local tanning enterprises, limited in size by government legislation and local tradition, traditionally employ vegetable tanning agents that are derived from forest vegetation. Deforestation has driven up the cost of these agents and has encouraged a shift to faster chemical tanning agents. Particularly favored is chrome tanning, which employs over 200 chemical agents including cadmium and arsenic (Bowonder and Ramana 1986: 3). Limited seasonal stream flow exists to transport these agents from the local environment. The region's red loamy soil is highly susceptible to infiltration, so during the monsoon large quantities of contaminants percolate into the groundwater. Clean groundwater is one of the essential ingredients for a successful tanning industry, as well as a necessity for domestic consumption. The unrestricted dumping of polluted effluent threatens an unregulated common property resource, the groundwater, as well as the viability of the basis of the local economy, the tanning industry, and the health of the local population.

Similar impacts on water quality characterize the industrial history of most "developed" countries. To cite but one example, the Blackstone River in southeastern New England, the water power sites along which were privatized during the early stages of the industrial revolution and provided the basis for major industrial growth, became one of the most severely polluted rivers or streams in the United States. This occurred because no control was exerted over the industrial effluent dumped into the river by each mill complex within the river basin. The river pollution caused by industrial development has taken a heavy toll on the marine resources and water quality of Narragansett Bay into which the Blackstone empties (Lewis and Brubaker 1989). Although municipal wastewater treatment plants along the river have succeeded in reducing the discharge of suspended solids by 41 percent since 1983 (Kadri 1991: 1), the level of pollutants in the river remains dangerously high. Heavy metals, including cadmium and copper, and PCBs exceed the chronic levels deemed safe by the Environmental Protection Agency (EPA) and at some times and in some places along the river these pollutants exceed acute levels. The result is a waterway that is not considered safe for boating, swimming, and the consumption of any fish able to survive in the Blackstone's waters, a condition mirrored in the pollution encountered in other aquatic environments (plate 6.1).

One reason for this continuing high level of pollution is the presence of large amounts of heavy metals in the sediments trapped behind the river's many mill dams. Despite the deindustrialization that has taken place with the collapse of the region's traditional

Plate 6.1
Pollution in Massachusetts Bay

textile, electroplating, and shoe industries (James 1929; Dunwell 1978; Reynolds and Myers 1991), their historic pollutants remain. Each storm stirs up these sediments, and each breached dam of the abandoned mill factories initiates a new cycle of erosion and sends an unwanted contribution downstream and ultimately into the sea. The result is a legacy of degradation that has been inflicted on the current generation, imposing ruinous costs if the entire river course and its discharge is to be cleaned up sufficiently to reach the federal government's EPA-mandated safety standards.

Not only has the pollution severely affected water quality for humans, but it has also reduced riverine fish populations. The water quality issue is compounded by the dams created during the industrial epoch. These structures have presented an impenetrable barrier to migratory fish, who once ascended the river to spawn, providing Native Americans and early Anglo-Saxon settlers with an important source of food. Atlantic salmon, American shad, alewife, and smelt, anadromous fish once native to and common in the river (Stolgitis 1991), are today conspicuous by their absence. Nor is there much likelihood that they will return in the near future. For even if the Blackstone's water quality was improved to the point at which environmental conditions were suitable, the barrier effect of dams without fish ladders would remain both difficult and costly to overcome.

The Blackstone is a small stream, only 45 miles (72 km) in length, the management of which is made more difficult by the problems of coordinating the activities of two state governments (Massachusetts and Rhode Island). The difficulties of managing a larger water body with multiple political jurisdictions can be enormous. The Baltic Sea provides an example of the problems inherent in managing a large-scale CPR. Nine independent nation–states surround the Baltic, making common policy on its management a difficult task. Fishing is an important economic activity in the Baltic, and industrial and agricultural changes in the countries surrounding the Baltic have had a major impact on the sea's marine environment. The major changes that have occurred in the past 50 years are associated with eutrophication. Nitrogen, herbicides, and pesticide increases in land runoff (Fleischer et al. 1987), a fivefold increase in metal concentrations in the past 50 years (Hallberg 1991), industrial wastes from the pulp and paper industry (Hansson 1987) among others, and locally significant nutrient discharges from a small but growing fish farming industry (Ackefors and Enell 1990) have created conditions for major change in the Baltic marine environment. Eutrophication has been the logical consequence of these new and increasing inputs to the Baltic, and many changes in the species composition of the flora and fauna have occurred.

The increased nutrient matter in the Baltic has promoted more primary production among the phytoplankton and zooplankton populations, and this increase in available food has sparked an apparent increase in the size of the fish population, primarily cod,

herring, and sprat, that is of interest to fishermen. In coastal areas
there are reports that pikeperch have increased in abundance while
whitefish have declined (Hansson and Rudstam 1990). Overall,
despite some fluctuation in the species composition of fish population, the Baltic fish catch has increased more than tenfold in the past
50 years (Elmgren 1989).

Superficially it would appear that eutrophication represents an
act of *creative destruction* in which increased nutrient loading
results in more primary production and larger fish stocks.
However, the picture is more complicated than this optimistic
assessment would suggest. Some of the increase in fish production
is associated with more intensified fishing in the twentieth century
(Hansson and Rudstam 1990), coupled with a limited knowledge of
the size of Baltic fish stocks earlier in the century. The lack of
satisfactory time series data may mean that fish population
increases attributed to the greater food stocks produced by eutrophication are illusory. In part, the greater success of human predation on fish is likely to be a result of the dramatic decline in the
population of three species of seals and one small whale that were
major competitors with humans for Baltic fish. These marine animals captured 5 percent of the primary production at the turn of the
century, but today vestigial remnants of the seal population only
account for 1 percent (Elmgren 1989). Moreover, it is now clear
that fish ladders, once considered a successful device to insure the
survival of Atlantic salmon (*Salmo salar*), have not overcome the
obstacle posed by hydroelectric dams. By cutting off the spring
freshwater pulse that signals the existence of food and nutrient
conditions suitable for spawning, the hydroelectric industry has
had a substantial negative impact on salmon. Combined with intensified capture of salmon by offshore fishing, the pulse-removal
impact of dams has reduced the freshwater catch of migrating
salmon by 50–90 percent compared to the 1950s (Jansson and
Jansson 1988: 134). Moreover, eutrophication, by creating favorable conditions for heightening the growth of algae and other
organisms, decreases the oxygen supply. This, in turn, can have
serious impacts on the composition of marine species, the availability of food supply for animals at higher trophic levels in local food
chains, and the conditions needed for successful reproduction. Cod
in particular are susceptible to these changes (Hansson and
Rudstam 1990: 125), but other commercially valuable species
such as herring are also likely to experience negative impacts.
Thus, the balance between positive impacts that promote increased
primary production and negative conditions that lead to sudden
decline is often delicately poised. The result can be a sudden shift
of the system into a *destructive creation* mode, with an abrupt
collapse in the common property resource being managed.

Wild animals

In most countries, wild animals are a common property resource. In
many industrialized countries those wild animals that are valued for

recreational hunting purposes are managed relatively carefully with a view to controlling total population numbers and insuring adequate hunting during the designated hunting season. Some wild animals, such as elk, whose antlers are valued in Asia for their medicinal properties, are valuable enough as a renewable resource to have become the basis for a thriving ranching industry (Williams 1992), although the practices of this growing industry have excited the opposition of conservationists and animal rights advocates. However, most wild animals fall outside either category.

Elephant hunting in Zaire is an example of an animal resource that has undergone several stages of exploitation (Kisangani 1986). Before European colonization, elephants were an important meat resource for the BaMbuti (pygmy) forest people, who were able to capture the animals using either nets or bow and arrows, and to exclude potential competing groups from frequent access to the elephant resource, because these groups lacked familiarity with the forest environment and feared becoming lost if they went deep into the trees. Technological limitations prevented the BaMbuti from overexploiting the elephants. Similar overexploitation was avoided during the Belgian colonial era because – by Draconian measures – colonial authorities were able to prevent the natives from hunting in national parks and reserved lands and successfully to limit the number of licensed hunters. Hunting was also largely restricted to elephants older than 55 years of age. Given the long gestation period of the elephant and the eight to ten years required before reaching sexual maturity, restricted hunting of younger age cohorts preserved the reproductive structure of the elephant population (Kisangani 1986: 150). Increased off-take from younger age cohorts only occurred when economic recovery pressures after World War II and late-1950s independence processes relaxed government control. After independence, efforts to establish small indigenous cooperative hunting groups failed as the demand for ivory in the international marketplace increased. African elephants were particularly vulnerable to ivory hunters because both males and females have tusks, thus exposing the mature breeding population to potentially disastrous pressure. The use of poisoned water and fruit to kill elephants indiscriminantly, regardless of age and sex, became a common practice in many parts of Zaire and East Africa, as opposed to the more selective reliance on firearms to kill mature males, which characterized the colonial era. At independence, Zaire possessed an elephant herd of approximately 150,000 individuals, distributed throughout the country (Kisangani 1986: 156). While figures on the size of the contemporary herd do not exist, elephants have disappeared from all but a few isolated parts of the country. In this rapid decline in its elephant population, Zaire mirrors the situation in much of sub-Saharan Africa.

A contrasting example of an animal that has not received management attention is provided by coral reefs. Marine animals that are common property resources, coral reefs are a vital element in many tropical ecosystems. Coral is a critical component in preventing coastal erosion in many tropical areas. Storm waves expend most of their destructive energy on the coral reef and not the shoreline. When coral reefs are destroyed, coasts experience erosion, a direct form of land degradation. Corals are also important marine ecosystems because of the diversity of species that they help to sustain (Kuhlmann 1988). In Southeast Asia, major portions of human protein intake come from coral reefs (McManus 1988). The algae that coral reefs produce sustains many economically important fish in the open ocean, pearl mussels are often harbored by coral reefs, and tourists are attracted to the aesthetics and biodiversity of coral habitats. All of these features make coral reefs an extremely important habitat.

Yet corals are highly sensitive to fluctuations in temperature and to changes in water quality, especially sediment. Corals throughout the tropics are at risk to degradation. Deforestation in nearby terrestrial watersheds increases sediment deposition, which can kill coral. Dredging operations around harbors or for mineral extraction can increase turbidity to dangerously high levels and reduce light. Without adequate sunlight, reef building cannot continue. Pesticides and herbicides from agricultural operations, industrial effluent, insect control spraying, blast fishing, physical damage and oil discharge by passing ships, and removal of coral parts for souvenirs for the tourist industry or the growing hobby of marine aquariums, all have a negative impact on coral survival (Williams and Williams 1988; Shinn 1989; Ward 1990). The impact is analogous to poaching of wild animals, such as elephants (Kisangani 1986), whereby the short-term users benefit while the larger community suffers.

Management of small-scale CPRs

Small-scale CPRs have a long history in much of the world. They have been the primary way in which common property resources have been managed over time. It is a fundamental difference in scale, and therefore in the immediacy of local control, that distinguishes small-scale CPRs from the larger-scale CPRs discussed in the previous section. Small-scale CPRs are generally managed successfully because the community that manages the resource and the beneficiaries of the management are synonymous. Any violations of use rules and any negative change in the quality of the resource being exploited are observable to everyone. Action can be taken quickly either to exclude mismanagers from access to the resource or to lighten the pressure being placed on the resource by establishing temporal or quantity limits on use of the resource.

There are many examples of small-scale CPRs (NRC 1986; Berkes 1989). Draz (1990) examined the *hema* (protection) concept as applied to grazing, fodder, and forest resources in the Arabian Peninsula and found that approximately 3,000 contemporary or historical examples existed in Saudi Arabia. Most of these *hema* zones were organized at the village or tribal level and relied on local sanctions and leadership to enforce compliance with community rules. Hobbs (1989) reports the existence of a strong environmental ethic among the Khushmaan bedouin of the Egyptian Red Sea Hills that has developed out of long familiarity with – and sometimes abuse of – their native habitat. In Syria, the traditional concept of protection has been resurrected after decades of neglect to form the basis of contemporary rangeland management (Shoup 1990).

In Morocco, there exists an institution, the *agdal*, that regulates pasture access. An *agdal* is a pasture area that is subject to explicit rules governing access to the resource in terms of who can graze, determining when grazing can begin and when it must end, and setting limits on how many animals can be placed on the pasture. In the western High Atlas, at Oukaimedene, an *agdal* has been in operation since the seventeenth century under conditions of environmental stability and with remarkably little conflict (Gilles, Hammoudi, and Mahdi 1986). The *agdal* is especially important in controlling grazing in mountain environments. Both the traditional pattern of nomadic pastoralism and village-based transhumance depend on the existence of productive high-altitude pastures upon which herds can graze at some point during the summer dry season. Management of an *agdal* is by consensus, and is exercised by a controlling council and one or two particularly knowledgeable local resource managers. As long as community solidarity is maintained, their decisions are easy to enforce. However, whenever consensus breaks down, conflicts emerge that can undermine the viability of the *agdal* system (Artz, Norton, and O'Rourke 1986).

Often, these disputes lead to the intervention of central governmental agencies. At other times these non-local institutions intervene in pursuit of a differing definition of the common good. For these institutions, the nation–state represents the best interests of all its citizens, and so should control the use and management of common property resources. In North Africa, this led colonial powers to replace tribal order with colonial disorder (Bedrani 1991) as traditional CPR institutions were undermined in favor of European settlers and local privileged groups (Bencherifa and Johnson 1991). Lands alienated from community control placed particularly severe pressure on those areas that continued to be managed as CPRs. Few local CPR groups were as successful in resisting the alienation of control over and access to local resources as the Bontoc and Kalinga rice farmers of northern Luzon, who managed to stave off efforts by the Philippines government to construct a series of hydroelectric dams in their territory that would have flooded their fields (Malayang 1991). Most local CPR groups have experienced difficulties because their tenure rights to resources were not formally documented, and so they had no legal status with the colonial or post-independence governments. Even where such rules were codified, as with the *dina* laws regulating the interactions of herders, agropastoralists, farmers, agrofishermen, and fishing groups in the inland delta of the Niger River in Mali, post-independence government intervention has undermined the traditional local authority system of masters of land and water who monitored use and settled disputes (Moorehead 1989). These management institutions were already under stress from more than a decade of drought, but government institutional intervention has contributed to substantial environmental decay. In all such situations, in which a central government is unable to replace successfully a functioning local CPR system with a management regime of its own, the result is a scramble for access to the resource, which results in severe and rapid land degradation.

In India, disputes over the legal status of village forest councils have placed in jeopardy small-scale CPRs that were effectively managed. These local groups were successful because they were able to limit the amount of wood taken, restrict hunting and collecting in the forest to appropriate seasons, protect young animals and plants until they could reach harvestable size, forbid the cutting of "keystone resources" such as fig species and *Prosopis cinerarea* that were valuable food sources to wildlife and domestic stock, and completely protect special places that could serve as genetic reservoirs for forest regeneration (Gadgil and Iyer 1989: 247). Income from user fees was reinvested in employing a watchman to protect the CPR. The role of spiritual values in nourishing successful management of CPRs in this and other instances is particularly important, since these values create a moral imperative that justifies individual sacrifice in favor of a common good. Gadgil and Iyer (1989: 241–2) also argue that Indian caste society played a major role in successful CPR management, because these tightly knit groups were linked to the exploitation of particular ecological niches, yet were connected in multicaste villages into a network of symbiotic relationships. Each group was small enough to achieve consensus in the use of its particular CPR, but was not able to survive if it failed to respect the CPRs of other groups whose specialized production was essential to community well-being.

Conclusions

The management of large-scale common property resources is fraught with difficulty. Typically, mismanagement occurs as a function of scale. When the common resource is defined in very large terms, it becomes increasingly difficult to restrict access to the resource. If all of the American public owns the national park system in common, it becomes very hard to restrict access to the resource without inequity and the emergence of political complications. Yet too many people in the park can destroy the basic amenities that visitors seek. Similarly, the difficulty of reaching agreement between sovereign states makes it hard to control undesirable impacts on common resources, such as air or water, that are considered to be free goods with equal rights of access for all.

However, the existence of CPRs need not result in a "tragedy of the commons" as postulated by Hardin. There are three primary conditions that must be met in order to avoid degrading the commons. Historically these conditions have been best fulfilled in small-scale CPRs. First, many small-scale CPRs are not managed as open-access resources upon which everyone has an equal claim. Historically, rights to use common resources were restricted to particular uses, at specified times, by a limited array of individuals (Cox 1985). These chartered limitations were matched also by, and often based upon, physical constraints. Thus, in medieval England, farmers who had the right to pasture animals on the uncultivated "waste" in the summer were unable to keep more animals than they could sustain in winter on the post-harvest residues and hay that their tenancies or shares in common fields could produce (Cox 1985: 54). The critical feature of small-scale CPRs is that their use is not the product of a free-for-all, but rather is replete with limitations. And the primary restriction is that only certain people have a legal right to use the common resource.

Second, access rights and usage limitations are carefully monitored in all successful CPR systems. At the scale of the village or clan, violations of community norms in the use of CPRs are quickly known. In consequence, social pressure can be brought to bear on violators of the rules of use (Berkes et al. 1989: 93). When this community pressure is backed up by legal sanctions, it is very effective in controlling access to the CPR and limiting degradation. In many cases (NRC 1986; Berkes 1989) local groups are able voluntarily to impose limitations on their present use of a CPR in the interest of long-term sustainability. When this group is not only small and based on common kinship but also has interacted over a long time – preferably for generations – it is likely to possess the shared value system that is needed in order to distribute resources equitably, thus reinforcing support for sustainable resource management.

Third, this pattern of voluntary restraint is very much dependent upon local control of the CPR. Not only does this keep management of the common resource local, where action and impact can be readily observed and corrected, but also it insures that the products extracted from the CPR, or the resources generated from the sale of these products, will be available locally. In this way both benefits and costs are spatially associated. Local control insures that the group that uses and manages the resource is kept small. A small user group insures that commonality of purpose and community social control are maintained. Fragmenting resource use into a multiplicity of user communities, argue Gadgil and Iyer (1989), is the way in which Indian society has insured sustainable management of common property resources. Caste, craft, and community exploit a diversity of separate ecological niches. Each activity group has a vested interest in the sustainability of the CPR being exploited. Among pastoral groups this interest in sustainability encourages a flexible definition of resources and allows other groups to increase their security by insuring access to neighboring tribal territories under conditions of interannual scarcity (Perevolotsky 1987). In highly fragmented social structures or variable environmental settings, a premium is placed on tying all groups together spatially and socially by a loose structure of mutual obligation, need, and reciprocal exchange that minimizes competition and conflict. The result is a *mosaic principle*, one in which each group focuses on management of its particular niche, and the complete picture is only available when one steps back from the individual fragment far enough so that the entire composition is visible.

Urban Growth

One important repercussion of the interaction between the industrial and agricultural revolutions has been rapid population growth, especially with regard to urban areas. In 1800, only one city, Beijing, exceeded a population of 1,000,000. Worldwide, only 74 cities had populations greater than 100,000 and, with the exceptions of Mexico City and Cairo, all were located in either Europe or Asia. Today, at least 199 cities have populations in excess of 1,000,000, with every inhabited continent having a share in this increase (Chandler 1987). This urban growth is associated with the surge in overall population growth, made possible in part by the increase in food production as well as increasing industrial and commercial activities, by-products of industrialization.

Originally, urban growth was primarily concentrated in the industrial countries. Today, it is a worldwide phenomenon, with some of the highest rates taking place in the developing world. Land scarcity in rural areas and the search for economic opportunities are two causes that contribute to the explosive growth of cities in these nations. No matter what the driving forces causing urbanization may be, as populations increase urban areas expand areally. Urban sprawl, associated with twentieth-century urbanization, consumes significant acreage of formerly highly productive agricultural lands in order to house increasing populations and to provide land for other urban activities. As land use is metamorphosed from rural to urban, vegetated land cover is to a high proportion replaced by sterile concrete and asphalt; productive wetlands are filled and replaced by urban structures; and pollution associated with many urban activities results in a spectrum of chemical and biological changes that often degrade the affected areas (Flawn 1970: 119).

During the establishment of older urban centers, a critical criterion that determined the location was proximity to fertile land (Mumford 1961). With the continuing urban explosion, this juxtaposed highly productive farmland is being destroyed even in countries that have a shortage of fertile lands. Urbanization, with all of its positive and negative feedbacks, is truly a worldwide phenomenon. It is one of the most critical events – including both anthropocentric and nature-driven processes – shaping the Earth's surface. Running the whole gamut from pre-Columbian *chinampa* agriculture in the Basin of Mexico and ancient agricultural lands in the lower Nile Valley to the modern citrus groves near Orlando (Florida), productive agricultural lands are being consumed by urban sprawl. Today, the highly productive *chinampa* agriculture (Xochimilco's floating gardens) occupies less than one-hundredth of the area that it did during the Aztec Empire (Oterbridge 1987). The continuing expansion of Mexico City is likely shortly to lead to the complete disappearance of these agricultural lands (Ezcurra 1990) and a significant percentage of all agricultural lands in the Basin of Mexico. Similarly, with the growth of Disney World and other recreational and entertainment activities, orange groves are becoming rare in the immediate vicinity of Orlando.

From the strictly economic perspective, in most cases the conversion of land to urban uses is extremely rational. Yet, with the reduction in biomass production inherent in the conversion from rural to urban land use, urbanization can be viewed as directly resulting in the degradation of land. The consumed lands required for urban functions become a sacrifice zone due to urban sprawl. Furthermore, as urban areas grow, they require the development of ever longer supply lines to satisfy cities' energy, food, water, and material needs. In Arizona, in the dry American Southwest, rapid urban growth is placing increasing demands on the area's limited water supply. Water rights are being purchased from farmers by cities to meet their water craving. This reallocation of water from irrigation to urban use results in the contraction of productive rural lands in order that urban growth can continue unabated. Not only are the lands adjacent to urban areas affected, but also, in order to satisfy the energy, water, and material requirements of cities, additional sacrifice zones often result. These sacrificed areas frequently are not directly connected to, and sometimes are quite distant from, the urban area that caused their occurrence.

Surficial land changes associated with urban areas

To exist, cities require construction. Construction materials move in large quantities into any active community. For example, in the Los Angeles area during its period of rapid growth, per capita consumption of sand and gravel was in excess of four tons per year (Reining 1967). The sand and gravel pits supplying these materials usually experience various degrees of degradation as the minerals are removed. Often, these disturbed areas characteristically result in a pockmarked landscape. These pitted areas are examples of sacrifice zones that, while spatially detached from the urban area, represent land degradation resulting directly from urban activities (plate 6.2).

Two ubiquitous by-products of urban construction in general are that: (1) topography is greatly altered; and (2) the permeability of the area, especially in nonresidential areas, is greatly reduced. Cut-and-fill operations are an inherent property of shaping landscapes

Plate 6.2
A gravel pit at Northboro,
Massachusetts

to meet the demands of urban activities (Flawn 1970). Exemplary of the topographic alterations needed to meet the needs of an urbanized society, between 1803 and 1988, 1,604 hectares (3,960 acres) of land were reclaimed from the immediate waters surrounding Boston as a way of meeting both the space needs of the growing population and requirements of the infrastructure, especially airport expansion (figure 6.1). Cut operations were also widespread throughout this period. Originally, the central core of Boston had a hilly topography. Most of the hills were leveled both to use the material for the filling up of the Back Bay, ponds, and wetlands as well as to facilitate construction (Wilkie and Tager 1990). A remnant of Beacon Hill, the site of the seat of state government for Massachusetts, today rises only slightly above the broadly leveled downtown landscape of modern Boston, which bears slight resemblance to the original wetland and hilly terrain of the area. In general, the removal of earth materials in cut operations removes an area's soils and brings less fertile materials to the surface. This represents a permanent degradation in the potential of the land for biomass production. This degradation is reflected in most landscaping operations for modern office buildings built in any city. To establish gardens for aesthetic reasons, during the construction phase topsoil has usually to be transported to the building sites and placed on the grounds before landscaping with any significant vegetation growth is possible.

Not only are topographic changes and decreases in fertility of surface materials features of urbanization, but a significant percentage of urban surface materials bear little resemblance to their natural properties. Almost any modern urban center has had a large proportion of both its vegetation and soil replaced by concrete, asphalt, and other building materials. With the exception of

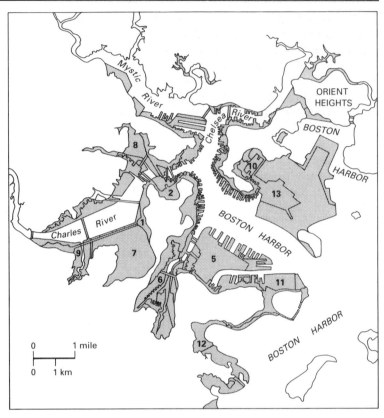

Figure 6.1
*Land reclaimed from Boston
Harbor*

	Location	Approximate acreage	Dates		Location	Approximate acreage	Dates
1	West Cove	80	1803–1863	8	Charlestown	416	1860–1896
2	Mill Pond	70	1804–1835	9	Fenway	322	1878–1890
3	South Cove	86	1806–1843	10	East Boston	370	1880–1988
4	East Cove	112	1823–1874	11	Marine Park	57	1883–1900
5	South Boston	714	1836–1988	12	Columbus Park	265	1890–1901
6	South Bay	138	1850–1988	13	Logan Airport	750	1922–1988
7	Back Bay	580	1857–1894				

parks, close to 100 percent of the surface of modern cities' commercial cores is completely occupied by buildings, sidewalks, and streets. Situated in the heart of downtown Boston, Boston Common, where once it was permissible for livestock and horses to graze, is now surrounded by a densely populated urban conglomeration that is largely inhospitable to most living creatures except the urbanized human. Residential areas undergo similar changes, but at a lower intensity. The urban landscape, in contrast to rural areas, is significantly covered with impermeable materials, largely sterile with regard to plant growth. With the exception of animals and plants directly linked to humankind (pets, shrubbery, weeds, rodents, and so on), the potential for biomass production is curtailed drastically in urbanized areas. Thus, without even considering derelict urban land (Kivell, Parsons, and Dawson 1989), the process of urbanization today almost always results in a form of land degradation. Few, if any, modern cities have not undergone drastic alterations in their vegetation and topographic situations similar to those of Boston. These changes are an intrinsic property of twentieth-century urbanization. They are not solely concentrated in the industrialized developed world, but are worldwide in their distribution (Lewis and Berry 1988).

Satisfying the needs of urban areas

As cities increase in population, a number of changes take place in the relationship between the urban area and the environment. As urban populations grow, a lower percentage of urban demands with regard to energy and water supply, building materials, food, transportation, and waste disposal can be provided from local resources. Not only do the impacts of each of these activities on the environment increase in magnitude with urban growth, but they often occur in distant places and in areas with relatively low populations. Because of this spatial discordance, urban populations are often shielded from the negative impacts of their activities and the demands that are placed on the land resource. This acts to dampen any immediate corrective responses to negative environmental impacts driven by urbanization.

Energy

Electrical energy, a necessity for any urban area, primarily is generated by either hydro, coal, gas, oil, or nuclear fission. Each of these power sources has impacts on the land resource. Reservoirs required for hydroelectric generation flood lowlands. In the process hectares of fertile lands and areas of high biodiversity are often destroyed (Barrow 1991). These flooded areas become sacrifice zones for the well-being of a distant urban area. Associated with the mining of coal for thermal power plants are a variety of environmental consequences that have direct bearing on the land degradation problem. Spoil, a by-product of mining operations, covers large tracts of land and renders them largely useless for any beneficial land use (plate 6.3). In the past and in many contemporary areas, these spoils are areas that generate acidic waters. The acidic runoff from the overburden degrades both surface and ground waters, often killing vegetation. Without proper pollution control on power-generating plants that utilize coal, sulfur dioxide emissions released into the atmosphere can destroy or degrade the surrounding area's vegetation cover, which in turn initiates soil erosion. Massive land degradation resulting from both the mining of coal and its utilization in power generation to meet urban and industrial needs has occurred in Saxony (Germany), Silesia (Poland), and Bohemia (Czech Republic). The land degradation resulting from these activities is of such a magnitude that this part of central Europe has been referred to as a "Bermuda triangle of

Plate 6.3
Coal mine tailings near Hirwaun, Wales, UK

pollution" (Ministry for Economic Policy and Development of the Czech Republic 1991). The other mineral energy sources – oil, gas, and nuclear fuels – likewise can have deleterious effects on the land resource. Ramifications of the extraction of these mineral fuels are discussed in following sections.

In many Third World cities, the major source of energy is wood, often in the form of charcoal. With the rapid growth of urban populations in Third World nations, the demand for charcoal has increased drastically. For most urban centers in Africa, the demand for wood and charcoal is growing much faster than the forests can reproduce. Around a significant number of towns and cities, the environmental impact of this phenomenon is intense. For example, since 1960, the zone of wood exploitation for charcoal has moved more than 300 km (185 miles) south of the Khartoum market (figure 6.2). Today, it continues to move southward. In this semiarid setting, the removal of tree cover has been a contributing factor to the desertification occurring in some parts of the region (Berry 1983).

Building materials
Construction materials (e.g., sand, stone, and clay) are needed in large quantities to meet the various needs of urban areas. Besides the buildings themselves, all forms of transport infrastructure need construction materials if supplies are to flow both toward and out of cities. Modern highways cover about 10 hectares of land for every kilometer segment (40 acre/mile) (Flawn 1970). Each hectare of land covered by roads, vital to the viability of the urban areas that they connect, represents a sacrifice zone for urban success. Furthermore, as with other forms of mineral extraction, construction materials needed to build and maintain the highways scar the land and add dust to the air. Unlike some forms of mineral extrac-

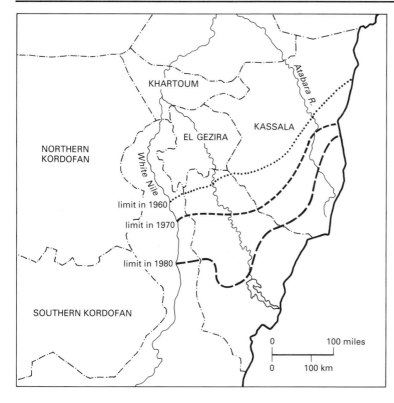

Figure 6.2
The expanding wood and charcoal exploitation zone south of Khartoum, Sudan

tion, which produce toxins that kill vegetation, the production of construction aggregates is relatively benign: but it often leaves behind a degraded, pockmarked landscape devoid of soil cover. In New England, many of the former gravel pits, a primary source of construction materials, upon exhaustion of their sands and gravels, became waste disposal sites (sanitary landfills) for the urban area that had been the demand catalyst for the extraction of construction materials in the first place. Prior to government controls, which today require the protection of groundwater supplies, both household and industrial wastes were often buried in these construction pits without proper safeguards. One result of this past practice is that today there is widespread groundwater pollution throughout many parts of New England, rendering this subsurface water useless as a water supply in the affected areas. Yet, when managed correctly, these landfill areas can be restored for some limited types of urban development, such as playing fields. However, the land-use options are more restricted than if they had not been mined. From this perspective, they represent a clearly degraded condition. Additional relations of energy and mineral production with land degradation will be examined in greater detail later in this chapter.

Water supply

Securing a sufficient and reliable water supply has always been a critical resource need for urban areas. From ancient Jerusalem to contemporary Los Angeles, urban growth has required public works that attempt to offset the variability and limitations of the natural supply of water to the area. To attempt to meet the needs of population growth, plus the urban requirements of industrial, recreational, commercial, and sanitary activities, most modern cities, even in humid areas, must import water from distant areas to supplement local supplies. Whenever water transfers occur, a set of changes take place in the affected hydrologic system (Schneider, Richert, and Speiker 1973). More often than not, these alterations result in land degradation.

Dams Where feasible, dam construction during the past 100 years has been an almost ubiquitous strategy for providing a reliable water supply for urban areas. The reservoir that is created behind the dam collects and stores water during periods of abundant supply or low demand, for usage when the natural, unregulated supply would be deficient. It also allows for the collection of water from a large area (catchment), often quite distant and in an area of low water demand and low population density, to be transferred to a high consuming urban area. Given the water demands of modern cities, unless an urban area is situated adjacent to a large source of fresh water – such as Chicago on Lake Michigan or Pittsburgh astride the Ohio, Allegheny, and Monongahela Rivers – water transfers from areas of low water use to urban areas are a necessity. Contemporary urban areas require water supplies in excess of the potential of water production directly derived from the lands occupied by the urban area.

Even though Boston is located in a humid climatic area, by the 1920s additional water supplies had to be secured outside the immediate urbanized area, due to the 60 percent population growth that the city had experienced since 1890. To meet the needs of metropolitan Boston, construction of the Quabbin Reservoir was initiated in 1928 in rural west-central Massachusetts, approximately 90 km (60 miles) from the urban area that it was to supply with water. If the Boston metropolitan water district did not have the ability to transfer water from distant river basins, severe water shortages would continuously exist. The only alternative to inter-basin water transfers would be to develop entirely new water use and water supply strategies. While local rivers exist in closer proximity to Boston than the distant Quabbin watershed, their ability to meet the water demands of the metropolitan area were constrained by two factors. First, their water quality was already low in the 1920s due to pollution that resulted largely from industrial wastes. Second, it was impossible to build reservoirs on these river systems since the lowlands that would be flooded by dam construction were already occupied by densely populated urban communities.

An inherent property of all reservoir systems is that they require a sacrifice zone. The lands flooded by the reservoir for water storage represent an absolute loss of the land resource. The sacrifice zone attribute is further amplified when reservoirs are to be utilized primarily as drinking water sources. In this case, multipurpose uses of the reservoir's waters, such as recreation, are often limited in many cases as a strategy to protect the water quality for the urban area. The Quabbin is the world's largest artificial lake used solely as a domestic water supply. Its construction required 311 km^2 (120 sq. miles) to be expropriated for its management, of which 101 km^2 (39 sq. miles) is flooded by the lake. With regard to the Quabbin Reservoir, not only were four towns and several small villages obliterated, but also parts of a forest and agricultural ecosystem were destroyed (figure 6.3). All future options for land use of the flooded lands were removed for the benefit of a distant, more politically powerful metropolitan area. Land-use controls on the immediate catchment of the reservoir have resulted in a managed forest where only a few tree species are permitted to grow and any active land use is prohibited. For example, access into the forest for picnics and hunting is rarely permitted. These restrictions place some limits on both the biomass production of the area and its utility for the local inhabitants.

The Los Angeles/southern California area further exemplifies the almost insatiable demands that urban areas can place on regional water resources. This is an especially dramatic example because southern California is situated in a dry climatic zone that has experienced an explosive population growth. Seventy percent of all stream flow in California occurs north of Sacramento, while 80 percent of California's population lives in metropolitan areas from Sacramento south to the Mexican border. To supplement the limited ground and surface waters that could be captured in Los Angeles' local reservoirs, as early as 1905 plans were initiated to transfer waters from the Sierra Nevada (Owens Valley) to southern California. With southern California's continuous development, by the 1920s it was clear that if population growth was to continue in this region, additional water supplies would be needed. These water demands culminated in the construction of three major aqueducts to meet both the urban and agricultural needs of southern California. Today the resulting complex hydrological system transfers water from river systems over 320 km (200 miles) away. Runoff from snow falls in the Sierra Nevada and stream flow directly pumped from the Colorado River (river discharge derived primarily from snowfall and precipitation occurring in the Rocky Mountains over 1,000 km (600 miles) away), are diverted and transported by this aqueduct system to bring life and prosperity to arid southern California's thirsty farms and urban areas.

The Owens Valley, prior to the diversion of its waters, was once occupied by fertile verdant land fed by the meltwaters flowing off the nearby Sierra Nevada Mountains. Today it is a degraded, dry dust bowl. For all practical purposes, all of its waters have been funneled to meet the urban needs of Los Angeles (Hundley 1992). The contemporary conditions of the arid, degraded Owens Valley illustrate the concept of a sacrifice zone *par excellence*. This competition between traditional agricultural uses of water and growing urban and industrial demand for an increasingly scarce water resource (Englebert 1982) dominates discussion of water policy in the American West. Water-use competition is likely to force out of production those agricultural districts that are unable to achieve

Figure 6.3
Areas flooded by the Quabbin Reservoir

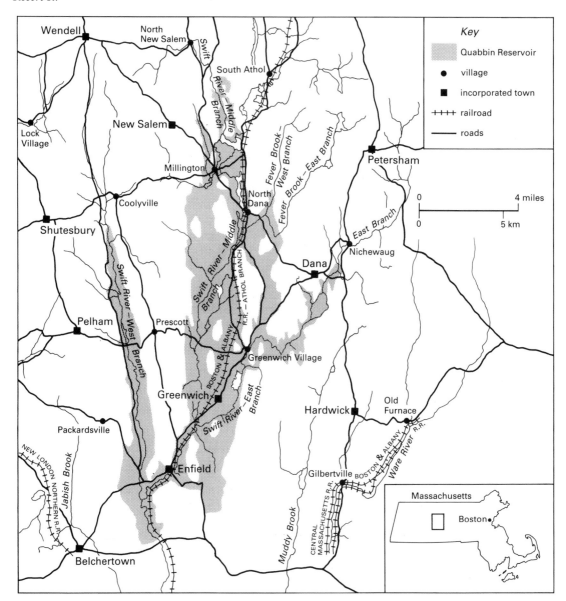

major water-use efficiencies (Englebert 1984), and thus increase the spatial extent of a type of land degradation analogous to that of the Owens Valley. Similarly, the diversion of the Colorado's waters, to both California and Arizona, is of such a magnitude that during most years it no longer has any significant freshwater discharges into the Gulf of California (Gulf of Cortez). The result is a very degraded Colorado Delta at the river's mouth in Mexico (Babbitt 1991).

The building of reservoirs to meet the water needs of urban areas always results in the degradation of the land in the areas directly impacted by the dam. Other changes, such as water table alterations and salinity changes in river flows, can result in an additional degrading land resource. Changes in the hydrologic balances associated with dam construction and water diversions are particularly prone to result in negative impacts in sensitive areas such as arid lands. Los Angeles and surrounding cities and towns exemplify some of the problems resulting from urban demands and the approaches that historically were developed to satisfy them (Hundley 1992).

In southern California, where local aquifers existed, communities often have over-pumped them as another means of satisfying short-term water demands. The continuous pumping of the groundwater to meet the needs of many coastal urban centers, such as Beverly Hills and Long Beach, has resulted in saltwater intrusions (Miller 1993: 235). By the 1940s the salt/fresh groundwater boundary had progressed up to 3 km (2 miles) inland in the vicinity of Los Angeles. The resulting saltwater intrusions have caused a loss of future options for the use of local groundwater supplies. While this does not *per se* represent land degradation, the decline of the groundwater resource has altered some inland areas where former freshwater springs have become saline.

In summary, to meet the water demands of a growing population in the dry populated areas of southern California, distant river flows have been drastically reduced by the diversion of their waters into the aqueduct systems feeding this area. In response to the reshaping of their region's natural water distribution to satisfy human demands, a series of other changes have upset the hydrologic system's *status quo*. The curtailment of sediment reaching the oceans has resulted in beach and coastal erosion (see chapter 5). The drastic reduction of the Colorado River's discharge throughout the lower portion of its basin has resulted in salinization problems as well as a general degradation of its delta, especially in the Mexican portion of the system. In 1922 Aldo Leopold described the Colorado Delta as a "milk and honey wilderness." The surface was a series of green lagoons rich in wildlife and vegetation. Today, with the Colorado flows almost completely diverted for both agricultural production to feed urban areas or for direct urban water utilization, the delta ecosystem shows little resemblance to its natural state. What water remains in the river is so saline (the average is more than 700 ppm) that only degraded vegetation that can exist on highly alkaline soils is widespread. Desert sands are found in areas that once were highly productive (Babbitt 1991).

Groundwater   Over-utilization of groundwater supplies to meet urban water demands, as well as lower surface water infiltration due to new surface materials associated with urbanization, can result in land degradation. Three examples, Long Island – a part of the New York Metropolitan area – Houston, Texas, and Tokyo, Japan will be briefly presented to illustrate further some linkages between urbanization and unintentional land degradation under two different environmental situations.

The bedrock of *Long Island* (figure 6.4) is overlain with unconsolidated deposits, up to a maximum of 650 m (2,000 ft) thick, of largely glacial and alluvial origin. The majority of these deposits are

Figure 6.4
The location of salty groundwater in southwestern Long Island (after Swarzenski 1963)

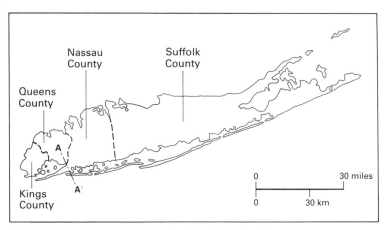

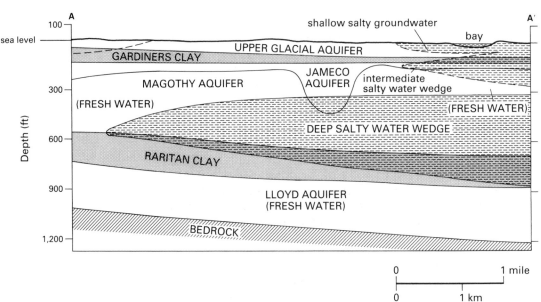

permeable and permit a rapid infiltration of precipitation. With an average annual precipitation of slightly more than 1,000 mm (40 in) and an evapotranspiration rate under 500 mm (20 in), a significant fresh groundwater reservoir of between 37–75 trillion liters (10–20 trillion gallons) existed under the island prior to pumping (Heath, Fosworthy, and Cohen 1966). Under natural conditions (pre-development), an overall equilibrium existed between groundwater recharge and discharge. This hydrologic balance resulted in a stable interface between fresh and salty water that was situated either at the coast or offshore depending on the nature of the underlying surface deposits.

As the New York metropolitan area expanded onto Long Island, a number of changes in the groundwater system occurred. First, cesspool and industrial pollution infiltrated into the shallow portions of the water table. Second, increasing construction resulted in an increasingly impermeable surface. This, along with the development of sewer systems that steered the runoff directly to the ocean, lowered the recharge of the groundwater. Third, increased pumping of the underground reservoir further contributed to the lowering of the water table. The result of these changes was that the saltwater/ freshwater interface began to migrate landward in those zones where pumping exceeded recharge. Moreover, pollution in specific areas, especially along the western portion of the island that was part of New York City, reached sufficient levels as to prevent the groundwater from being used as a water source (Heath, Fosworthy, and Cohen 1966). In those areas where either polluted or saline waters now discharge into streams and ponds, the affected areas have clearly degraded. The degradation of the groundwater in the western portion of the island has resulted in the need to supply these areas from the New York municipal water system, which imports waters originating north of the city.

In an attempt to reverse or at least stabilize the depletion of the groundwater reservoir in both the suburban and the few remaining rural areas of Long Island, a number of policies were implemented. First, pumping of water is monitored to attempt to keep it in balance with recharge. Second, treated waste waters are returned to the shallow glacial deposits to recharge the aquifer. Third, few storm sewers now empty directly into the coastal waters: instead, a series of pits throughout the island (plate 6.4) have been constructed as catchments for storm runoff. These pits (ponds when full of water) act as groundwater recharge zones. However, because the function of the pits is solely to recharge the aquifers, these are areas that have degraded in terms of actual biomass production. Grasses and scrub saplings now cover the land. With moisture conditions in these pits ranging from complete saturation (ponds) when filled with the runoff to scorched during dry, hot summer periods, a degraded land condition exists in response to attempts to meet the water demands of this ever increasing urbanized area. This degradation is reflected by lower real estate values for the house lots that abut these "sinks." These pits clearly represent a sacrifice

Plate 6.4
*A Long Island storm runoff
catchment pit*

zone that has been established to save a degrading water supply. All of these policies have acted to arrest the decline of the groundwater reservoir, but they do not have the ability to restore the island's groundwater system to its original state.

Subsidence in urban land levels due to the withdrawal of liquids is a common phenomenon in areas underlain with poorly consolidated sediments. In the majority of settings, when land settles due to the extraction of fluids, the major problems are usually restricted to damage to structures, such as buildings, roads, and pipes. Such is the case in Mexico City, where drainage of the former bed of Lake Texcoco has resulted in major settling problems. An example of this process is the slow sinking of the massive Palace of Fine Arts, the first floor of which is now below street level. While this type of settling results in economic costs, it does not result in land degradation. However, when subsidence occurs in areas situated close to sea level, land degradation does occur.

The lowering of land levels, usually associated with the pumping of groundwater or petroleum, results from the deliquefying of poorly consolidated sediments. The removal of the liquid from the sediments subtracts an integral part of the sediment's mass. This causes a decrease in the buoyant pressure that the liquid had contributed to the sediments. As a result, the sediments can no longer support the pressure resulting from the overburden and they begin to compact. Once compaction occurs, it is almost impossible to restore the land to its former condition, even if an effort were made to inject liquids back into the sediments.

Houston, Texas is built over the poorly consolidated sediments of the Gulf Coast Aquifer. Furthermore, it is situated in a flat, low-lying coastal area. Due to the withdrawal of large amounts of water from the aquifer, 1–2 m (6 ft) of subsidence has occurred in the Houston–Galveston area (Flawn 1970). As a result, some portions along the coastal zone have become submerged below sea level. However, perhaps the most serious problems of this nature occur in Japan.

Groundwater removal has resulted in urban sections of *Tokyo and Osaka*, where over $2\frac{1}{2}$ million people live, subsiding below the high tide level. In Tokyo, parts of the city are subsiding at rates up to 500 mm/yr (20 in), while parts of Osaka have rates of 76 mm/yr (3 in) (Goudie 1981). With subsidence up to 4 m (13 ft) in Tokyo and 2.6 m (9 ft) in Osaka, without offsetting engineering works, these areas would now be susceptible to daily flooding (Forrester 1978).

While not directly connected to urbanization but partly a result of satisfying the energy needs of urban areas, similar ground subsidence results from oil and gas extraction. In southern California, where many of the oilfields are situated in urban areas, the subsidence has direct impacts on the built environment. At Long Beach, California, extraction of oil from the Wilmington oilfield has caused coastal areas to sink up to 9 m (29 ft) (Flawn 1970). Since a large proportion of the area affected by this was only 1–3 m (5–10 ft) above sea level, over 1,215 hectares (3,000 acres) of land has been lost (degraded) as it became submerged below sea level. Damage to harbor facilities and other infrastructure due to this subsidence has been enormous.

As is evident from the examples in this section, satisfying an urban area's water demands can result in unintentional land degradation. The submergence of coastal lands due to the pumping of ground-water results in the submerged areas becoming sacrifice zones. The need to set aside land in reserve for storm water catchment pits on Long Island to insure sufficient groundwater recharge, due to the alteration in surface permeability associated with the change from a rural to a built-up area, results in the lands within the pits being susceptible to frequent flooding. This limits the biomass productivity and utility of the lands within these depressions. Land forever flooded behind dams in order to supply urban areas with ample water results in the reservoir areas becoming a sacrifice area for a distant urban area. These illustrations outline one by-product of urban growth: to insure the viability of an urban area often necessitates the sacrificing of a component of the land resource for the benefit of the overall urban system. The resulting land degradation occurs because the overriding consideration in all the cases is to satisfy the water demands of urban communities. The protection and utilization of one resource, water, was achieved at the expense of another resource, land.

Energy Production and Other Raw Materials

Mining

The industrial revolution required natural resources both for energy to drive the machinery that had been substituted for human labor as well as to acquire the materials required for manufactured products. Clearly, increased mineral utilization was a decisive foundation of the industrial revolution. In response to this need, the exploitation of the Earth's mineral resources has increased rapidly since the 1700s. Between 1750 and 1900 mineral use increased tenfold. Since 1900 mineral use has increased an additional 13-fold (Young 1992). As minerals are part of the environment, production (mining) of them without modification of the environment is in most cases unrealistic.

However, the modifications need not result in land degradation. New habitats can be created during mining operations by altering drainage and topography in harmony with the geomorphic and climatic situations (Goudie 1981: 85; Barrow 1991: 229). Nonetheless, to prevent land degradation from occurring, mining operations usually need to be controlled by strong legislation. This is because, in general, higher costs are associated with the extraction of the mineral if environmental damage is to be prevented. Since mineral resources are where you find them, not where you would like to find them, prevention of deleterious effects associated with their removal must be planned around, not planned for (Flawn 1970): that is, mining operations must take place in the locations where the minerals are located. They cannot be assigned to areas that satisfy the whims of society.

Today, in some countries that have a strong environmental lobby, when even with good planning the environmental consequences are deemed too costly to society, mineral extraction may be prevented. Such has been the case in some areas of Alaska, California, and coastal New England where, to date, exploitation of potential offshore petroleum has been prevented. But the overall reality from the worldwide perspective is that, in most cases, the short-term economic benefits of mineral extraction are all powerful and they are actively mined.

As minerals are part of the environment, their removal must upset preexisting conditions. Environmental damage is associated with most mining operations. There are two reasons why land degradation generally results from mineral extraction. First, industrial development remains the highest priority of most nations. "The American Colossus was fiercely intent on appropriating and exploiting the riches of the richest of all continents, grasping with both hands, reaping where he had not sown, wasting what he thought would last forever" (Pinchot 1947). The viewpoint expressed by Pinchot is still widely followed by many nations in their pursuit of short-term economic rewards. A second reason, somewhat related to the previous one, is that short-term economic benefits, such as reaching production goals and employment, delegate environmental considerations to the back burner (Ministry for Economic Policy and Development of the Czech Republic 1991). To insure that the deleterious aspects of mining operations are minimized requires, in almost all cases, conservation strategies that would increase the overall cost of the mineral extraction. Worldwide, prices of most mined minerals usually reflect only the direct immediate costs such as fuel, equipment, and labor. Rarely have provisions been made to include the costs of destroyed landscapes in the price determination of minerals

Direct impacts of mining on the land resource
Mineral extraction by mining takes place either as a surface or underground operation. Both methods can alter the preexisting environment to a sufficient magnitude that land degradation results. With over 500,000 hectares (1.24×10^6 acres) of land directly disturbed by mining activities each year (Young 1992), many areas in close proximity to mining operations are absorbing the environmental costs of mining for the users of the minerals. Some implications of these land alterations are discussed in the following sections.

Underground mining Subsurface mining is an ancient activity. The salt mines in the vicinity of Hallein (Austria) and Bad Reichenhall (Germany) have been active since prehistoric times. The oldest coal mine in Glamorgan (Wales) began operation prior to 1400. By 1600, Wales was already exporting coal to France (White 1991). With the industrial revolution, mining operations have become more extensive to satisfy both the demands for materials (e.g., iron and lead) and energy (e.g., coal). Regardless of the mining method, all subsurface mining moves materials from beneath the surface to the surface itself. In mining areas, where groundwater is superimposed upon the mineral vein being mined, pumping to lower the water table also is required. All of this extracted material contributes to the support of the surface. Regardless of the mining method and type of support provided for the roof of a mine, surface subsidence and collapse are common environmental adjustments to the removal of underground materials in most areas that have been actively mined. Subsidence alters drainage patterns and damages surface infrastructure (e.g., roads and utilities) and buildings in built-up areas. Collapse drastically alters the surface configuration of the affected areas and clearly degrades the immediately affected areas. Often, mining operations result in surface waters infiltrating into the subsurface voids, especially after the mine is exhausted and pumping ceases. In some areas of Appalachia, large underground impoundments of water have filtered into coal mines. These waters have become very acidic and, when they are returned via pumping (in active mines) or by subsurface flows to surface waters, their low pH value devastates the aquatic systems impacted by them.

Another aspect of underground mining is that waste material is a by-product of any mining operation. Gaining access to the vein or seam of the desired mineral, as well as transporting the mineral to the surface, requires large quantities of waste materials to be removed to the surface. These tailings are often piled up in large mounds in close proximity to the mine. The composition of many tailings can contain toxic minerals such as mercury or iron sulfide. Water percolating through these waste materials often produces water quality problems downstream from the tailings similar to those associated with subsurface water flows from within the mines.

Tailings dating from lead mining operations in Roman times are still poorly re-vegetated (Barrow 1991). It would be a safe bet that the mountains of slag and shale (see figure 6.4) in Wales, if left to natural processes, would take a thousand years before the evidence of the past mining activities was covered over by vegetation and forgotten (White 1991). Yet today a massive and expensive reclamation of these coal tailings throughout Wales is well under way (plate 6.5a and b). The coal tips are being leveled and the land is being made useful, at least for some human activities.

This reversal of land degradation is important because coal tailings, with their highly acidic properties, not only result in largely sterile conditions on the mounds themselves, but also rain water

Plate 6.5
Coal mine tailings being reclaimed near Abercarn, Wales, UK

running off of these tailings is often so acidic as to kill vegetation both in the immediately affected lands as well as aquatic life in streams and rivers receiving these waters. Many lands and streams within the Appalachian coalfield areas of western Pennsylvania, West Virginia, eastern Kentucky, and eastern Ohio are devastated by the acidic waters resulting from coal mining operations. The enactment of environmental legislation limits the damage associated with active mining operations; but the land degradation associated with past mining has left a filthy legacy of degraded landscapes dispersed throughout Appalachia.

Surface mining When possible, surface mining is usually favored over underground mining. The reasons for this are primarily economic. Compared to underground mining, the cost per unit of production is considerably less in surface mining operations. Additional advantages are: (1) the larger percentage of mineral recovery in surface mining, often close to 100 percent; and (2) the greater safety of mining operations that are relatively free of accidental explosions and mine collapse. Some minerals with a low value per unit volume, such as sands, gravels, and phosphates, are only surface mined. In stark contrast to these short-term economic advantages, since these types of operations always significantly disturb the surface, large-scale land degradation has resulted from these activities in almost every nation.

The Oceanic nation of Nauru perhaps best exemplifies the economic benefits of surface mining as well as the environmental devastation that all too often accompanies this type of mining. Through the mining of rich phosphate deposits found on the island, Nauru has obtained one of the highest per capita incomes in the world (approximately \$20,000). Yet, in around 1995, when it is estimated that the phosphate deposits will be completely depleted, the island's interior, 80 percent of the island's total area, will have become a wasteland of coral pinnacles, representing a classic example of land degradation (Graves 1990).

In the United States, surface mining has disturbed over 23,000 km^2 (5.6×10^6 acres). Coal mining accounts for 41 percent of this figure, while sand and gravel removal is responsible for 26 percent (US Department of the Interior 1967; Tank 1983). With the development of efficient earth-moving equipment, such as dippers on large power shovels that can hold over 95 m^3 (125 cu. yd) of earth and rock, deep mining continues to decline relative to surface mining (Barrow 1991).

It is difficult to prevent land degradation when surface mining occurs. First, in some operations huge depressions result. The Kennecott Copper Mine in Bingham, Utah is over 700 m deep (1/2 mile), and covers more than 7 km^2 (2.8 sq. miles). Second, the overburden (soil, subsoil, and unconsolidated earth and rocks) must be stored and then replaced systematically in their original order after the mineral is removed. Even under optimal conditions, which rarely occurs, restoration usually results in a landscape that is less productive than it was prior to mining. Subsurface drainage is always disturbed and re-vegetation is often slow (Jordan, Gilpin, and Aber 1988). The restoration of land is always more difficult when toxic materials are involved or leaching of the overburden occurs during its storage. These conditions often occur in mining operations associated with coal, lead, copper, and gold, which together comprise a significant percent of surface mining operations.

Because of the critical role of minerals in development, mining operations are inevitable. How could any urban area exist without the quarried stone, sand, and gravels needed for construction or the minerals needed for industry and energy? However, the destruction of the land resource, with soils blighted and water fouled, needs to be counted against the expediency of economic short-term gain (Simons 1992).

Nuclear energy production

One of the authors clearly remembers from his childhood the commissioning of the first US commercial nuclear power plant in Shippingsport, Pennsylvania. Among other advantages proclaimed for atomic power during this early stage of nuclear energy was that it would provide a clean alternative fuel (environmentally benevolent) to the region's traditional polluting energy source, coal. Yet, the raw materials required for nuclear fuels result in the same disturbances of the landscape as other mined minerals. In addition, the hazardous nature of the spent fuels creates another set of problems with regard to finding a safe site to store them. During the infancy of atomic power, little public awareness existed to the "new" environmental problems that would be an inevitable by-product of this proclaimed clean energy.

Today, it is clear, even if one only considers the issue from the perspective of land degradation, that both the hazardous wastes associated with nuclear power generation and the radioactive impacts of nuclear accidents make nuclear energy environmentally risky. Because of these inherent risks, Denmark has never permitted atomic power plants to be constructed within its boundaries, and Sweden has a policy of decommissioning all of its existing plants. In the United States, it has become so costly due to required environmental safeguards and the inevitable litigation of nuclear opponents that few plants are currently under construction. Additionally, in most parts of the United States, it is so politically risky that no new plans for nuclear power plant construction are currently in existence.

Power plant accidents

In March 1979, when the Three Mile Island (Pennsylvania) atomic power plant sustained a near disastrous event, the general public in

Table 6.1 A comparison of estimated releases (in percent) for the Chernobyl and Three Mile Island nuclear power plant failures

| | Three Mile Island | | Chernobyl |
Isotopes	Containment building	Released to the environment	Released to the environment
Noble gases	48	1	100
Iodine	25	0.00003	20
Cesium	53	Not detected	10–15
Ruthenium	0.5	Not detected	2.9
Cerium	None	Not detected	2.3–2.8

Source: modified from Collier and Davies (1986).

the United States became more alert to the potential for acute environmental problems associated with energy generated from nuclear power. However, as most of the radiological impacts of the Three Mile Island accident were confined to the immediate area of the power plant, minimal land degradation is associated with this event (table 6.1).

In contrast to the fortunately relatively mild Three Mile Island incident, on April 26, 1986, another nuclear power plant accident occurred at Chernobyl (Ukraine). This thermal explosion released radioactive isotopes far beyond the confines of the reactor building. This event has resulted in a multitude of major environmental consequences, including land degradation that continues to affect this area of the Ukraine (Gittus et al. 1988).

The radionuclides released from the Chernobyl plant were transported almost worldwide. Outside of the former Soviet Union and Eastern Europe, Austria and Scandinavia experienced some of the largest amounts of fallout on their lands (OECD 1987). While crops and milk products were contaminated in many European countries for a short period of time (OECD 1987; Mould 1988), with regard to land degradation, the damage resulting from Chernobyl was confined to the Ukraine, Belarus, and Russia. Significant contamination was concentrated over an area of approximately 1,000 km^2 (400 sq. miles) radiating outward from the destroyed reactor. Contamination has affected both plant and animal life within this zone, and greatly limits land-use options for the foreseeable future (Mould 1988). While it is beyond the scope of this book to examine the possible long-term effects of Chernobyl on land degradation, in the immediate surrounding areas of the plant the decontamination activities themselves, which were required to prevent pollutants from entering the area's river and water systems, resulted in significant land degradation. In the vicinity of Pripyat, the town closest to the plant, decontamination actions left only one tree standing (Mould 1988). The spared tree was a memorial to Ukrainians killed during World War II in this area. Workers carrying out decontamination strategies in close proximity to the nuclear plant cut down all of the trees in this previously forested area and buried them. The soil in the vicinity of the power plant was removed to depths between 1 and 1.5 m (3–5 ft) and buried elsewhere (Mould 1988).

Because of the intensity of the radioactivity in proximity to the plant, a 30 km wide evacuation zone centered on the power plant has been closed to human activities. This zone is now set aside for the investigation of the effects of the accident on the environment. While only future research will be able to present a complete picture, by any criterion the contamination of the lands in the vicinity of the plant and the required decontamination procedures represent a new form of land degradation resulting from some of the technological advances of modern societies.

Hazardous radioactive wastes

While less dramatic than acute nuclear accidents such as at Chernobyl, Three Mile Island, Kyshtm (Russia), and Windscale (UK), the chronic accumulation of hazardous wastes from nuclear facilities also has land degradation impacts. In the United States, federal nuclear activities are not required to adhere to the same standards as civilian installations. Only occasionally do the negative impacts of the management of governmental installations become part of the public record. One problem brought to the public's attention with regard to governmental operations concerned the Savannah River Nuclear Facility.

Located on a 777 km² (300 sq. miles) parcel of land in South Carolina, bordering the Savannah River, is a major US Government nuclear facility. It includes five nuclear reactors and several large processing installations. Because of the activities carried out in this area, large quantities of radioactive and nonradioactive wastes, byproducts of the plants' activities, are stored at various locations on its grounds (USGAO 1984). In 1981, the public became cognizant that some of the stored hazardous wastes were leaking into the groundwater below the Savannah River facility. If these waste materials were to enter the nearby Tuscaloosa Aquifer, a major water supply for the Southeast would have been threatened. Clean-up operations were initiated to rectify this problem, and the Tuscaloosa Aquifer appears to have been protected (USGAO 1984).

If radioactivity had entered the Dnieper River system from the Chernobyl accident or the Tuscaloosa Aquifer from the Savannah River's hazardous wastes, the ramifications for the bioproductivity of the affected land and water systems would have been momentous. Land degradation problems associated with radioactivity are a negative ramification of progress in the industrial/technological sector. Because of the wide spectrum and the nature of the negative impacts that can result from the misuse of nuclear energy, prevention of the degradation from ever occurring is clearly a superior option to decontamination of an area after an accident has transpired.

Wood and other biomass utilization

Besides mineral utilization, vegetal materials make significant contributions to the industrial demands of energy and matter. Wood provides approximately 20 percent of all energy in Asia and Latin America. In Africa, 50 percent of all energy is provided by the burning of wood and charcoal (Arnold 1992). In theory, unlike most minerals, vegetal materials should be treated as renewable resources in that new growth should replace the harvested plants. In reality, human management of growth and harvest cycles often are deficient. When this occurs a degrading environment and a lower production in biomass results after a number of harvests. Through a brief examination of two vegetal commodities, trees and cotton, some relations between land degradation and biomass utilization will be explored.

Trees

Comparison of a map of natural vegetation to a map of actual groundcover reveals a major discrepancy between areas of natural forests and current land use. Approximately 33 percent of the world's forests have been cleared for agriculture, grazing, and urbanization (Postel and Ryan 1991). Of the remaining 4.2 billion hectares (10.4 billion acres) of forests and woodlands, only approximately 25 percent are primary forest, the vegetation type presented on maps of natural vegetation. The remaining 75 percent are either secondary or managed forests that differ significantly from the primary tree cover that they replaced. Most of the world's remaining primary forest is either tropical rainforest or taiga in Alaska, Canada, and Russia (Siberia). Almost all of Europe's natural forests are gone, while in the United States, excluding Alaska, less than 4 percent remain (table 6.2).

Environmentally, trees play a multitude of roles. They affect climate, are a carbon sink, minimize soil erosion, contribute to a complex biodiversity, and strongly influence groundwater/surface water flows. In many areas they are a source of animal feed and

Table 6.2 Estimated forest cover (millions of hectares)

Location	Original forest	Current forest	Existing primary forest
Europe	n.a.	157	<1
United States[a]	384	244	13
Brazil	286	220	180
Other	5,530	3,623	1,320
Total	6,200	4,244	1,519

[a] Excludes Alaska.
Source: data modified from Postel and Ryan (1991).

provide supplements to the human diet (Antonsson-Ogle 1990). By intercepting rainfall and providing litter, they affect groundwater recharge and hence have a significant effect on stream discharge (Pereira 1973). Additionally, trees can improve soil fertility by adding nitrogen to the soil and/or reducing salt concentrations. Under current demographic, economic, and political conditions, in many parts of the world, current management practices are increasingly failing to renew existing forests, let alone restore those woods already sacrificed. Through the early 1980s, it was estimated that forest cover was decreasing at a rate of 11 million hectares per year (WHO 1992). Recent estimates indicate that deforestation is now occurring at a rate over 50 percent higher, 17×10^6 hectares (4.2×10^7 acres) per year (World Resources Institute 1992; and see table 6.3). Reforestation and afforestation only offset these losses by 10 percent.

The greatest deforestation today is occurring in Africa, Latin America, and Southeast Asia. Contemporary pressures, such as landlessness, on the remaining tropical forests could result in the same magnitude of change as has occurred in mid-latitude areas unless international efforts to control tropical deforestation are successful (Postel 1988). However, because the environmental situation in many of these tropical areas is more fragile than in the former mid-latitude forested areas, unlike the southern Appalachian area discussed in chapter 5, these lands could experience a more intense land degradation that would be more difficult to reverse. In Indonesia, a 1988 estimate indicates that 4 percent of previous forested lands (48×10^6 hectares, or 119×10^6 acres) already had been permanently lost due to resettlement programs (Gradwohl and Greenberg 1988). Much of this land had been cleared by timber operations that did not include a tree planting component. It is questionable if the small farms created on these former forested lands will be able to prevent land degradation, given both the physical environment of intense rainfall and the limited economic and human resources available to the settlers.

Originally, widespread wood fuel shortages in the developing world were viewed as the significant cause of tropical deforestation and, in many places, desertification (Eckholm 1975). Later it was theorized that widespread tree cutting was largely a result of agricultural expansion (O'Keefe, Raskin, and Bernow 1984) and urban energy demands (Munslow et al. 1987; Lewis and Berry 1988).

Table 6.3 Deforestation rates, 1990

Region	Area forested (10^3 hectares)	Deforestation (10^3 hectares)	Rate (%)
Africa	241,800	4,800	1.7%
Central/South America	753,000	7,300	0.9%
Asia	287,500	4,700	1.4%

Source: modified from UNEP (1991).

Today, it is felt that except in extremely local areas, fuelwood demands are not a catalyst for deforestation (Leach and Mearns 1990). Excluding land clearing for agriculture/pasture and urbanization, the major catalyst for forest cutting is wood, either for lumber, paper pulp, or other products such as plywood. If the remaining primary forests are to survive, a significant effort needs to be undertaken to curtail the world's demand for wood. According to the World Resources Institute (1992), over $2 billion of tropical woods are used yearly to make the forms for poured concrete used in construction. These forms are used generally up to a maximum of three times and then destroyed. The US government estimates that at least 10 percent of lumber and plywood/chipboard used in construction could be saved by changing standard, traditional house-building techniques without any sacrifice to quality. A simple change, such as increasing the distance between wall beams to 60 cm (24 in), is but one example.

Curtailing waste related to convenience is another area that would reduce demand for wood. As one example, The World Resources Institute estimates that over 25 billion pairs of disposable chopsticks are produced annually (World Resources Institute 1992). Not only should demand be curtailed through more rational use of wood, but afforestation, which is rarely practiced in most of the areas where tropical woods are cut, should be an integral component associated with tree cutting. If reforestation expenses were included in wood costs, not only would waste be likely lowered, but reforestation would contribute to preventing the forested areas from degrading.

Timber harvests vary annually depending on the state of the world economy. However, in the past 25 years they have increased by 50 percent. In 1988, 1.66 billion m^3 (2.17×10^9 yd^3) of timber were reported to be cut (FAO 1990). The United States, the former USSR (mostly Russia), and Canada produced 53 percent of this total. Much of the timber cut by industrial companies comes from plantations or forests where tree planting occurs shortly after tree cutting. From the land degradation perspective, secondary and managed forests can be successful in minimizing soil erosion and preserving water resources. However, they never maintain the biological diversity found in primary forests (Postel and Ryan 1991).

With regard to the ability to support a diversity of species, prevent soil erosion, and enhance water resource preservation, secondary forests often fall short of their natural counterparts. In the national forests of the United States, most logging operations in recent decades utilized clear-cutting. This is a strategy in which generally all of the trees are cut at the same time on large tracts that include from ten to several hundred 16–24 hectare (40–60 acre) plots. The result is that many of the national forests today resemble a checkered mosaic of highly fragmented plots of mature trees, separated by lumber roads and a limited number of recently replanted commercial tree species. Some biologists feel that the practice of slicing up the forest into such small blocks and the replacement of diverse mature forests with a limited number of commercial species threatens the overall health of the remaining forests.

Only 10 percent of the 10×10^6 hectare (25×10^6 acre) mid-latitude rainforest in the Pacific Northwest remains intact (Miller 1993: 343–4). The Northwest forests, which previously masked the terrain completely, have been replaced by a checkered mosaic composed of 16–24 hectare (40–60 acre) plots of bare ground, recently replanted commercial saplings, and some mature forests. This continuing change in groundcover threatens the habitat of many plants and animals. Furthermore, current replanting strategies only account for a regrowth of 64 percent of the current volume being cut (Sonner 1992). The result of current logging practices in the national forests of the Pacific Northwest is a reduction in the diversity of species along with accelerated soil erosion resulting from the poorer groundcover. These conditions represent a chronically degrading environmental condition. If policy changes on logging are not implemented, large areas of forested hilly lands in California, Oregon, and Washington eventually could evolve into large areas of severe land degradation. The Forest Service, in partial response to public pressure resulting from the 1992 Rio Earth Summit, has altered the former clear-cutting strategy. Now some mature trees will be left standing with the hope of encouraging a more natural regeneration of forest lands. Whether or not this is sufficient to reverse current trends remains to be seen.

Throughout the tropics, a similar pattern of replacing numerous species with a few favored ones has been the recent pattern. In particular, eucalyptus has been preferred over existing local species. It has been planted throughout the tropics as a substitute for indigenous species because it is fast growing and, when cut for coppicing, its shoots quickly develop into new tree growth. However, the oil in its leaves results in a ground litter that curtails the development of a good groundcover. As a result, accelerated eroded soils under eucalyptus trees are a common occurrence. Generally, reforestation utilizing eucalyptus trees is not environmentally friendly. Thirty percent of soil degradation is directly related to deforestation alone (World Resources Institute 1992: 114). Thus, good reforestation programs and more rational use of timber would significantly contribute to a decline in contemporary land degradation.

Cotton

Cotton fabric is one of the most desired materials for textiles. A tropical/semitropical plant, cotton requires a long warm growing season and ample soil moisture for successful cultivation. Within the former Soviet Union, very few areas experience the climate and soil conditions needed for commercial cotton production. To meet the needs of their domestic market as well as to increase their exports of cotton products during the 1950s and 1960s, "cotton independence" was a major goal of the USSR. "Produce millions of tons of cotton at any cost, and fulfill The Plan, at any cost" became a central planning goal of the national government (Precoda 1991: 114). To create the environmental conditions needed for successful cotton production and other agricultural crops in Central Asia, especially Uzbekistan, major water diversions were undertaken to increase irrigation (Rutkowski 1991).

The diversion of waters within the Aral Sea Basin for irrigation expansion has initiated and is continuing one of the most significant contemporary ecological disasters. Widespread land degradation, still expanding, is one result of the diversion of river waters onto cotton fields. The Aral Sea, an ancient body of water, is likely to cease to exist in the near future. Already, the reduction in its size has degraded a complex land/water ecosystem. There is good evidence that the changes set in motion by this attempt to control nature are not only resulting in unintentional destructive changes in the immediate areas of cotton irrigation and the Aral Sea, but have set into motion a chain of events that is resulting in negative regional – and perhaps global – climatic change as well.

The Aral Sea before water diversions   Prior to the 1950s, before the diversion of rivers entering the sea began, the Aral Sea had been essentially stable for hundreds of years. Its surface generally fluctuated between 50 and 53 m (164–174 ft) above sea level (Precoda 1991). Including the islands within the sea, its surface area was around 67,000 km^2 and its volume was over 1,000 km^3. While having no surface outlet, its waters were saline, but not excessively so. These conditions resulted from the generally prevailing equilibrium among freshwater surface and groundwater inflows, evaporation from its surface, transpiration from the wetlands surrounding a significant proportion of its coastline, and outflows via groundwater (table 6.4).

These "natural" conditions resulted in a highly productive ecosystem within a climatologically dry area. The Aral Sea provided a diverse setting for a multitude of commercial activities, including shipping, animal trapping, fishing, and paper production. A rich

Table 6.4 Characteristics of the Aral Sea, 1960-89

Year	Height above sea level (m)	Area (1,000 km^2)	Volume (km^3)	Mineral content (g/l)
1960	53.3	67.9	1,090	10.0
1965	52.5	63.9	1,030	10.5
1970	51.6	60.4	970	11.1
1975	49.4	57.2	840	13.7
1980	46.2	52.4	670	16.5
1985	42.0	44.4	470	23.5
1989	39.0	37.0	340	28.0

Source: Kotlyakov (1991)

aquatic life existed within the sea. Highly commercial species such as pike and sturgeon accounted for over 45,000 tons of fish annually caught from the sea and its tributary rivers. The coastlines, rimmed with endless reeds, provided a rich habitat for birds and animals. Immediately inland were over 250,000 hectares (620,000 acres) of forests and shrubs. The Amu Delta, along the sea's southern shore (figure 6.5), provided highly productive year-round pasture (Precoda 1991).

The Amu and Syr Rivers, the major sources of water entering the Aral, had been utilized by local inhabitants as critical water sources both for irrigation and drinking water for centuries. However, unlike the changes introduced by the massive water diversions beginning in the 1960s, these older uses of the water system did not profoundly upset the hydrologic balance of the sea. The traditional irrigated areas were confined to the Amu and Syr floodplains. On these irrigated lands, drainage had been improved and heavily consumptive moisture-demanding natural vegetation, such as reeds, had been removed. The inhabitants primarily cultivated grains and fruits in these areas. These crops had lower water demands than the natural vegetation that they replaced. The net effect was that the overall hydrologic balance of the sea was not upset by this traditional activity.

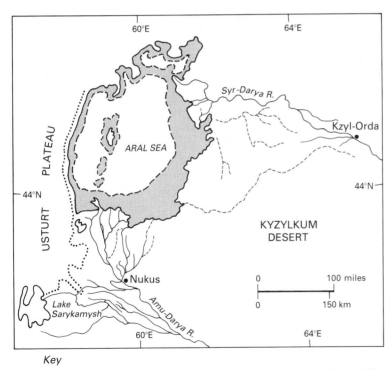

Key

•••••••• eastern border of the Usturt Plateau   ——— previous boundary of the Aral Sea

‑ ‑ ‑ ‑ ‑ ancient river beds   ▨ dried-up Aral Sea bed

— — — present boundary of the Aral Sea

Figure 6.5
The shrinking Aral Sea (after Kust 1992)

The Aral Sea since 1960   According to plans, the Amu and Syr
waters used in the basin's irrigation schemes were eventually to
result in an increase of 1.5 million tons in annual cotton production (Precoda 1991). The massive expansion of irrigation to reach
this goal began in 1956, when some of the Amu's waters began
to be diverted into the Kara Kum canal. In 1960, because of the
increasing diversion of freshwater discharges into the irrigation
projects, river flows into the Aral Sea decreased to levels that
were less than the evapotranspiration and groundwater losses
from it. At this time, the sea began to contract. Since then, the
shrinking of the sea has continued unabated. This reflects the continuing imbalance of excess annual evaporation over the decreased yearly inflows into the sea. By 1988 less than half of the
35 km^3 of river discharges needed to prevent further shrinkage of
the sea was actually reaching it (Ellis 1990). Diversion of significant volumes of the Amu and Syr's discharges for the consumptive use of irrigation continues (table 6.4). If current trends
continue, it is estimated that the sea will not exist at all by 2010
(Sigalov 1987). As desiccation of the sea continues, a multitude of
deleterious changes throughout the basin are under way. These
changes are of such a magnitude that they threaten the production of cotton, the primary reason why the diversions occurred in
the first place.

Destructive creation, resulting from the river diversions, has pro-
liferated in the region of the Aral Sea. The sea itself has ceased to be
a productive habitat. Of the more than 20 species of fish formerly
found in the sea, increasing salinity and dried up coastal spawning
grounds have caused the disappearance of almost all forms of
higher aquatic life (Micklin 1991). The sea's fishing industry has
been completely destroyed.

With vast areas >30,000 km^2 (11,600 sq. miles) of former sea bed
now dry, the exposed precipitated salts covering the previous sea
bottom are now a major source of windborne pollution. According
to recent data, 100 million tons of fine dust and salts are transported annually by winds blowing over the exposed portions of
former sea bed (Precoda 1991). This problem will continue to
grow in magnitude as the sea recedes and ever increasing salt
deposits on the former sea floor become exposed. Some of the
windblown salts are being deposited in the irrigated agricultural
areas. By raising the soil's pH, these deposits are lowering yields.
Estimates indicate that 5×10^6 hectares (12×10^7 acres) of irrigated land will eventually become saline from these aerial deposits,
thereby decreasing the soils' fertility and preventing most forms of
agriculture, including cotton production. Since the harvesting of
cotton was the major reason why the region's river waters were
diverted from the Aral in the first place, this negative, land-degrading feedback seems particularly ironic and destructive. The salt-
laden air also is responsible for a multitude of respiratory and
other health problems for the inhabitants of the basin (Ellis
1990). In the Karakalpak region of Uzbekistan (just south of the

sea), the incidence of throat cancer and respiratory and eye diseases has soared. One likely contributing cause has been the increased levels of airborne dust and salt. Other likely factors are pesticides and insecticides associated with the cotton production made possible by the diversion projects.

The exposed sea beds are not only a source area for dust and salt pollution in the lowland areas around the sea, but also windborne dust and salts are being carried to and deposited on the distant glaciers and snowfields along the Afghanistan/Tajikistan/Kyrgyzstan/China borders (e.g., the Pamir Mountains). Some of these areas are within the catchment of the Aral basin. These windborne deposits appear to be contributing to a significant acceleration in the ablation of these areas' glaciers and snowfields (Precoda 1991). The long-term effect of these changes on the rivers flowing out of these areas is unknown. Yet if any significant changes do occur, they will cause further alterations in the hydrologic conditions of the basin, as well as in other river systems that have headwaters in the Pamir Mountains.

Prior to the shrinking of the Aral, the frost-free period in the surrounding area was around 200 days. Through its maritime influence, the sea protected large areas of Uzbekistan from the harsh winter winds that blow from the northeast. As these cold winds blew over the sea, moisture was evaporated from its waters and their temperatures moderated. A proportion of the moisture evaporated from the sea eventually precipitated as snow in the distant mountains to the south. These are the very areas that today are likely to experience accelerated ablation due to the increase in dust and salt and the decrease in moisture carried by these winds. As the surface of the Aral becomes smaller, the ability of the sea to ameliorate the cold winds decreases. Frosts are increasingly coming earlier in the autumn and later in the spring. The frost-free period is now as short as 170 days (Kotlyakov 1991). Thus climatic conditions for cotton are becoming ever more marginal. According to Moroz (1988), "Climatologists think that for cotton-growing, the Central Asian region will be lost permanently." If this occurs, it will be the final irony in this example of land degradation. In the attempt to modify nature by increasing the production of cotton, the Aral Sea Basin will have undergone significant land degradation and will no longer have climatic conditions that permit cotton cultivation, even if ample water exists and the soil resource can be preserved.

In the case of the Aral, some of the degradational processes set into motion by the massive diversion schemes were predicted; but they were ignored due to the critical national goal of increasing cotton production. No matter what strategies are decided upon in the future, the Aral Basin will never recover to its former self. Even if cotton production is systematically lowered and water waste is curtailed, many of the degradational events that have been set into motion by the water diversion projects will be irreversible. The salinization of some of the irrigated lands, the decreased productivity of the sea itself, the climatic change to a more severe winter and hotter summer, and the changes in the distant snow-fields and glaciers illustrate some of the complex linkages and interactions between the sea and its surrounding environment. These linkages have been drastically altered by the single-dimensional plan to increase production through the utilization of this region's water resources without considering the feedbacks between the water, land, and wind systems.

Transportation

As land, water, and air transport evolved, each improvement generally has placed increasing demands on the environment. Ships have become larger, requiring deeper and larger harbors. Horses, donkeys, llamas, and other beasts of burden have been replaced by railroads, automobiles, and trucks. Compared to animal transport, these mechanized forms require gentler road gradients. As these vehicles have become larger, faster, and more powerful, curves in the roadbeds have been reduced or eliminated, road dimensions increased, and natural surfaces replaced by man-made materials so that these land transport carriers can function efficiently. Airplanes, like their water- and land-based counterparts, have increased both in speed and size. Runways have necessarily become longer, wider, and – as with highways – natural materials have been replaced by either asphalt or concrete. The grass or dirt airstrips of yesteryear are anachronisms.

Older transportation systems largely were situated in topographic settings that naturally satisfied the demands of the conveyance. With advances in earth-moving machinery and more effective use of explosives, topography is now often "created" to meet the demands of the specific transportation equipment. Harbors and navigable rivers were once restricted to locales in which natural conditions were suitable. Today rivers are dredged and dammed and canals are built to expand shipping possibilities, as well as to prevent rivers from shifting their courses. Through canal and dam construction, river traffic can now navigate from the Danube to the Rhine. This permits direct river transport from the Black Sea to the heart of Western Europe. Similarly, canals, dredging, and other engineering improvements on the Great Lakes/St Lawrence waterway enable Buffalo, Toronto, and numerous other former inland ports to become gateways for ocean shipping. On the Mississippi, meanders have been cut to shorten the shipping distance on the lower Mississippi by hundreds of kilometers. The river's shipping channel in the delta has been "stabilized" through continuous river control operations that prevent the Mississippi from shifting its main delta outlet 160 km (100 miles) westward along the route occupied by the Atchafalaya River (McPhee 1989). Without human intervention that has prevented the Mississippi from shifting its outlet, New Orleans would not be able to function as a major seaport. Unlike Bruges (Brugge, Belgium), where silting of its harbor prevented it from remaining a major port, through continuous public work expenditures and technological advancements, New Orleans so far has avoided suffering the same fate. While the reasons would be different (silting versus river shift), without engineering works and continuous maintenance of the Mississippi's main channel that prevent its westward shift, New Orleans would have lost its port functions.

Modifying the topography for surface transport is common to most societies. Earth-moving equipment is an integral component of modern railroad, highway, and airport construction. Where trucks and trains previously had to climb steep hills laboriously, terrain gradients have been reduced through cut, fill, and tunnelling operations. The narrow and steep winding horseshoe curves of yesterday's highways and railroads have been replaced by modern multilane expressways and straight gentler railbeds that permit the maintenance of high speeds in hilly settings. Airport locations, which once had to be situated on flat and open terrain, are now far less restricted. Hills are removed and wet areas filled, if needed, when the priority for airport construction is high. Expansion of Logan Airport (Boston) occurred by extending its runways into a portion of Boston Harbor through massive landfill operations. The new airport for Hong Kong requires extensive alterations in topography in order for it to meet the needs of modern aircraft.

From the perspective of attempting to satisfy the immediate needs of specific transportation goals, modification of the Earth's surface is completely logical. Nevertheless, in the process of meeting these needs, the alterations often set into motion land degradation. From the bioproductivity perspective, just replacing the natural surficial earth materials with concrete and asphalt for roads in the United States alone accounts for 130,000 km^2 (50,200 sq. miles) of land degradation (Barrow 1991). In the remainder of this section other reasons for causal relations between transportation and land degradation will be explored.

Highways

Wherever humans go, roads are built. With modern earth-moving machinery and the use of explosives, primary roads that are intended to provide fast linkages between distant places, in particular, create anthropogenic environments. They alter areas by changing vegetative cover, land morphology, surface permeability, and surface and subsurface hydrology (Detwyler 1971). Slope stability problems are common along roads due to alterations in slope and drainage characteristics associated with road construction (Parizek 1971). This is because changes resulting from road building usually increase the potential for erosion and mass movements (Reed 1980).

In particular, three ramifications of road construction are the primary catalysts for the instability of lands juxtaposed to major

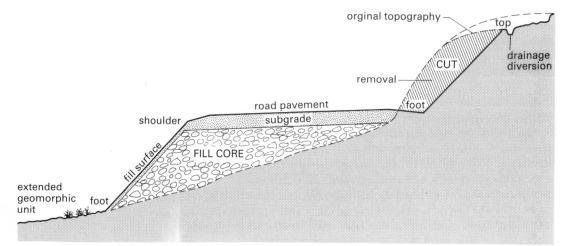

Figure 6.6
Typical cut and fill highway construction (from Molinelli 1984)

roads and railroads. These are: (1) the creation of cut slope areas, where terrain is excavated; (2) the establishment of fill slope areas to raise the preexisting ground to maintain road grade or, in areas susceptible to flooding, to raise the paved road above highwater; and (3) the immediate areas adjacent to the road that are directly affected by the drainage and other processes originating from the highway construction (Molinelli 1984; and see figure 6.6).

Worldwide, modern expressways crisscross through diverse landscapes as relatively straight ribbons with gentle to moderate gradients. Construction of this form, which permits consistently high vehicle speeds to be maintained regardless of the neighboring terrain, requires road cuts through hills and road fills along valley bottoms. As local relief increases in the zone that the highway traverses, the cuts and fills become ever more massive in order to meet the specifications of modern road building.

While the purpose of cuts and fills is to minimize steep road gradients, the land slopes on the cuts and fill themselves are often steeper than their natural angle of stability. There are both economic and environmental reasons for the steep slope characteristics of cut and fill. As the slope of a cut becomes steeper, the slope, the area of the preexisting terrain altered in road construction, decreases (figure 6.7). This both reduces the amount of land that must be purchased for road rights of way, and it directly disturbs a smaller area of the existing terrain. The flip side of this common engineering strategy of relatively steep slopes on the flanks of cuts

Figure 6.7
The erosion potential of highway cut slopes of different gradients

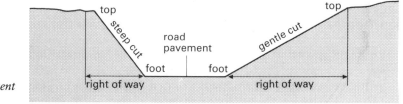

and fills is that erosional processes and rapid mass movements are more probable as the slope increases. This is especially true when the materials involved are not consolidated. While road design considers the physical and engineering properties of the materials involved, the fact that erosion and mass movement rates are generally higher in proximity to highways indicates that road slopes are often underdesigned (Skempton and Hutchinson 1969; Younkin 1974). The areas that experience either landslides, slumps, or excessive erosion clearly become degraded. Road building to increase the accessibility of the Alps is one of the reasons for the occurrence of landslides and areas of degradation in parts of Switzerland.

One method of minimizing slope instability, and hence degradation, along roads is the planting of vegetation. Kudzu, a fast growing broadleaf creeper introduced into the United States from Japan, was considered an ideal plant to provide a dense groundcover along road slopes and therefore to encourage slope stability. For this reason, it was planted along many roads cuts and fills. While its rapid growth did provide a good groundcover on highway rights of way, it has become a scourge in many areas of the southeastern United States. It is difficult to restrict kudzu's growth to where it was planted and supposed to grow. Kudzu has become a weed, spreading beyond the highways. This "miracle" vine has climbed and covered trees in the forests paralleling many southern highways. When the infestations are severe, tree growth is greatly curtailed and, in extreme cases, kudzu can kill the covered trees (*Forests Industries* 1984; Watson 1989). While this is not strictly a form of land degradation, it illustrates how an attempt to minimize land degradation has resulted in other negative feedbacks.

Lands immediately beyond road rights of way are often directly impacted negatively by a road's presence. In particular, water runoff is a major culprit. When poorly designed, drainage outlets, built to channel precipitation off the road surface, are responsible for gully erosion. In Kenya, highway construction in some parts of the highly productive Central Highlands has resulted in the formation of gullies 2 m (6 ft) in depth, over 1 m (3 ft) in width, and over 100 m (328 ft) in length at road drainage outlets. In the process, portions of farmers' fields have been washed away. The phenomenon of gully formation along drainage works associated with roadways is widespread. However, it is particularly disastrous in settings similar to Kenya, where the runoff has been diverted to flow over deep, unconsolidated earth materials. When bedrock is close to the surface, such degradation is minimized. Problems similar to those just presented are associated with both railroad and airport construction.

Harbors and coasts

From a geomorphic perspective, coastal areas are extremely dynamic areas. Change is normal in coastal zones. Energy associated with storms and longshore currents continuously moves materials along coasts and from the land to the sea. On the East Yorkshire coast in the UK, near the mouth of the River Humber, the coast has retreated 4 km ($2\frac{1}{2}$ miles) since Roman times. Twenty-nine villages and their surrounding lands have been lost to the sea by normal geomorphic processes (Sheppard 1912). On Cape Cod's eastward-facing shoreline, in the town of Wellfleet, the coast retreats 1–3 m (3–10 ft) per year in response to winter storms and occasional hurricanes. The Marconi Station, the site of the first trans-Atlantic wireless transmission, has already been lost to the sea at this location. Rapid coastal erosion in parts of the Cape continues to result in sand encroachment onto developed lands (plate 6.6), with numerous houses lost whenever severe winter storms occur.

Harbors are also potentially in a dynamic geomorphic environment. Two coastal phenomena, sedimentation and storm waves, have often required human interventions to keep a port's waters from becoming too shallow and to protect ships that are anchored. By definition, a harbor is a sheltered body of water protected from strong currents and waves. The very property that makes a harbor sought after as a haven for ships, namely quiet water, makes them ideal sediment traps. When rivers with heavy sediment loads or offshore currents transport materials into harbors, they likely will be deposited in the port. Over time, the harbor silts up and becomes too shallow for shipping. Ephesus, Ostia, and Bruges are examples of ancient ports that lost their harbor status due to siltation. In order to remain open, many contemporary ports require constant dredging operations. New Orleans and Buenos Aires are two examples of modern ports that, without dredging, would suffer the same fate as their ancient counterparts.

Plate 6.6
Coastal dwellings threatened by beach erosion

Dredging

Even where siltation is not a problem, the depth requirements of contemporary ships mean that most ports have to undertake dredging operations. The dredged channels have been deepened to such depths as to create unstable conditions that require constant maintenance along the channel slopes. In New York Harbor alone, 4.6×10^6 kiloliters (6 million cu. yd) of sediment is dredged annually (Fisher 1992). Today the competition among seaports to receive the largest and most economical ships requires channels up to 50 ft (15 m) in depth. This has initiated a new cycle of channel construction in a multitude of major harbors. For example, along the US East Coast, while New York (including all of its shipping berths, such as Port Elizabeth, Newark, and Staten Island) remains the largest port, its main rivals, Baltimore and Norfolk, already have deepened their channels to 50 ft (15 m). This has necessitated the New York/New Jersey Port Authority to undertake major dredging operations to deepen their shipping channels which, without dredging, would only be 5.5 m (18–19 ft) deep in many areas.

By preventing siltation in coastal situations where new highly productive land areas would otherwise form naturally, dredging curtails some aspects of bioproductivity. But by itself, dredging cannot be considered a cause of land degradation in most instances. However, as harborsides are often the scene of intense industrial activity and large urban conglomerations, many of their waters and sediments have become polluted. This has degraded the lands and waters in the vicinity of these activities. For example, many parts of Boston Harbor have essentially become dead bodies of water. Often the water is so polluted that potential endeavors on the lands bordering it are severely constrained. All forms of aquatic activity are negatively affected. A year rarely passes during which many of the beaches and coastal lands in proximity to Boston Harbor are not closed, during at least for part of the year, due to problems associated with pollution (see figure 6.1). The magnitude of the problem is so severe that major human interventions are required to reverse the ecological situation. Through a massive and very expensive wastewater treatment project (costing hundreds of millions of dollars) now under construction, it is hoped to reverse this degradation of Boston's land and water environs.

In New York Harbor, for years pollutants have found their way into its waters and contaminated its sediments. This is not a unique situation; the same situation exists in many ports. With dredging under way to deepen its channels, toxins such as dioxin have been found in some of New York Harbor's sediments (Fisher 1992). If buried elsewhere in the region, these toxic sediments will likely degrade outlying landfill areas. If the 460,000 contaminated cubic kiloliters (600,000 cu. yd) of silt are dumped in the ocean, the toxins could enter the food chain and degrade a productive offshore area. Thus, while dredging by itself does not result in land degradation, getting rid of the sediments can result in degraded environments.

Engineering structures

Ports need to be havens for ships during stormy weather. To improve the natural sheltering properties of harbors, as well as to prevent siltation in many cases, engineering works such as jetties, groins, breakwaters, and sea walls have been constructed. Some ports, because of a lack of natural harbors in the vicinity, are largely the result of engineering works. Vera Cruz, a major port for Mexico, is an example of an artificial anchorage. Because of the inherent dynamic nature of coastal and harbor areas situated at the interface of terrestrial and aquatic environments, anthropocentric changes in either milieu usually result in modifications in both.

Equilibria along coastal areas are inherently dynamic. Any structure that alters littoral drifts often upsets many properties along a coastal segment. Jetties and groins built to protect the harbor from waves and siltation are intended to trap materials moving against the barriers and alter their flow away from the harbor toward deeper waters. In response to the decrease of sediment transport in the currents resulting from these protrusions, the shore currents will erode sediments down current from the jetties (Pilkey and Neal 1992). The accelerated erosion of undefended coastal lands, as a result of engineering works intended to protect another segment of a shore, represents another manifestation of land degradation.

Modern transportation systems, no matter whether they are land-, water-, or air-based, usually require alterations in the topographic areas that are directly impacted by them. Harbor improvement is no exception to this rule. Because feedbacks exist in nature, when the previous balances that existed in the area are upset by construction activities, change almost always is set into motion. In the extreme, land degradation can directly occur from the "improvement." However, land degradation problems associated with transportation systems are more often than not the result of both the direct impacts of the transportation system and the associated activities that utilize the transport. The presence of dioxin in harbor sediments and the removal of these sediments for channel improvement is one example of the concomitant responses among land resources, activities, and transportation systems.

Summary

Destructive creation is an extremely widespread feature of human use of the Earth. The processes of change that degrade environmental quality and productivity are exceedingly complex in origin and are often indirect and counterintuitive in impact. However, two generic features characterize the examples presented in this chapter. The first is the change in scale from small to large systems that is made possible by powerful modern technologies. This shift in scale overwhelms local systems of knowledge and environmental management that have evolved over long periods of time. The pace at which change occurs makes it difficult for small-scale systems to adapt rapidly enough to the new conditions. The scale at which large-scale systems operate makes it easy to export costs onto more distant parts of the environment that are outside management control and to ignore those negative impacts. As in the case of the Aral Sea, when the negative consequences begin to be recognized, it is often too late to salvage the situation – and the problem is too vast to be tackled successfully with available resources. In many instances, a misplaced faith in the power of modern technology to solve all problems, and a lack of respect for traditional approaches to environmental management as a basic building block for developing new and improved sustainable systems, unleashes destructive forces that exceed the coping ability of existing management systems and technologies. Whether the issue is overgrazing, disposal of nuclear waste, the development of irrigated cotton, or urban growth, increasingly large-scale systems unleash a Pandora's Box of forces unforeseen by the initiators of new developments.

A second generic characteristic of destructive creation is the failure to think and plan in regional and holistic terms. Each activity is a world unto itself. Agricultural development takes little cognizance of the impact that its expansion has on pastoral operations, let alone urban activities. Common property resources are degraded because no one is accountable for the impact that their activities have on "free" goods, and traditional constraints to use are thoughtlessly removed. Good intentions based on inadequate knowledge of local environments and cultures generate unexpected feedback mechanisms that undermine the viability of the original project. Wastes are created for which there is no safe disposal mechanism. All of these problems, and many more, bespeak a fragmented and individualistic process of environmental management that creates long-term sacrifice zones and unsustainable development in the interest of short-term gains. The result is a process of creation that fails to achieve sustainability and drifts perilously closer to large-scale environmental destruction.

Yet this pattern of land degradation and lost development options is not inevitable. Models exist that offer guidelines for sustainable development. These successful examples of creative destruction are able to withstand sustained and increasingly intensive use. As a counterpoint to the instances of destructive creation noted in the last two chapters, instances of viable creative destruction are examined in the following chapter.

SEVEN

Creative Destruction

Introduction

Like the ancient Roman god Janus, represented by two opposite faces of dramatically different character, human use of the Earth has a dual visage. One face is creative destruction, the process by which the natural world is modified and sustainable land-use systems are developed. The opposite process is destructive creation, a condition characterized by the failure to achieve long-term sustainability and by the initiation of progressively more serious patterns of land degradation. Creation and destruction are closely linked for, in the interest of sustaining human population, one must often first destroy in order to create. Cultivation of most crops desired by humans, for example, requires the removal of preexisting natural vegetation in order to create the conditions that are most conducive to crop growth. Despite the evident risks involved in this strategy, from a reduction in species diversity to increased prospects for land degradation, simplification of the natural world in the pursuit of human sustenance is the resource management approach most frequently employed by humankind.

A successful process of *creative destruction* is characterized by several factors. Of primary importance is the subtle application of the *genius loci principle*. This principle stresses the special set of conditions found in every locale and the need to know these system states in detail. Each place is characterized by particular attributes of soil, climate, vegetation, culture, and system behavior. Without an intimate knowledge of these conditions, it is relatively easy to attempt to erect stable resource use systems but, instead, to generate systems that are vulnerable to collapse or that are poorly adjusted to local resource constraints and opportunities. Time is required in order to comprehend the basic dynamics of any place and to develop livelihood systems that are closely attuned to these rhythms. This is why many traditional resource-use systems provide important insights into how to erect sustainable systems.

Intimate knowledge of place makes it possible to identify the *critical zones* that are essential for system stability. These crucial resources are the ones that must be maintained and enhanced if a sustainable system is to be generated. Increasing the productivity of a critical zone often is linked to the corresponding decay of a *sacrifice zone*. Sacrifice zones are resource areas that are degraded in order to transfer the productivity of that sacrificial space to the critical zone. The movement of soil and water from hillslope catchment areas and their concentration in valley bottom terraces by the Nabateans, cited in chapter 2, is an example of the use of sacrifice zones to improve the productivity of critical resources. In other instances, there is no direct transfer of productivity to the critical zone from a sacrifice area. Rather, the changes in the critical zone can involve linked changes in other areas that reduce the productivity of those areas and make them sacrificial offerings, the degradation of which is accepted as a necessary cost of promoting progress and productivity elsewhere. Enduring creative destruction is sensitive to the changes that take place in both types of sacrifice zones, since otherwise degradation in the sacrifice zone can have negative feedbacks that diminish productivity in the critical zone.

In addition, there exist two other attributes of creative destruction. The first is *waste capture*, which converts a potentially negative output from one source into a positive resource. The more that wastes are internalized within a system and become input resources for other parts of the system, the more likely it is that the system will attain long-term stability. Chinese fish ponds are an apt illustration of this principle, since manure and other organic wastes from terrestrial activities are a basic resource for the carp cultivated in the pond catchment. Equally important is *cost export avoidance*, an objective that strives to prevent imposing negative impacts upon distant areas and ecosystems and converting these affected areas into sacrifice zones. This is a vital concern, because failure to diminish or prevent distant negative impacts can produce *unintended sacrifice zones*. Unexpected impacts, spatially connected to distant activites, are an important contributing factor to many contemporary land degradation problems. Because causation in these instances is far removed from the immediate setting of the negative change, it is usually difficult to rectify the situation until often irreversible damage has been done. It is one thing to sacrifice deliberately the productivity of one zone for the gain of another when this has been carefully considered and, in so far as is possible, the consequences have been meticulously evaluated. It is quite another matter to trash an environment heedlessly without compensating gains in stable productivity elsewhere. It is particularly in settings in which *cost export avoidance* is ignored that the often delicate balance between creation and destruction is upset and degradation occurs. Only when critical zones are preserved and the larger ecumene is protected from the export of undesirable costs can the creative destruction process bring into existence resource-use systems that possess the stability to recover quickly from disturbance and the resilience to absorb change without negative consequences.

In chapters 5 and 6 we examined some of the myriad ways in which destructive creation occurs in land, water, vegetation, common property resources, and urban systems. This chapter explores the operation of creative destruction in human systems. These examples are grouped into low-, moderate-, and high-intensity systems depending on the capital, labor, and technological investment that each system makes in extracting sustenance from its habitat.

Low-intensity Systems

Resource-use systems that are spatially extensive, employ a long fallow cycle, and invest relatively little human labor, capital, and technology in exploiting the total area to which they have access, are low-intensity systems. Hunting and gathering, swidden (slash and burn) cultivation, and nomadic pastoralism exemplify low-intensity resource management systems.

In low-intensity systems, a fundamental factor in their success is mobility. This element allows a band, village, or herding group to shift to new locations whenever seasonal conditions or local resource scarcity require. A linked variable is the size of the territory exploited by a particular group, which is often very substantial compared to the population supported. Thus, low population densities are typical in low-intensity systems. Long lag times characterize the periodicity of resource use, and it may be years before a specific site exploited in one year is revisited. At the same time, there are often seasonal periods of quite intense use of spatially limited resources, and these episodes of concentrated use, coupled with alternating periods of dispersion (Ingold 1987: 184–7), establish both the rhythm of seasonal activities and the base constraints that condition population dynamics and resource use. Population growth in excess of what can be maintained during the most constrained period requires either a change in location or an altered resource exploitation strategy, and often signals a shift to more intensive systems of resource use.

In a significant fashion, low-intensity resource-use systems exhibit few of the characteristics of creative destruction. In particular, they are able to avoid frequent use of *sacrifice zones* because the overall intensity of resource use is sufficiently low and the pressure on limited resources is sufficiently diffuse to make the construction of a sacrifice zone unnecessary. Moreover, low-intensity systems are able to protect the *critical zones* and resources that are essential to sustained exploitation. Departures from the practices employed by these systems result either in intensification and creative destruction or land degradation. In this sense, low-intensity systems are the base setting from which more intensive, creative destruction modes of livelihood may develop. A brief examination of hunting, shifting cultivation, and herding follow in subsequent sections as a prelude to examination of the more ubiquitous intensive examples of creative destruction that exist in most populated regions of the contemporary world.

Hunting and gathering

The oldest human mode of food production is the combination of hunting and gathering activities (Lee and DeVore 1968). This system of resource management encompasses all but the past 10,000 years of human existence. It is the ultimate base from which all subsequent livelihood systems evolved. Although hunter–gatherers possessed a sophisticated hunting technology and had a deep knowledge of both plant and animal characteristics in their local area, they were not able to support large human population densities. Generally, hunter–gatherers have utilized a variety of population control devices (prolonged lactation, infanticide, post-partum taboos, and so on) that discourage population growth (Neel 1970). Today's hunter–gatherer cultures only survive in the most marginal and remote settings, environments not able to support many people. In the past, hunter–gatherers would have lived in more abundant habitats that were capable of sustaining larger populations. Undoubtedly, it was from these better locations that the experimentation leading to agricultural domestication took place. Also, it is from these better sites that hunter–gatherers have gradually been pushed until they were restricted to their present isolated locations. Even in these limited habitats, replete with constraints, most hunter–gatherer groups are able to enjoy an adequate diet and to experience significant leisure (Lee 1968). They are able to accomplish this by virtue of values such as sharing, gender-based division of labor in which men hunt animals and women collect plant material, and cultural practices that emphasize seasonal mobility. The success of hunter–gatherer cultures in sustaining themselves throughout most of human history is a product of two factors: the modest concentrated pressure that they brought to bear on the resources in their territory; and the modicum of creative destruction that figured in their livelihood system.

The bulk of the hunter–gatherer diet is composed of plant material, not meat. The women are primarily responsible for these food resources, although there are some communities in which the women also figure as prominent hunters (Estioko-Griffin and Griffin 1981). Depending upon the abundance of the habitat, the range of plant resources collected can be quite large. Among the !Kung, for example, 200 plants are identified, and some 85 edible plants are gathered seasonally in an environment that can hardly be considered lush (Lee 1969: 59). The palatability of these items varies, and there is a seasonal cycle in their collection. Thus, as the dry season increases among the !Kung, it becomes necessary to shift from the more desirable species and consume the less tasty. As collecting continues, it also becomes increasingly difficult to find food within easy walking distance of the base camp. Lee (1969: 31) indicated that a 6 mile (10 km) collecting radius is the maximum that people can conveniently go from their base camp in a day in order to hunt and collect. This "day prism" sets a limit to the time–space resources of a band (Carlstein 1982; and see figure

7.1). When available resources begin to decline, a band must either
shift to poorer resources or move on to another site. In drier places,
there is always a movement away from permanent water sources
during the wet season in order to conserve the resources of this
more bountiful area for the dry season. In more moist habitats,
bands must still move in order to avoid overexploiting local
resources. Failure to move combined with excessive local extraction would invariably result in the starvation of the band, a result
that has undoubtedly occurred on occasion in the past. Such localized population collapses allowed for recovery of the resource base
before any future cycle of human use. Movement was also mandated because significant storage was not practiced by hunter–
gatherer bands. Without stored reserves to carry the population
through a long "empty" period, movement to new districts and/
or consumption of less palatable species was mandated.

Within this framework of selective and relatively short-term use
of resources, hunter–gatherer groups had relatively little impact on
their environment. There were three ways in which gatherers may
have had an impact. First, by creating disturbed conditions around
their camp sites and along their favorite trails, gatherers may have
produced conditions that were suitable for the growth of some of
the species that they collected. Fire was the major tool for environmental modification employed by hunter–gatherer groups. As such,
it was the second mechanism used by gatherers, probably as much
by accident as by design, to favor some of the food plants that
played a role in their diet. Use of fire by !Kung collectors to
enhance the growth of plant foods is reported by Wiessner (1982:
65), although overall it is not likely to have had a great deal of

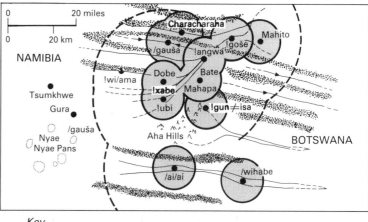

Figure 7.1
*The collecting territory of the
!Kung Bushmen: the intensity
of land utilization as a function
of distance from permanent
water holes (after Lee, in
Vayda 1969)*

Key

⬤ within 6 mile radius of a permanent waterhole

◯ within 6 mile radius of a large summer waterhole

⌒ within 20 mile radius of a permanent waterhole

▓ sand dunes

➝ ephemeral streams

impact on productivity. By virtue of their collecting activities, gatherers expanded the range and frequency of the food plants that they preferred. However, except for the use of fire to alter species composition in favor of plants that they regularly consume and at the expense of others, gatherers have little creative destruction impact upon their environment. Indeed, among the !Kung, who derive more than one-half of their vegetable food from one source, the mongongo nut (*Ricinodendron rautanenii*), millions of nuts are left uncollected on the ground each year (Lee 1969: 59). Even with a basic dietary staple, the !Kung did not begin to interfere with the reproduction of mongongo nut by natural means; no efforts to achieve domestication occurred.

As is the case with most collectors, the !Kung's best creative destruction activity is a negative one. It is found in the mobility that characterizes their seasonal pattern and the social network that enables them to leave their home territory and to stay with friends and relatives in distant bands during periods of scarcity. By temporarily reducing carrying capacity under adverse conditions, hunter–gatherers avoid doing serious harm to their environment. This permits their habitat to recover so that it can once again sustain their population and its extractive activities.

If the evidence for deliberate creative destruction is weak among gatherers, it is also problematic among hunters. By far the most important tool used by hunters was fire, which was employed in two ways. In the first instance, fire was used as an aid in hunts in order to drive animals over cliffs, into marshes, or into other settings where they could more easily be killed. In the second case, fire was used to favor the growth of grass or browse that would both attract game animals to sites closer to human settlement and provide more abundant and nutritious fodder for animals. Simmons (1989: 78–9) believes that fire, especially in wooded areas, can improve the quality of fodder by a factor of ten. This occurs because fire mineralizes the surface litter and makes nutrients more readily available to the soil and ultimately to the fodder plants that grow in it. Benefits not only include more desirable browse with higher leaf protein content, but also greater concentrations of game animals in smaller areas, so that hunters have to invest less energy in their capture. Given the attentiveness that characterizes hunter and gatherer relations with their home territory, it is likely that in many instances they readily recognized the utility of fire. Consciously employing fire to modify vegetation in a direction that is conducive to human sustenance is, of course, the essence of creative destruction. At the same time that some species will find burnt areas attractive, other species will find the altered environment unsuitable for their use; their numbers will decline, at least locally, as a consequence. Fire used in hunting would have that same general effect, although the purpose of its use was not always directly aimed at environmental manipulation. Fire was used by !Kung to attract game (Wiessner 1982: 65) and burning was a common practice in the woodlands of eastern North America in order, among a multiplicity of uses, to provide browse for deer (Day 1935; Cronon 1983: 147–51). Thus, it seems clear that some creative destruction was practiced by hunters to increase both grassland and browse, although it is unclear how really extensive this practice was.

More destructive creation could have been initiated by hunters directly by predation on animals themselves. By selecting key species upon which to concentrate their efforts, and by failing to spare pregnant females and adolescents, hunters could place specific species and age sets at risk (Simmons 1989). These activities may have gradually altered the population structure of the preferred animals and reduced the number of older individuals present. The result would then have been a larger and much younger population with a rapid growth to hunting size, which would have used less energy in support of aged individuals. With a population pyramid much closer to a herd of domesticated cattle, the antelope, seals, or deer being hunted would have channeled more energy into the diet of the hunting community. More excessive demands than this that are made upon animal populations by their human predators are generally a product of contact between contemporary hunters and modern markets, with dangerously excessive off-take rates being the result (Simmons 1989: 71). Thus, it is the demands of an urban market that commercialized the hunting of green turtle by the Misquito Indians, rather than their own consumptive demands, and has resulted in drastic overexploitation of this important marine resource (Parsons 1962; Carr 1967; Nietschmann 1972, 1973).

Some evidence has been presented to suggest that hunters entering North America at the end of the last glaciation were responsible for hunting to extinction a number of genera of megafauna (Martin 1973). Certainly, the arrival of hunters synchronous with the loss of a number of species of large game was a suggestive coincidence. Yet there are few examples of hunters running amok and exploiting their prey out of existence in pre-European contact conditions. Thus, it is unlikely that human predation was the primary causal mechanism in the loss of North America's large fauna. Nonetheless, it is possible that heavy hunting combined with the very severe environmental changes that accompanied the end of glaciation could have had a catastrophic impact on fauna that was already under very severe stress (Butzer 1971: 503–15). In at least one case, that of Maori immigrants to New Zealand, a hunting culture was capable of driving a number of flightless birds to extinction (Simmons 1989: 79). However, such instances appear to be isolated and the product of very special circumstances. Traditional hunting technology was sufficiently limited, despite its long evolutionary development and considerable sophistication (Oswalt 1973), that hunters lacked the ability to decimate the animals that were their preferred food species.

Indeed, most hunter–gatherer groups possessed a value system that promoted respect for the game that was an important part of their subsistence. This respect for and sense of kinship with the main object of predation is quite different from the attitude that develops when physical and psychic distance between humans and wild animals increases. In the classical world of Rome and Greece (Anderson 1985), this turned hunting into a sport for the elite, a spectacle of ritualized mass killing in the arena for the enjoyment of the mob, and a setting for the display of manly courage.

A striking example of the differing impact of hunters upon their prey is afforded by Kemp's (1971) energetic analysis of the Baffin Island Eskimo, whose adoption of the rifle and motorboat, among other technological devices, changed their relationship with the seals that they hunted (figure 7.2). Under traditional conditions, the hunter took only as many animals as were needed, and captured those animals only after a long, arduous wait near the seal's air holes or by difficult stalking efforts by boat on the open water and by foot on the floe ice. Believing that Eskimo and their prey were joined in a common realm of spirituality, when possible the hunter tried to avoid offending the spirit of the animals that he captured by killing no more than he needed. Almost every seal harpooned was retrieved because a harpoon formed a firm attachment that prevented the carcass from sinking. With the introduction of the gun, the ease and range of capture was increased. However, the price of this increase in labor efficiency was a dramatic increase in the number of seals killed but lost by sinking before they could be

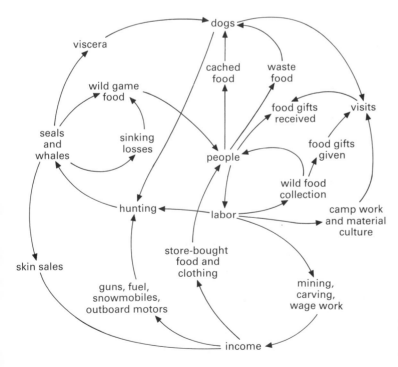

Figure 7.2
The structure of an Arctic hunting society

retrieved by the hunter. Losses of up to 60 percent of the animals shot were observed by Kemp. Moreover, the seals have become cautious about approaching people, a disruption in intimacy – rooted in a technological change – that has profound potential implications for the wise maintenance of the basic animal resource.

Hunter–gatherer bands exhibit little conscious creative destruction behavior. On another level, however, their livelihood system reflects a stability that is usually only exhibited after creative destruction has taken place and sustainability has been attained in modern systems. While there is little firm evidence to indicate a strong tendency for hunter–gatherers to overexploit their food sources, there is evidence that in special circumstances this may have occurred. Certainly, some habitat and species modification has been promoted, especially through the use of fire, to the benefit of human sustenance without initiating significant land degradation.

Shifting cultivation

Shifting cultivation is an agricultural system in which crop production moves to new locations whenever yields decrease to the point at which continued labor investment in the crop and the field environment is no longer worthwhile. Fields so abandoned are usually left fallow for a generation or more (Richards 1985: 50) rather than for a few years. The objective is to allow natural vegetation to recolonize the cleared field. In this way, plant roots tap deeper mineral layers and leaf litter (at least to some degree) contributes nutrients and organic matter to the soil without humans having to engage in expensive and time-consuming rehabilitation efforts. In most instances, practitioners of shifting cultivation return to plots that they have cleared on a previous occasion. Diminished clearing effort in new-growth forest is likely the primary reason for this practice. Carter (1969), in studying the "slash and burn" cultivation systems of lowland Guatemala, estimated that clearing an old-growth climax forest required 368 man–hours of labor to prepare a 2.8 hectare (7 acre) field. In contrast, a similar size plot on which saplings were growing required only 113 man–hours to clear, while but 84 man–hours needed to be invested in clearing dense low-growth vegetation. Thus a cyclical return to formerly utilized plots avoids the arduous work involved in clearing old-growth timber. However, the temptation to return to previously cleared areas too frequently must be avoided at all costs. Too frequent return reduces the accumulation of nutrients in the soil and can shorten the period within which farmers can anticipate receiving adequate yields. For this reason, maintaining the length of the fallow cycle is essential to the success of shifting cultivation. This is also why extremely long fallow agricultural systems such as shifting cultivation are land extensive, support low population densities, and make little investment in infrastructure.

Creative destruction is an integral part of any shifting cultivation system. The primary creative destructive act is the removal of forest cover in the area in which cultivation is to take place. Large trees are girdled, small trees are cut, and material not needed for housing or fencing is piled in heaps. Often branches and small saplings are collected from a larger forest area, transferred to the field site, and added to the vegetation piles (Manshard 1974: 58). This transfer of primary productivity in the form of burnable trash treats the uncut forest as a mild form of sacrifice zone, since organic material that should be available via decomposition for plant growth in one area is moved to and concentrated upon an adjacent zone. Once dried, the vegetation mass is burned and the ash residue functions as a fertilizer source for the newly established field. This nutrient flush plays an important role in the initial yields generated from the cultivated plot. The use of fire to liberate nutrients in the ash also heats the upper soil horizon and plays an important role in destroying the seeds of some weeds (Manshard 1974: 58). Reduction in the local availability of competing weedy vegetation is important to the initial high yields of fields, just as the increase in weeds over time contributes to greater maintenance costs and ultimately to the decision to abandon the plot.

Vegetation removed and altered by cutting and burning is essential for the success of the shifting agricultural enterprise; the forest in both its morphology and its species composition is sacrificed for the benefit of the agricultural population. Distant sacrifice zones are not created; rather, one component of a given unit of space, the forest vegetation, is sacrificed in order to create conditions in that same space that are conducive to the crops that people desire. In most instances, these crops would not be able to grow in sufficient abundance under natural conditions to support a small, semisedentary village population living for several years in the same place. The presence of a closed canopy forest would sufficiently shade many potential crops to inhibit growth. While light holes would always exist in the forest as a result of natural events, edible plants growing in these locations would be most likely to be used by mobile hunters and gatherers. Forest removal is the essential creative act mandated for the success of shifting agriculture. Clearing the forest is a human device for enlarging and enhancing the conditions that occur naturally, but are found too sporadically to sustain an agricultural, semisedentary existence.

The success of shifting cultivation depends on four factors. The first is the design of the gardens themselves. Particularly in the tropics, gardens are not monocultural; rather, they contain an abundance of different plants, the "high diversity index" of which mirrors the variety of plants found in the forest (Ruddle 1974: 5). Plants cultivated in shifting cultivation gardens are intercropped. Not only are only small numbers of any one plant present at any particular point in the field, but also plants of different heights are intermixed. The leaf architecture of this complex structure insures that raindrops do not fall directly onto the soil surface. Instead,

raindrops cascade more gently toward the surface through an inter-
mediary series of layers. As a consequence, soil splash is greatly
reduced (Richards 1985: 60) and erosion is diminished. Richards
also points out that soil conservation is encouraged by employing
minimal tillage in high-rainfall areas in order to minimize soil disturbance, and by keeping some uncut tall trees and the stumps of
cut trees within the field. The cohesive action of the intact root mass
enhances soil stability. To a degree, these isolated tall trees, together
with the diverse heights of the densely packed cultivated vegetation,
replicate the structure of the forest that the crop habitat replaces
(Geertz 1966: 16–25). In similar fashion, the small cultivated clearings reproduce an analogous scale opening to the natural light holes
produced by blowdowns and other natural processes, which encourage rapid growth of the crops by permitting sunlight to reach the
vegetation. Small-scale fields such as these also contribute to erosion control (figure 7.3; and see Watters 1971: 7). By mimicking on
a small scale the same organizational principles as the forest that it
replaces, the shifting cultivation field represents a stable act of
creative destruction.

A second adaptive feature of shifting cultivation is the way in
which forest recolonization is encouraged. When fields are
cleared, individual trees are always left standing in the fields.
These trees are never so large or so numerous as to represent a
shade problem for the crops. However, the existence of a number
of trees in the field with well-established root systems means that
some of the minerals leached from the surface soil layers will be
captured by the tree. The presence of individual trees is also a secure
source of seeds for forest regrowth once the plot is abandoned
(plate 7.1); so too are trees along the forest-field edge, provided
that the plot is not too large. Thus, built into forest clearance is
the mechanism for forest reestablishment.

Nonetheless, periodic forest removal is guaranteed to have an
influence on the species composition of the regrowth forest.
Because tropical rainforest trees in particular are widely scattered,
there is no guarantee that the regrowth forest plot will reflect the
same species composition as did the originally cleared plot. This is
even more unlikely when one considers the tendency for shifting
cultivators to retain within plots those trees that have some useful
purpose in their dietary regime. To be sure, regardless of species,
very large specimens are readily ignored if there is no pressing
reason for removing them. However, all things being equal, it
makes sense to retain scattered individual trees that are helpful to
the local community. In parts of Latin America not only are valuable crop trees retained and protected within agricultural clearings,
but also fruit trees are planted within fields when they begin to lose
their fertility for crops. These sites can be identified long after
agricultural use has been abandoned by virtue of their unusual
concentration of beneficial trees (Watters 1971).

As a result of this selective pressure, the structure, composition,
and morphology of the forest that is the raw material for the shifting

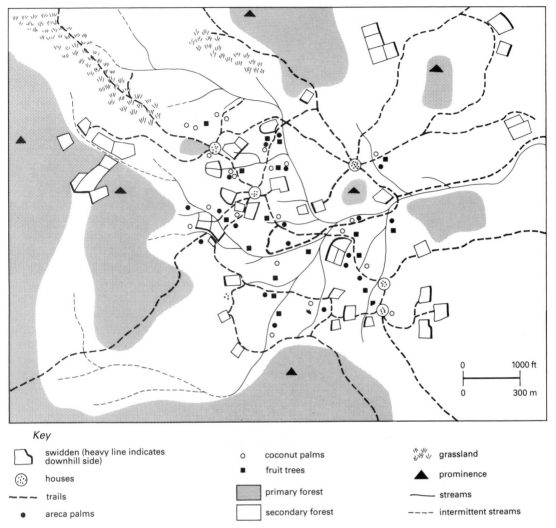

Key

⬠ swidden (heavy line indicates downhill side)

⊙ houses

--- trails

● areca palms

○ coconut palms

■ fruit trees

▨ primary forest

▢ secondary forest

🌿 grassland

▲ prominence

— streams

---- intermittent streams

Figure 7.3
Land use in a shifting agricultural society: the Yagaw area in 1953 (after Conklin 1954)

cultivation system can be expected to undergo change over time. Hecht, Anderson, and May (1988) have described a successional forest in the intermediate ecological conditions between the Amazonian rainforest and the semiarid northeast region of Brazil. This forest is a *babassu (Orbignya phalerata)* palm forest, which is the basis for much more than the shifting cultivation system of the region's inhabitants (figure 7.4). From this particular new-growth forest are extracted items such as thatch and other construction materials, fiber, animal feed, palmetto, and charcoal. All of these items are of vital importance to the income of the region's impoverished population, and all of them come from an early successional stage in forest redevelopment. This is a stage that local folk would probably be happy to see permanently maintained, were it not for the need to remove the *babassu* and other species episodically in order to plant crops. Most of the shifting cultivationists who collect

Plate 7.1
Shifting cultivation fields in
East Kalamatan (courtesy of
D. Kummer)

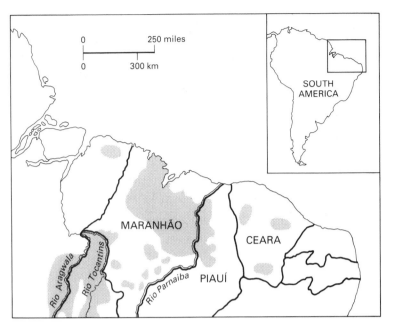

Figure 7.4
Secondary forest in northeast Brazil dominated by babassu *palm (after Hecht, Anderson, and May 1988)*

the products of the *babassu* palm do not have legal tenure to the areas from which they derive their livelihood. Therein lies a major problem. The expansion of commercial exploitation into the same land areas, in particular ranching, mechanized cultivation, and plantation production of *babassu* fruit for industrial uses, creates a situation of progressive impoverishment for already poverty-stricken people who have few development alternatives other than shifting cultivation and *babassu* exploitation (Kates and Haarmann 1992). The existence of the *babassu* palm forest is itself an artifact, a product of an altered successional sequence initiated by shifting cultivation.

Fertility maintenance is a third feature of shifting cultivation's creative structure. This is primarily achieved through the ash produced by burning vegetation accumulated on the field. Burning increases the phosphorus available to cultivated plants and the ash raises the soil pH into the intermediate range (Richards 1985: 56). While these effects are temporary, they are critically important for decent crop yields. Beckerman (1987: 81) reports that the Bari of Venezuela obtain yields of 18 tons/hectare (7 tons/acre) per year from their swidden fields, a yield that is twice that which more sedentary farmers are able to obtain from the same habitat. Beckerman regards this figure as the norm for shifting agriculturalists throughout South America's tropical forests. The productivity of swidden in Southeast Asia is more variable and is generally lower, ranging from 500 to 3,000 kg/hectare, although yields of 1,000 kg/hectare are common (Kunstadter 1987: 146). In the latter case, these yields compare favorably with rainfed fields in the same habitat, but are modest in comparison with the yields obtained from irrigated fields. Nonetheless, the productivity of swidden fields in the short-use/long-fallow system is considerable, and the fertility of soils and the productivity of the system is a direct result of the long fallow cycle that allows for an adequate period of recovery from use. Attempts are also made by swidden agricultur-alists to extract the greatest possible return from fields and to pro-long their use to the maximum extent once the effort has been made to clear the plot. Richards (1985: 57–8) reports widespread use of intercropping as a fertility and likely pest control management device. In these situations, farmers develop pragmatic combina-tions of crops in order to fix nitrogen for nitrogen-demanding plants as well as to maximize returns from microscale variations in soil type. Crude rotation schemes in which hardier crops are planted in successive years are another fertility management device that is designed to extend the life of cleared fields as much as possible.

Episodic mobility is the fourth and final factor that contributes to the stability of mature shifting cultivation systems. It is particularly important that the period in which a field is not used be a long one. For this reason, the uncleared portion of a community's territory is as much a part of the total resource base as the cleared field. The situation is analogous to that of the large catchment area attached to each small Nabatean field. Without the extensive forest fallow area, the small "slash and burn" plots would not be productive. Moving the field site, and often the village as well, is essential in order to permit forest regrowth, which restores local fertility. Without this natural regenerative process, and without the use of fire to unlock the nitrates and phosphorus contained in the vegeta-tion, a rapid shift into land degradation would occur. However, as long as the fallow cycle is maintained, the shifting cultivation system continues to employ constructively the principles of creative destruction.

Pastoral nomadism

A third type of low-intensity system that has traditionally employed creative destruction is pastoral nomadism. Pastoral nomadism is an extensive land-use system that is based on the herding of animals. This economy is most commonly found in areas that lack sufficient reliable water to sustain significant agricultural activity. Instead, people rely on moving animals to seasonal pastures where ade-quate grass and water can be found. It is an especially well-suited adaption that permits nomads to take full advantage of the highly variable nature of rainfall in semiarid areas. To accomplish this, pastoral nomads must develop a lifestyle that emphasizes mobility and that subordinates their material culture to the need to move many times during a year. A tent, rather than a permanent house, is their main dwelling, and their most valuable possessions (in addi-tion to their animals) are easily moveable goods such as rugs and jewelry. However, despite their focus on animals and their mobile lifestyle, pastoral nomads are hardly self-sufficient. Whenever practical, they plant crops themselves, and their diet is very much dependent on agricultural products. In most parts of the world, this has meant that nomads and farmers have engaged in a number of economic interactions in which their respective products are exchanged to their mutual benefit.

Like hunting and gathering and shifting cultivation, prior to modern conditions, traditional pastoral nomadism has a very limited impact on its physical environment. This is due to the large territories within which nomads traditionally operate, the small population densities supported by pastoral nomadism, and the frequent movements in which pastoralists engage. The pattern of movement is a very extensive rotational grazing system that involves oscillating between summer and winter pastures (Johnson 1969; Ingold 1987: 179–95). This movement pattern is an ecologically conditioned practice that is adjusted to the seasonal availability of water and grass in a given district at a specific time of the year (plate 7.2). The objective is to protect critical pasture zones from overuse by moving to new areas seasonally. Thus, movement to high-altitude summer pastures has the advantage not only of reaching fodder and water resources that cannot be exploited in the winter, but also of reducing pressure on the crucial lowland dry season pastures. In terrain in which an altitudinal variation in pasture resources does not exist, herders move their animals to distant districts that receive only occasional rainfall in the wetter season in order to conserve grazing possibilities close to reliable water resources in the dry season (Galaty and Johnson 1990: 33).

Little adverse impact is brought to bear on the pastoral environment by these mobile herds, because under pre-modern conditions animal numbers seldom reached sufficient size to place local resources under prolonged pressure. While herd numbers would increase during a sequence of good years, the corresponding decrease in herd size under extended drought conditions, or as a result of exposure to an epizootic epidemic, would result in a dramatic decline in grazing pressure. As long as herd numbers did not fall below a critical minimum, the viability of the pastoral nomadic community was not threatened. Every effort was made during a run of good rainfall years to increase the herd size to a point at which it could withstand severe reduction under adverse conditions and still provide minimal support for the dependent human population. As a consequence, catastrophic failure seldom occurred or affected more than localized groups.

The only sacrifice zone that ever temporarily emerged in traditional pastoral nomadic conditions was found around dry season water sources. O'Leary (1984), in discussing the pastoral patterns of the contemporary Rendille of northern Kenya, argues that herders take the easy option in determining their dry season grazing location and in providing security. This leads the Rendille to locate near springs, shallow water holes where the effort required to raise water to the surface is minimal, and settlements with secure water

Plate 7.2
*Seasonal grazing in Morocco
and the Sudan*

supplies. Security concerns encourage herders to locate in areas in which raiding is unlikely, and only after alternative pastures are depleted will they move to more vulnerable sites. Extrapolation into the past of the present-day pastoralist's preference for secure water source sites, where modern technology makes it possible for deep groundwater resources to be exploited and adjacent vegetation resources to be overexploited, is unreasonable. Around surface springs and shallow groundwater sources that could be reached by traditional technology, there is a more reasonable case to be made for possible overgrazing. Repetitive, concentrated grazing in dry season sites tended to reduce the frequency of the most favored, palatable forage species as animals selectively rummaged in their habitat. This selective pressure, in turn, encouraged the expansion of species that possess thorns, unpalatable leaves, or harmful exudes (Simmons 1989: 146). Despite this selective pressure, the limited water resources of most shallow, hand-dug wells make it very unlikely that unpalatable vegetation was substituted for desirable fodder species over extensive areas. Overgrazing did occur in specific places, as large herds were temporarily attracted to available water resources, but the scale of the disturbance always was small.

Severe damage at these sites was controlled by local tribal authorities who activated political power and status distinctions between local tribal segments in order to expel client groups from threatened resources (Peters 1968; Bernus 1990). Differential access rights to water and grazing resources were mechanisms that enabled nomadic pastoralists to lighten pressure on critical fodder resources and to manipulate the stocking ratio in a given habitat in order to insure the survival of some herders and their animals. This stratagem was always a last resort, and a variety of institutions existed in order to distribute pressure on the land resource more equitably before attempts were made to drive low-status groups away from resources. For example, in periods of abundance when herds were growing, surplus animals were transferred to the care of "stock friends" living in adjacent districts, as a device to spread the owner's risk and lower local stocking levels to manageable proportions. Unlucky friends, kinsmen, and neighbors with inadequate herds were lent animals in order to spread stock more widely in the social system, and at the same time disperse livestock more widely in space. Ties of kinship and marriage were maintained with other groups in a complex web of social relations. This network of social debts and obligations enabled herders to request access to the temporarily more favored areas of other groups should natural conditions on their own tribal turf prove unfavorable in any given year. In these ways excessive concentrations of animals and people that might threaten the viability of critical grazing zones were avoided and the intensity of use was diminished. Detailed local knowledge of the carrying capacity of the local resource base made it possible for herders to use the local environment intelligently (Hobbs 1989). In short, compared to modern pastoral conditions, traditional pastoral nomads used their environment lightly and had only a modest adverse impact upon their resource base.

Fire was the major tool of creative destruction employed by traditional nomadic pastoralists, just as most contemporary pastoral groups use fire at some point in their seasonal cycle for a multiplicity of purposes. Burning grassland just before the beginning of the rainy season encourages many grasses to sprout. These new shoots provide a welcome source of fresh fodder at the end of a long period of poor grazing. Regular and repetitive burning creates a more open habitat – a setting in which pastoralists can more easily see their flocks over large distances – and eliminates protective cover that might shelter potential predators. Fire also manipulates the species composition of the vegetation. The primary impact of fire is upon woody vegetation, since trees and shrubs that lack a thick insulating bark are adversely affected by the heat of the fire.

Grasses are favored in these circumstances. The removal of the shade dominance of trees and shrubs creates more niche openings for other plants, thus increasing the diversity of the grassland flora that is of primary interest to the herder (Ruddle and Manshard 1981: 123). A denser sod cover also reduces erosion. When combined with light grazing, fire removes dead stems and old growth and promotes an increase in the vigor and growth rate of many desirable fodder plants (Warren and Maizels 1977). Increased availability of grass means that more livestock can be supported on a given area. More livestock means greater manure deposition and a higher rate of nitrogen return to the soil. Since domestic stock have definite consumption preferences, the plants that they favor tend to become more widely distributed as grazing animals spread the seeds of preferred plants about the landscape as an accidental by-product of defecation. The key ingredient in this process of coevolution is the combination of burning and grazing at intensity levels that are not excessive. Light, episodic impacts have profoundly beneficial long-term effects.

Not only does fire improve the quality and quantity of the forage available to domesticated livestock, but also in many places the removal of moist, shady habitat reduces the breeding opportunities for insects that carry disease. In sub-Saharan Africa, the tsetse fly, the intermediate vector of trypanosomiasis (sleeping sickness), is controlled by burning that removes its breeding habitat. Cessation of managed burning leads to an increase in brush and a decrease in the safe habitat within which nomads can herd their animals. Lacking sophisticated chemicals to use in an anti-tsetse campaign, materials that pose their own set of environmental problems when employed in large concentrations (Linear 1985), traditional pastoral nomads have had fire as their only tool for instituting large-scale environmental modification. That tsetse abatement has been an accidental result of burning for other purposes in no way diminishes its importance as an act of creative destruction (Lewis and Berry 1988). By burning, nomads have created a grassland habitat that is conducive to their grazing animals at the expense of forest and shrublands.

Despite the reality of traditional pastoral nomads causing almost no land degradation (McCabe 1990), herders have a bad reputation for following destructive land management practices (Wikjman and Timberlake 1985; Hjort af Ornas 1990). This image is a product of failure on the part of central planners to understand the dynamic nature of nomadic pastoralism (Ruddle and Manshard 1981: 124) and the tendency of settled farming societies to export environmental costs onto pastoral areas (Simmons 1989: 264–71). Nomadic groups that once were able to control their own destiny have lost power and territory to an expanding agricultural economy. This agricultural expansion invariably takes the best pastureland, and leaves the herder with the less desirable land; the more arid the district in which this process takes place, the more destructive and the more rapid is the impact on the surrounding habitat (Falloux and Mukendi 1987). Confined to increasingly more marginal habitats, nomadic herders have had to maintain their animals for longer periods in rangeland that once was only used seasonally. The inevitable result is land degradation in the excessively pressured districts (Olsson and Rapp 1991).

In many river floodplains, nomads were able to graze their animals on fallow land, riparian forest, and post-harvest crop residues once the agricultural season was completed. Conversion of these seasonal agricultural zones to perennial irrigation has cut the nomad off from vital dry season pasture (Adams 1989). Not only does the farmer lose the manure formerly deposited by the nomad's herds, thus rupturing the manuring cycle that once tied farmer to herder and captured waste resources, but also herders lose seasonal use rights that reduced pressure on rainy season pasture. In effect, part of the costs of floodplain irrigation development are exported onto the rangeland. The remaining grazing areas are made the sacrifice zone for prosperity in the agricultural sector.

Many pastoralists have become sedentarized, both on their own initiative and in response to pressure brought to bear by nation–states for political and ideological reasons. The result is an excessive concentration of people and animals, usually around sites with decent water resources (plate 7.3), and the outcome is degradation in the surrounding vegetation and soil resources (Bedrani 1983; Janzen 1983; al-Ibrahim 1991). War and civil conflict have also restricted nomadic movements and placed more pressure on local habitats (Bascom 1990; Arkell 1991; Dahl 1991). Nationalization of the rangeland (Beck 1981; Bedrani 1991) has meant that nomads have lost the ability to control their resource base. The shift of political power to governmental agencies has usually proven to be ineffectual, because these institutions seldom possess the staff or the local support to implement decisions about resource management. Land degradation is the inevitable consequence of the resulting scramble to extract something useful from the rangeland resource base before access to the rangeland is forever lost to the pastoralist.

Plate 7.3
*Undisturbed steppe compared
to an overgrazed zone near El
Khuwei, Kordofan, Sudan*

In a spatially extensive, long-fallow, low-intensity ecological system such as nomadic pastoralism, few, if any, sacrifice zones exist. When they do occur, these sacrifice zones are limited in spatial scale and temporal duration. Creative destruction in traditional nomadic pastoral systems is a common process, but it takes place slowly through the gradual expansion of rangeland habitats. Fire is the primary tool employed in this positive process of habitat modification. When substantial land degradation occurs in pastoral environments, it is almost invariably linked to the export of costs from other sectors of economy and society onto the pastoral community. This process of land degradation impacts vulnerable pastoral people who are much more victims than they are villains in the development of rangeland degradation. For it is when cost export avoidance fails in other habitats and economic sectors, largely because linked cultural ecological systems are treated in disaggregated fashion, that serious land degradation occurs in rangelands exploited by pastoral nomadism.

Moderate-intensity Systems

The agricultural revolution, whereby humans began consciously to manipulate and control plants and animals, altered the relationship of humankind to its environment. Unlike the low-intensity systems described in the previous section, which today are relic mobile livelihood systems existing on the margins of sedentary civilization, moderate-intensity systems represent a fundamental change in the nature–society relationship. Moderate-intensity systems demonstrate an enhanced ability to produce significant impacts that transform concentrated spaces, last for substantial periods of time, and affect distant ecosystems. More sophisticated and numerous technological inputs characterize moderate-intensity systems when compared to low-intensity systems. One major difference between the two is the greater ability of moderate-intensity systems to shorten, and in some cases even eliminate, the fallow period.

Positive use of creative destruction underlies this ability to use land resources more intensively. Much larger populations are supported in a smaller territory than was the case in low-intensity systems. As a result, human labor can be invested in greater amounts to transform local space. It is equally significant that, unlike low-intensity systems, the actions of moderate-intensity systems begin to affect areas far beyond the boundaries of their immediate application.

By massive use of human labor in some cases and/or the selective application of technology – including the use of machinery – in other situations, constraints of the natural environment that restrict production are ameliorated. As is typical of most successful examples of creative destruction, moderate-intensity systems evolve slowly and are rooted in an intimate understanding of local habitat conditions. New areas of sustainable use on lands previously avoided due to environmental constraints, and drastically shortened fallow cycles on established lands that require careful management of system inputs and outputs, are two hallmarks of moderate-intensity systems. Both types of improvement result in the ability to sustain large populations on limited areas for unlimited periods of time.

An important difference between low- and moderate-intensity systems is exemplified by the ability of moderate-intensity systems to expand food-producing activities despite often poor natural conditions. This expansion is the result of a direct modification of the land, initiated to meet the desired utilization. This characteristic is completely unlike low-intensity systems, in which vegetation cover is almost exclusively the product of the direct alteration that occurs due to human activities and in which conscious modification of terrain features is unknown. Low-intensity systems modify their habitat very little, and that only slowly over long periods. Moderate-intensity systems accelerate this process often through direct alterations of critical components in the existing topographic and/or hydrologic systems.

Prior to the development of hybrids and bioengineered plant varieties, expansion of plant varieties beyond the environmental settings that were naturally conducive to their growth was primarily accomplished by altering the locale's constraining component. In particular, drainage and/or terrain morphological changes often were undertaken to meet the specific demands of the crops that the inhabitants wanted to grow. This ability to transform radically the environment of nature and convert that environment into habitats that are the product of human conceptualization, and that are utterly dependent upon humans for their continuation, is the unique contribution of moderate-intensity systems. In these settings, nature and society no longer exist as separate categories, but rather are integral parts of common systems. Moderate-intensity systems take a giant stride down the road toward the domestication of nature, a process that succeeds only when sustainable agroecologies can be created.

In this section we consider several moderate-intensity systems that have successfully achieved a sustainable agroecology by the use of the principles of creative destruction. We begin with two examples where traditionally the use of large amounts of labor was essential in order to create – over centuries of evolutionary development and elaboration – sustainable and productive cultural ecological systems. These are the terrace systems that transform steep slopes into fields, and the integrated agricultural–aquacultural systems that have converted coastal and deltaic lowlands in many parts of Asia from forest and marsh into highly productive farmland.

Our second set of examples examines the American Midwest, a region that today is one of the world's premier rainfed agricultural areas. Throughout this agricultural heartland, technology, applied at varying levels of scale and importance, has played a critical role in creative destruction. In general, the example looks at the role of drainage in converting this region's large expanses of poorly drained, but good-quality soils into prime agricultural terrain. We look at how a simple technology, land tiles, made it possible to open to cultivation vast areas of the Midwest across which soils were wet enough to impede agriculture but not wet enough to be considered a swamp or marsh. We examine how farmers adjusted the needs of land drainage to an existing arbitrary land division system, the township/section survey system of the USA.

In the second part of this example, we explore the role played by larger-scale, machine-dominated drainage technology in partnership with labor, which transformed a substantial swamp in the heart of the American Midwest. The Kankakee Marsh, which remained a poorly settled island surrounded by productive farmland and the Chicago hinterland until the turn of the twentieth century, was altered (by creative destruction) from a diverse wetland ecosystem into productive farmland that better met the needs of the local population and the nation. Today, this area of northwestern Indiana and northeastern Illinois has evolved from lands where moderate-intensity agricultural interventions were applied at the turn of the century into a successful zone within the American Corn Belt, utilizing high-intensity interventions that permit large agricultural yields in a nonrotational, nonfallow, high-energy input, sustainable agricultural system.

Terracing and creative destruction

In most instances, agriculture on steep lands results in excessive surface water runoff. Two ramifications of this are: (1) excessive soil losses; and (2) relatively low infiltration of precipitation into the soil. Unless runoff is controlled on steep lands, eventually land degradation occurs from the loss of the upper fertile soil horizons and/or gully formation. One human response to curtail excessive soil erosion and to encourage infiltration of moisture into soils on steep agricultural lands is to terrace the steep slopes, a practice that topographically alters the landscape.

Terraces represent a widespread contemporary agricultural practice that dates back to antiquity, and has undergone an imperfectly understood pattern of evolutionary development and diffusion (Spencer and Hale 1961). The Nabateans (see chapters 1 and 2) utilized terraces as one component in their agricultural system by which to increase infiltration (soil moisture) in an extremely dry area. In China and Ethiopia (Westphal 1975; Deqi and Li 1990), stone terraces have a long history as a conservation practice by which to control soil erosion by water on both steep lands and easily eroded soils. In the Andes, Inca development of terracing was intended both to trap and control the runoff generated by orographic precipitation on the western slopes of the mountain range as well as to control soil loss (Pawluk, Sander and Tabor 1992).

Terraces are generally constructed by cutting into the soil on a slope segment and then using the material removed in the cut to add fill along the downslope portion of the terrace. When water conservation is a primary purpose of terraces, the terrace bench is either constructed level or slopes inward toward the hill (figure 7.5). This minimizes any downslope surface water runoff and encourages maximum infiltration. Where the major purpose of the terraces is to reduce the erosional potential of runoff water, the bench portion of the terrace is either level or outward-sloping. In these instances, the bench is constructed at an angle of slope that is less than the original hillslope (figure 7.5). Construction of erosional control terraces usually requires runoff ditches and/or conservation banks to disperse the runoff into thin, nonerosive flows over the terraced fields. The goal is to prevent runoff from concentrating into either rills or gullies.

By altering the configuration of hillslopes, terrace construction creates slope angles that are both less than (bench) and greater than

(a)

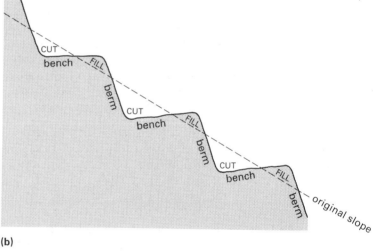

Figure 7.5
Water conservation and erosion reduction terraces.
(a) A water conserving terrace. The bench slopes inward to the hillslope, to save water. Drainage is from the back of one terrace to the next.
(b) An erosion reduction terrace. The bench is level or slightly out-sloping. The bench portion of the terrace is always gentler than the original slope. The berm portion of the terrace is always steeper than the original slope.
The natural hillslope is most stable, and so terraces are inherently unstable

(b)

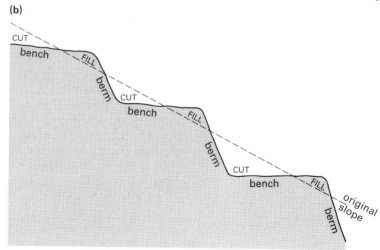

(terrace berm/wall) the original hillslope. Since the natural terrain likely represents the most stable morphology (Hack 1960) for the region, terracing inherently creates unstable conditions. This instability is found in two areas: (1) the bench areas, with their lower slopes, become zones that have a higher potential for deposition; and (2) the berm zones, with their steeper slopes, are prone to erosion.

Only through continuous maintenance will the vast majority of terraced lands remain stable. For example, terraces that for centuries were stable are now being destroyed in many parts of the world from Murcia, Spain to Yemen (Vogel 1987). Due to rural outmigration, farmers no longer have the available labor to maintain all of the terrace walls and drainage channels in these areas. Much of the workforce has left the countryside to pursue better economic opportunities elsewhere. Runoff that previously was controlled by continuous terrace maintenance now results in rill and gully formations. These phenomena, the result of uncontrolled sheet wash, are slowly destroying the terraces. The bench areas of the terraces are becoming steeper while the steeper terrace walls are gradually being destroyed. Both of these processes remove the area's topsoil, which is being deposited along the valley bottoms. Thus, once maintenance becomes slack, terraced land often leads to accelerated degradation along the affected hillslopes (plate 7.4). Ultimately, the destruction of the terraces will restore the hillslopes to an equilibrium condition that no longer requires the human input of maintenance. Albeit the new equilibrium condition will have a drastically altered soil, hydrological, and vegetational status compared to the conditions existing prior to the human intervention of terracing.

Terracing of lands requires major labor inputs or the use of heavy machinery in order to reconfigure the hillslope and construct the requisite drainage ditches. This is even true for the *fanya juu* terraces in Kenya (Lewis and Berry 1988) and the *fosses aveugles* in Rwanda, where the majority of the cut and fill required for terrace formation is done by erosion and deposition processes (figure 7.6). Similar methods of terrace construction are employed at these two sites. First, a drainage ditch is dug orthogonal to the direction of the downslope. The excavated soil from the ditch is placed immediately above and parallel to the ditch. Grass strips are planted in numerous rows along the low ridge formed by these excavated materials. Runoff from the field above the grass strips transports soil from the field to the grass strips. Contact with the plants decreases the surface water velocity, which encourages deposition in the grass strip zone. Over a number of years, the erosion on the field results in a proportion of this soil being deposited in the grass zone. This material, along with the continuous maintenance of the drainage ditch from which the farmer throws the soil deposited in it up into the grass zone, results in the formation of a berm along the grass strips.

If the *fanya juu/fosses aveugles* terraces are to be successful, care must be taken to insure that the spacing of the grass strips is close

Plate 7.4
Terrace degradation near
Tafraout, Morocco

enough in order that the soils eroded from the upper portion of the
bench will not bring infertile soil into the root zone of the crops. In
many areas of central Kenya, thick volcanic soils exist, and the
erosion and deposition of these fertile soils to form terraces has
made the *fanya juu* technique successful (Lewis and Berry 1988).
However, in parts of western Rwanda, where soils are both less
fertile and thinner, this method of terrace formation has contributed to the land degradation problem (Lewis 1992). Often distances
between grass strips were too great given the thickness of the soils
(figure 7.7). This has resulted in the highly acidic and infertile B
horizons found in many parts of this area being brought close to the
surface along the back portions of many of the terrace benches. One
outcome of this has been poor crop yields on the back portions of
these terraces. To increase food production on the back portions of
the terrace benches, during each growing season many of the farmers have begun to remove the better soils from the berm immediately above the zone of poor soils. This material is then spread
along the back bench and acts as a fertilizer. While improving
agricultural yields in the short term, this counters the long-term

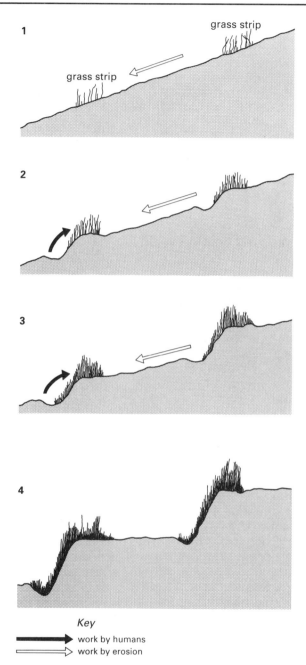

1

grass strip

grass strip

2

3

4

Key
→ work by humans
⇒ work by erosion

Figure 7.6
The transformation of grass strips into terraces (after Wenner 1980; Lewis and Berry 1988)

purpose of the terrace, namely reducing the downslope movement of soil. This human erosion (removal from the berm) and deposition (spread over the back portion of the bench) process results in the movement of approximately 62 ton/hectare/yr (25 ton/acre/yr) of material downslope (Lewis 1992). Thus terrace construction in some Rwandan areas represents another contributing factor to the widespread degrading condition of many highland areas in

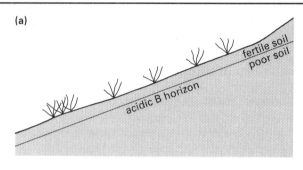

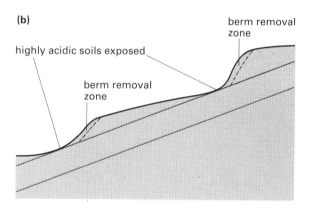

Figure 7.7
The acceleration of soil loss on Rwandan terraces (from Lewis 1992)
(a) Natural condition
(b) Terraced condition

western Rwanda (Lewis and Nyamulinda 1989). This is one example of how topographic changes on stable steep lands can promote inappropriate interventions that contribute to land degradation by encouraging the very unstable conditions that the terrace technology was intended to prevent.

When farmers terrace the land it is done with the intent of growing specific crops that likely would not be sustainable or possible under natural conditions. Therefore, as in any agricultural endeavor, terrace lands are covered by vegetation that has replaced the natural groundcover. Through this alteration, the previous ecosystem is greatly altered or destroyed and a new one is created. In northwestern Rwanda, as the foothills of the Virunga Mountains have been cleared and terraced for agriculture, the forest ecosystem has been destroyed. This created landscape has permitted the Rwandan people to increase agricultural production and meet their food needs as their population has grown. However, the destruction of the tree cover has also had major deleterious effects. It has destroyed a significant proportion of the habitat of the mountain gorilla, now a protected, but highly endangered, species within the country. Even though the remaining forested lands are within a national park, it remains to be seen if enough of the mountain gorilla habitat remains to insure their long-term survival. The hillslope terracing and forest removal process in northern Rwanda along the Ugandan/Zairan border illustrates an essential feature of creative destruction. It inherently results in both winners (those that benefit from the change, particular farmers) and losers (the mountain gorilla and other wildlife).

Creative destruction in the Chinese pond–dike integrated system

Sites at transition zones between major ecological systems often are productive habitats for humankind. No site is more conducive to access to diversified ecosystems than is a coastal location. Here both terrestrial and aquatic ecosystems can be exploited readily without engaging in substantial movement. Within the littoral zone, these advantages are maximized in delta environments where rivers empty into the ocean. Historically, these sites have attracted human settlement by virtue of their abundant plant, animal, and aquatic life. In the past two centuries, government-sponsored settlement schemes have transformed these habitats in dramatic fashion, and in no region more so than in coastal districts of south and eastern Asia (Richards 1990). Here most of the coastal wetland districts have been transformed into zones of peasant agriculture by forest clearance, drainage, and flood protection embankments.

Spurred by the desire to enhance agricultural productivity in support of a growing population, this process of land development destroys coastal vegetation that contains many valuable resources – particularly mangrove ecosystems (Hellier 1988). These terrestrial ecosystems are linked to offshore seagrass ecosystems that are important food and habitat resources for marine animals. Many species, such as the spiny lobster, pink shrimp, and various grunts and snappers, migrate between these systems during their life cycle (Fortes 1988: 211). Any reduction in the coastal vegetation results in an increase in erosion that smothers the seagrass beds that provide shelter and food for marine animals, at the same time that habitat destruction eliminates the breeding site for many species and the feeding setting during their juvenile stage. Thus terrestrial modifications constitute a threat to the primary productivity of aquatic ecosystems. This produces a classic case of creative destruction in which the transformation of mangrove and other coastal ecosystems into agricultural land turns the marine ecosystem into a sacrifice zone in return for increased terrestrial productivity.

Although accelerated by systematic government action in the past two centuries, this process of transformation has ancient roots that began with the development of paddy cultivation in the region (Lo 1990: 405; and see figure 7.8). In its most advanced state, coastal and delta modification follows the pattern described by Ruddle and his colleagues in their analysis of a rural commune in the Zhujiang (Pearl River) Delta in South China (Ruddle et al. 1983; Ruddle and Zhong 1988). Here conversion of the wetland soils of the delta environment has not halted with paddy rice development. Instead, there has evolved an agricultural system that tightly integrates

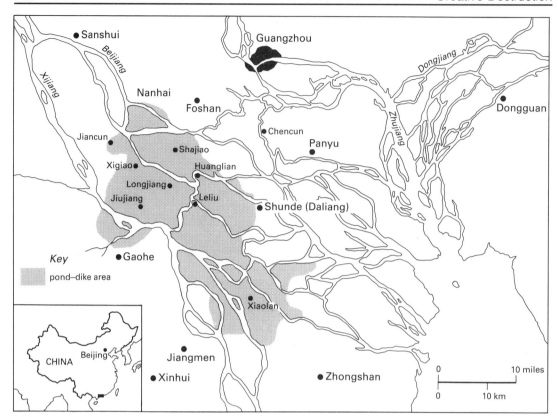

Figure 7.8
The pond–dike area in the
Zhujiang (Pearl River) Delta
(after Ruddle et al. 1983)

aquaculture with crop production on the dikes that separate the fish ponds. In this system, waste products from one subcomponent become the inputs to other segments (figure 7.9). Livestock, crops, and fish are treated as integral parts of one system rather than as separate entities managed by different production units. The result is sufficiently stable to be sustainable for centuries – yet resilient and flexible enough to accommodate change.

It has taken the pond–dike system over 1,000 years to evolve to its present state (Ruddle et al. 1983: 49). This time period is important in determining the character of the system. It means that it has grown organically from a rich base of empirical knowledge. This bedrock of indigenous knowledge is enormously sensitive to the peculiarities of the local environment, and as such demonstrates how the *genius loci principle* is applied. Because it is so rooted in place, the pond–dike system is not readily transferred elsewhere, although its management principles can inform land management in other locales.

The foundation of the system is the pond and dike subsystems and the complex manner in which they are woven together. The pond is typically an east–west-aligned rectangular water body, 2.5– 3 m (8–10 ft) deep and covering less than 0.5 hectares (1.2 acres). The ponds are shaped in a ratio of 6 : 4, an evolved configuration

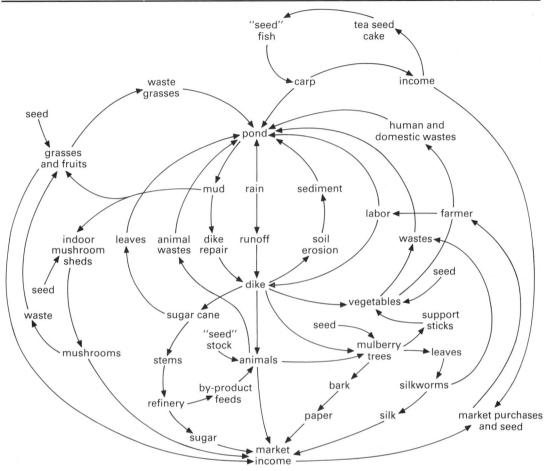

Figure 7.9
The structure of the pond–dike system (based on Ruddle et al. 1983)

which minimizes bank erosion due to wave action. The depth of the pond also reflects empirically derived folk knowledge, since digging deeper risks reducing sunlight to the lower depths and diminishing the productivity of the bottom-feeding species.

The beauty of the pond structure is that it is conceived as a total volume that can be exploited by polycultural techniques. This is achieved by raising a variety of different species of carp in the pond (Lo 1990: 410). Indigenous to China, carp have evolved specific feeding preferences. Some consume detritus on the pond bottom, some eat grass, others specialize in snails, and still others concentrate on algae and plankton. As a result, the entire pond volume can be filled with different species and ages of fish in varying stages of development, each of which utilize a different segment of the pond. The key to the entire system is the grass carp. This is a herbivorous species that lives in the top layer of the pond and consumes both aquatic plants in the pond and the grass clippings and crop residues that are raised on the dike and dumped into the pond. Grass carp are inefficient feeders that probably do not digest 50 percent of the food that they eat (Ruddle and Zhong 1988: 30).

Therefore, partly digested grass carp excrement becomes the raw material for plankton growth. Silver and bighead carp, phytoplankton and zooplankton consumers respectively, benefit most from this fertilizer input. Invertebrates living on the bottom of the pond feed on the plankton and unused detritus that rains down upon them from the levels above. These creatures in turn are the food source for black, mud, and common carp who prowl the pond floor. Common carp in particular are aggressive rooters in the bottom mud since they also feed directly on detritus, and by their actions they stir up material that enters the water volume and becomes available as food to other carp species at higher levels in the pond. In their mutual interdependence, by consuming each other's wastes and by-products, the carp mirror in the pond ecosystem the tight system integration that characterizes the entire pond–dike system.

The pond itself is kept constantly fertilized at a high level in order to produce the maximum amount of food for the fish. This high level of fertilization requires constant vigilance on the part of the pond operator in order to insure that the dissolved oxygen levels in the pond do not sink too low. Were this to happen, fish would become lethargic, growth would slow, and, if dissolved oxygen levels were to drop too low, fish survival would be threatened. The first sign of the emergence of these problems is the movement of the bottom feeders, the common and mud carp, to the surface (Ruddle et al. 1983: 55). Attention must also be paid to the turbidity of the water, since overfertilization can so cloud the water that sunlight penetration is reduced. Any loss of solar radiation reduces photosynthesis, and thus strikes at the primary productivity of the plankton and algae upon which the aquatic food chain depends. The source of the organic fertilizer used to fertilize the pond is partly the undigested wastes of the fish themselves, but primarily is derived from terrestrial sources. Terrestrial manure, in turn, is produced from the excrement of animals – especially pigs, humans, and silkworms – and the organic wastes from agricultural activities carried out on the dikes. Pig sties and latrines are commonly located over the pond so that manure will drop directly into the water, because these wastes are most valuable when fresh (Ruddle and Zhong 1988: 33). Wind and wave motion then redistributes the wastes more evenly throughout the pond. It is the bacteria and the protozoa in the manure that are the prime food source for bottom-feeding carp, and the fish faeces returned to the bottom are also available for recolonization by bacteria and subsequent reconsumption by fish in a continuous cycle (Ruddle and Zhong 1988: 32). Once established, the fish pond is capable of sustaining high stocking rates as long as manure inputs, water turbidity, and dissolved oxygen levels are continuously monitored in order to insure system stability.

Until recently, juvenile fish (fry) used to stock the ponds were captured from adjacent streams. This was necessitated by the difficulty of getting carp to breed in captivity. Today, river pollution places these natural sources of fry at risk. Fortunately, at approximately the same time as river pollution has become a serious problem, there have been dramatic improvements in nursery breeding and rearing of fish. To prevent in-breeding, however, fry and larvae are still extracted from natural sources (Ruddle and Zhong 1988: 34). Whether artificially bred or captured in nearby streams, fry are raised in specially prepared nursery ponds for about one month. When the fry reach fingerling size (ca.3 cm (1 in) long), they are transferred to separate ponds where they are "grown out" in a two-stage process (each stage using separate ponds) lasting about one month per stage. When the fingerlings reach approximately 12 cm (5 in) in length, they are transferred to the production ponds.

The second major component of the system is the dike. Dikes are artificial creations built up by the systematic excavation of the adjoining ponds. In this process of pond and dike creation, any semblance of natural vegetation and topography is destroyed and a completely humanized landscape is created. Despite the effort in pond design to minimize the impact of dike bank erosion, a gradual collapse of the dike edge and bank is inevitable, because none of these edge environments are constructed with hardened materials. The pond never fills in because two or three times each year mud is reexcavated from the pond bottom and placed on the bank (Ruddle et al. 1983: 57). Since the mud is rich in organic matter, it serves as the most important fertilizer source for the crops grown on the bank. The need to remove mud from the pond several times during the year is just one reason why the pond–dike system is very labor-dependent. It also demonstrates how dynamic the system is within a general structure of stability, and how important that basic dynamism is to the overall design and function of the pond–dike system.

Dikes are carefully calibrated structures and they can only exist within a narrow range of tolerances. The dike cannot be too narrow or it is impractical to cultivate: yet no dike can be too wide or there is not sufficient mud fertilizer to sustain year-round, continuous cultivation. Dikes that range in width between 6 and 20 m (20–66 ft) are typical (Ruddle and Zhong 1988: 55). Moreover, dikes cannot be too low or they become waterlogged. Conversely, if the dike is too high, its soils and plants are more readily affected adversely by drought. Thus, the dike exists within a rather narrow tolerance limit that must be carefully monitored and maintained.

The major crops grown on the dikes are sugar cane and mulberries. Less numerous are fruits, especially the litchi and longan for which the region is famous, vegetables, and oilseeds (mostly peanuts), which are intercultivated with the main crops or are cropped in rotation with them. Fruit trees were actually the first major commercial crop in the region when a production shift from paddy rice cultivation to commercial production began. However, after AD 1620 mulberry cultivation began to replace fruit as the price of silk rose substantially, and as the benefits of integrating pond and dike more closely began to be widely recognized (Ruddle et al. 1983: 49–50). In particular, farmers learned empirically that direct application of animal wastes to the plants harmed the mulberry leaves. However, the same manure dumped into the pond for the benefit of the managed aquatic food chain had only useful impacts when withdrawn after a period of pond fertilization and applied directly to the fields. The use of the pond as an intermediate fermenter converted harmful wastes into beneficial inputs and made the integrated land–water system possible.

Mulberry leaves are harvested seven or eight times between April and November and are fed to silkworms housed in sheds. Silkworm excrement, cocoon waste, waste water, dead larvae, and scraps of unconsumed mulberry leaves find their way into nearby ponds as fertilizer and fish food. So, too, does the waste from the silk spinning factory, the wastes from which are purchased by farmers as fertilizer for the fish pond. Feedback loops such as this silkworm to fish pond linkage are vital elements that complete the cycling of nutrients and waste products in the total system.

In the winter, when sheds can no longer be used to raise silkworms, they are used to cultivate mushrooms. Mud from the pond bottom is used to prepare the mushroom bed on the silkworm shed floor. Seed is purchased from off-farm sources, and most of the output of mushroom buttons is sold off-farm as well. When the mushrooms are harvested and the next cycle of silkworm cultivation is about to begin, the nutrient-rich mushroom mud is cleared from the floor and used to fertilize the vegetable gardens (Ruddle et al. 1983: 59). The sheds are then cleaned and disinfected, and the next cycle of silkworm cultivation begins.

Sugar cane is the other major dike crop, and has increased in importance during the past several decades to the point at which it surpasses mulberry trees in total acreage. Like all crops grown on the dike, the primary source of fertilizer for the cane is the mud dredged from the adjacent pond and deposited on the dike. The cane has several uses within the farm. In particular, the young leaves are a source of fish feed, while the old leaves provide shade for the gardens. The cane's major use is as a commercial crop. Stems, when harvested, are processed in off-farm refineries to produce sugar, which is both marketed nationally and consumed locally. Bagasse and other wastes from the refinery are returned as inputs to the farm operation and are important sources of animal feed.

Animals are a major component of the integrated pond–dike farm. Pigs are arguably the single most important animal, and are regarded as walking manure factories, the excrement from which is a vital input into the pond subsystem. In addition to their role as producers of manure for the fish ponds, large amounts of animal protein and cash income are derived by farmers from their pigs. Small numbers of water buffalo are also kept for traction power. Chickens, ducks, geese, and in some cases eels are raised by farmers in addition to the ubiquitous carp. All of these animals contribute to the subsistence and the cash income of the farmer, as well as returning excrement to the pond environment. The cycle is enriched by the growth of aquatic plants, particularly water hyacinth, in separate ponds as a food source for the animal population, especially the pigs. Supplemented by crop residues, kitchen wastes, and food material scavenged from accidental spills, these water plants provide the bulk of the food supply for the domestic animal population.

The pond–dike system is a tightly integrated example of creative destruction; it is a system joined together by a complex pattern of exchanges that only can be suggested in diagrammatic form. The integrated agriculture–aquaculture system leads to complete removal of the native vegetation and its replacement with a much simplified, human-manipulated complex of cultigens. Part of the success of the system is its ability to cut costs dramatically by finding most of its major inputs within the local environment (Ruddle and Zhong 1988: 152). Expensive imported inputs are kept to an essential minimum. This is accomplished by capturing as many wastes as possible from each subsystem and using these wastes as major inputs into other subsystems. Thus mud, which is a waste product filling the pond and threatening it with extinction, becomes a major source of fertilizer and a construction material for the dike. Similarly, the excrement of animals based on the dike becomes the primary energy input contributing to the controlled eutrophication of the fish pond. Carefully monitored and regulated, this process, which potentially could result in the death of the aquatic environment if dissolved oxygen levels were to decline too much, creates ideal growth conditions for the varieties of carp that inhabit the pond. The natural ecosystem of the delta and at least some of the productivity of offshore fisheries are sacrificed for the human-oriented productivity gains of the pond–dike farm. Because the farm system created thereby possesses high levels of both stability and resilience, and has been able to survive for centuries, this sacrifice seems defensible and a classic example of creative destruction.

Drainage and creative destruction in the American Midwest

The American Midwest, the nation's granary, reflects its recent glacial ice sheet heritage. Large areas of the region were poorly drained, as the river and stream systems had insufficient time to develop an integrated drainage by the time the major influx of European migration settled the area in the 1800s. To convert this region from the natural vegetation of woodlands, wetlands, and grasslands into prime agricultural land required major alterations in the drainage of vast areas as well as the removal of the prevailing forests. The alteration in the vegetation cover and drainage that rapidly changed this area from a region where hunting and trapping were replaced by farms is a prime example of massive areal creative destruction. The transformation of the lands from natural vegetation to a myriad of crops resulted in significant changes in the flora, fauna, and erosion/drainage characteristics of this immense region. The contemporary landscape, with farmland stretching from horizon to horizon, appears to the casual observer to represent a simple transition that humans have created in a natural setting just waiting to be cultivated. In actuality, it required major alterations in the natural environment. Today, it requires continuing technological inputs such as drainage maintenance, pesticides, fungicides, and fertilizers to keep sustainable the agricultural systems that replaced the natural ecosystem.

The thousands of square kilometers comprising the American Midwest, the country's agricultural heartland, possess numerous characteristics ideal for modern, high-technology farming. With a large percentage of its lands having slopes of less than two degrees, most Midwest topography can be described as ranging between flat and very gentle. Where relief is evident, it generally results from short and not particularly steep slopes. These terrain characteristics are ideally suited for modern farm implements. The soils that veneer most of this region are derived primarily from either glacial or alluvial deposits. They are relatively young, generally high in soluble mineral plant food, and most of the soils' nutrient deficiencies can be easily countered through application of fertilizer supplements, a critical requirement in the intense monoculture agriculture practiced today. From the agricultural perspective, the climate of long hot summer days and sufficient moisture complements the area's topographic and pedological advantages. Today, in many areas, highly productive farm fields stretch as far as the eye can see, from Ohio in the east to Iowa in the west. In the northern, cooler portions of this region, the flat checkerboard fields are covered by wheat, potatoes, barley, oats, sugar beets, or soybeans, while dairying is important in many of the hilly areas (Borchert 1989). In the southern, warmer sections of the Midwest, corn (maize) is king. In Illinois alone, 400,000 hectares (1×10^6 acres) generally are planted in corn (Horsley 1986). Soybean, wheat, oats, and livestock production follow in dominance.

That this bountiful area is the heartland of America's rainfed agricultural production is partially the result of its natural endowment. However, of paramount importance to its agricultural success have been the changes that its settlers have initiated and implemented on this landscape over the past 120 years. Farmers occupying former forest, prairie, and wetland areas reduced and eradicated multitudinous plant and animal species when creating the conditions needed for the successful introduction of livestock, pasture grasses, wheat, and other exotic species brought from Europe. Without the creative destruction of this region's enormously large areas of natural wetlands, a significant proportion of today's rich farmlands would not be cultivated. The wet prairies, swamps, and woodlands were largely avoided by the early pioneers during the first waves of settlement in favor of the healthier, drier lands. Yet, today, these former wetland areas contain some of the most productive farmland found anywhere within the Midwest. The implementation of vast artificial drainage schemes by both individuals and government organizations converted the thousands of square kilometers of poorly drained bogs, bottom-lands, meadows, and woods into the region's ubiquitous, highly productive, well drained farmlands (Meyer 1936; Hewes 1951; Kaatz 1955; Johnson 1976), the hallmark of the contemporary landscape.

Drainage throughout this area is not obvious to the casual obser-ver. Rarely are open drainage ditches utilized. Furthermore, the majority of the drainage ditches constructed in this region do not follow the region's natural gradient, where you would expect them to have been built. Their locations are determined not by topogra-phy but by land ownership. Property ownership and field pattern throughout this region are related to the location of the rectangular township and section lines, which are determined by survey, inde-pendent of topography. It is this survey system that gives the Midwest its checkerboard appearance (Pattison 1957; Johnson 1975). If, under this survey system, artificial drainage had been developed mirroring the land's subtle topography, it often would run diagonally across fields, a hindrance to both field preparation and harvesting. Furthermore, outlets to the ditches would have required many farmers to secure permission from their neighbors to cut their ditches through their property (Johnson 1976). This was something that most farmers would not appreciate. To overcome these shortfalls, tiling became the dominant method utilized for farmland drainage.

Tiling of the American Midwest began in the 1880s (Hewes and Frandson 1952). By 1910, thousands of kilometers of tiles had been laid (Johnson 1976). Drainage by underground tiles required farm-ers to dig a system of integrated ditches in their fields, place and connect the tiles in the ditches, and then to bury them. One rule of thumb used in a poorly drained soil was that the tile will draw water from one rod (5.03 m) on each side of the pipe for every foot (0.3 m) that it is buried (Hewes and Frandson 1952). Tiles throughout the Midwest are generally buried at depths of at least 1 m (3–4 ft). These depths permit both sufficient drainage at an economical cost for most soils as well as the minimization of land loss to drainage practices. A 1 m depth (3 ft) is usually sufficient to permit annual crops to be cultivated over the tiles without field preparation and harvesting activities damaging the buried tiles. Tiling was so successful in extending agriculture on lands pre-viously too poorly drained for cultivation that it is common for 40–60 percent of the land area in many Midwest counties to be tiled (Hewes and Frandson 1952). The Kankakee "Marsh" in northwestern Indiana and northeastern Illinois is illustrative of the role of artificial drainage and creative destruction that has resulted in the significant expansion of farmland throughout the Midwest. The creative destruction of thousands of square kilo-meters within this region is definitely one of the significant factors that has made the American Midwest one of the world's premier agricultural areas.

The Kankakee Marsh

As discussed in chapter 1, the development of long-term sustainable systems to meet the specific needs of humankind often requires major environmental changes in existing natural systems. The draining of wetlands throughout the Midwest for the expansion of agricultural activities is illustrative of creative destruction. The creation of the conditions needed for successful crop production in the poorly drained areas of the Midwest required widespread destruction – or at least massive alteration – of nature, with consequent changes in species diversity and habitat quality. The critical variable, the erection of a system that can withstand the test of time, appears to be met on these artificially drained lands. Corn (maize) yields averaging around 140 bushels per acre (8.8 ton/hectare) are common in northern Illinois and have remained high over a period of years (Illinois Department of Agriculture 1988). Agricultural yields from farms on these drained lands remain among the highest in the world (Horsley 1986).

Stretching for 135 km (85 miles) in northwestern Indiana and northeastern Illinois, the Kankakee wetlands were a poorly drained intermorainal area prior to the reclamation of its swamps and marshes for agricultural use. In its natural condition, the Kankakee River, with an average gradient of 79 mm/km (5 in/mile), meandered extensively and continuously shifted its course throughout the entire length of the wetland. This created a complex landscape with an intricate maze of poorly drained, abandoned channels, including oxbow lakes and sloughs (Meyer 1936). Generally, during 8–9 months of each year, water up to 1 m (3 ft) in depth covered a 5–8 km (3–5 mile) wide zone on each side of the river. Swamp vegetation occupied a 1.5–3 km (1–2 mile) wide band along the entire course of the river, with marsh vegetation dominating the other portions of this wetland. Both the American Indians and the early pioneers who settled in this general area largely viewed it as a CPR (see chapter 6), hunting and trapping the Kankakee's abundant wildlife. Thus, the early periods of utilization of this wetland were characterized by human adjustment to the natural environmental conditions of poor drainage (Meyer 1936).

Chicago, the major livestock and grain center of the United States at the turn of the century, is situated only 70 km north of the Kankakee area. With most lands in close proximity to Chicago already occupied by the middle of the nineteenth century, the lightly utilized Kankakee area was one of the last large blocks of land in the region that was available for agricultural expansion. Beginning around 1880, this land scarcity resulted in large herds of cattle being introduced into the "empty" Kankakee area despite its generally poor environmental situation. Initially, cattle ranged on the wild hay marshes that occupied the drier, higher lands along the marshes margins. However, shortly after the introduction of livestock (1884), steam dredging of the first ditch began to extend the cattle ranges areally and permit livestock grazing throughout the whole year (Reed 1920). The creative destruction of the "natural" Kankakee had begun. Engineering works, including dredging, ditching, and tiling, altered the environment to better meet human needs, first for grazing and eventually for farming. These landscape modifications ultimately resulted in over 203,000 hectares (500,000 acres) of good farmland being added to this portion of the "corn belt" (Meyer 1936). Today, the transformation of the Kankakee from a poorly drained wetland fit primarily for wildlife to well drained, highly productive modern farmland – with most of the original wildlife exterminated – is largely complete.

The meandering and wandering river has been replaced by a permanently dredged Kankakee River that is straight, with abrupt angular bends. By substituting a straight channel for the sinuous river, the river's gradient has been increased. The result is that today the river's velocity is swift compared to the slow current characteristic of the antecedent meandering river. This change in the geometry of the channel contributes to the draining of the Kankakee wetland by moving the waters through the area in a shorter time. Former marshes and swamps are now drained by the straight ditches and tiled fields that form artificial rectilinear drainage networks. These drainage works follow the surveyed boundaries of the township and range system, since these boundaries are the basis for property lines in the Midwest. The result of the ubiquitous drainage constructed throughout this former wetland is that today most of the Kankakee is covered by farmland. The previously dominant pasture, swamp and marsh vegetation is largely restricted to areas of former meander bends and sloughs, which remain close to the water table even with the artificial drainage (figure 7.10).

Channelization of rivers and streams and the construction of tiled and other forms of drainage were common strategies throughout the Midwest. This was a principal tactic to convert the region's poorly drained lands to the farmland found throughout this area today. Clearly, some negative aspects of this widespread alteration of drainage exist besides the immense changes that took place in the region's flora and fauna. The increased instability of some river channels due to the rapid runoff of precipitation and resulting land degradation is one such example (Simon and Hupp 1986). Likewise, areas of excessive soil erosion occur where good farming practices are not observed. However, in general, the conversion of the Midwest from its natural states of forest, swamp, marsh, and wet prairie to productive farmland is an example of creative destruction.

The Midwest's contemporary farming systems have evolved from the sustainable, but less intense, rotational systems of the first half of the century to the specialized and high energy demanding systems of today (Hart 1972). This trend toward greater crop intensity has evolved while largely maintaining sustainable patterns of land use. From the human perspective, clearly, the croplands that replaced

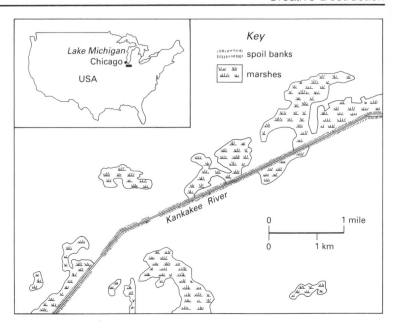

Figure 7.10
*Drainage of the Kankakee
Marsh near Hebron, Indiana*

the native vegetation and wildlife better meet both individual and
national needs than if the land system had been left in its natural
condition. Today, due to creative destruction in this region, carried
out by both governmental and private initiative, the area is characterized by some of the highest agricultural yields found anywhere
in the world. Its agricultural surpluses, largely the result of the
transformation of this zone into a well drained landscape, contribute significantly to the world production of grains. In the years to
come, farming practices will continue to evolve, and there is good
reason to assume that land degradation need not occur and that
sustainable agriculture will continue to be practiced throughout this
region.

Conclusions

The moderate-intensity systems discussed in this section possess six
common characteristics. First, all use labor in spatially much more
concentrated forms than is the case in low-intensity systems. The
existence of increasingly abundant and skilled labor, in turn, makes
it both necessary and possible to engage in permanent modification
of the natural habitats of the Earth. The increased use of technology
to assist human activities is a second characteristic. This technology
increasingly features larger and more powerful machines that supplement – and will continue to replace – human labor. Third, in
moderate-intensity systems humankind first begins to alter topography in a significant way. This begins with very simple, small-scale
changes and progresses to the modification of slope, hydrology, and
surface relief. In almost all such instances, however, the systems and
infrastructure created are dependent upon human vigilance and

support for their continued operation. Fourth, an artificial vegetation is substituted for the flora found in nature. In theory, this habitat modification is permanent, although withdrawal of the human energy subsidy would undoubtedly result in the collapse of the artificial system and the emergence of new, dramatically different ecosystems from the premodification habitat. A shift from long-fallow cycles to a much shortened fallow cycle, or even a pattern of continuous use, is the fifth distinctive feature of moderate-intensity systems. It is this ability to generate more production from more intensive use of the same plots that sustains the large labor supply needed to erect and maintain sustainable agricultural systems. Finally, with moderate-intensity systems the first examples emerge of sustained impacts upon the environment that have more than local implications. This change more than any other, because it often operates beyond the knowledge and control of local actors, increases the difficulty of creating sustainable agroecologies.

High-intensity Systems

Although the origins of high-intensity systems are found in the less intense production systems described in previous sections, in their use of capital, labor, and technology high-intensity systems differ from other production modes in more than just degree; for the change in degree represented by the emergence of high-intensity systems is sufficiently great that it becomes a difference in kind. Technological innovation is the primary feature of high-intensity systems. The lavish use of technology is focused on machines, which begin increasingly to substitute for rather than simply supplement human labor. Massive draglines used in strip mining, for example, make it possible to remove a volume of overburden in a matter of days that would take thousands of hand laborers years to accomplish. In its most profound state, a high-intensity system also may be able to substitute for the land component. This is accomplished, for example, in closed environment agricultural systems such as greenhouses, in which the soil, climate, nutrient, and temperature conditions are all controlled and manipulated in an absolutely artificial creation. Such systems can function without reference to soil quality and weather conditions because they can make up for nature's local deficiencies by controlling for the deficit ingredients. However, such systems are very dependent on inputs from outside their controlled environment and for this reason are inherently unstable. Because of this characteristic of larger systemic instability, we do not consider these controlled environment systems as meeting our sustainability criteria.

The use of machine technology as a labor substitute makes it possible to support large population densities in places distant from the primary production process. The worldwide shift of rural populations to urban centers is an artifact of high-intensity systems and the machines that sustain them. These systems also have profound impacts upon the natural world at great distances from the site of the intervention. As a consequence, managing high-intensity systems in order to promote creative destruction is difficult, since it is easy to initiate changes in one place that have negative, but unobserved or unassociated, consequences elsewhere. Thus, despite (or perhaps because of) its powerful technology, high-intensity systems have greater potential to induce destructive creation than do low- and moderate-intensity systems. Worse yet, the pace at which these changes occur makes it difficult to respond to adverse change in a time frame that promises successful results at a cost that is economically feasible.

Nonetheless, examples of creative destruction in high-intensity systems can be found. In this section, we examine four examples in which the forces of creative destruction have operated in high-intensity systems. Our first example deals with one case, the Broads of eastern England, in which an accident of history and environment transformed a wasted sacrifice zone into a critical resource zone that is itself now threatened by destructive creation. This is followed by an examination of the drained landscapes of the Zuider Zee polders, representative of similar systems worldwide, and the Scheldt estuary storm surge barriers that have met the needs of the Netherland's population for agricultural land, flood protection, and recreation. Finally, we consider how the introduction of irrigation technology into the Sonoran Desert created both the critical agricultural zones of the Imperial Valley and the sacrifice zones of the Salton Sea and the Colorado River estuary.

The Broads and accidental creative destruction

The Broads of Norfolk and Suffolk in eastern England (figure 7.11) are an example of inadvertent creative destruction. The Broads are a group of freshwater lakes of varying size that are linked in complex fashion to the region's partly tidal rivers through a system of lock-free channels (called dykes locally). These water bodies are connected to adjacent wetland, alder woodland, pasture, and arable fields to comprise an attractive rural landscape. Although evidence exists to support the belief that the general geomorphic and fluvial features of the region are the product of postglacial deposition and marine transgressions (Jennings 1952), an accumulating body of historical evidence indicates that many of the region's distinctive features are the product of human activity (Lambert 1960). In particular, the Broads themselves appear to be largely the relics of a medieval mining landscape (Moss 1979). Before AD 1450, peat mining met much of the local population's domestic energy needs. The pits were relatively deep, since the deeper, more consolidated peat layers constituted the most valuable source of energy. A rise in sea level, perhaps linked to climate change, led eventually to the flooding and abandonment of the peat mining operations.

In this instance, creative destruction, that is the formation of the Broads, is the accidental by-product of destructive creation. In the

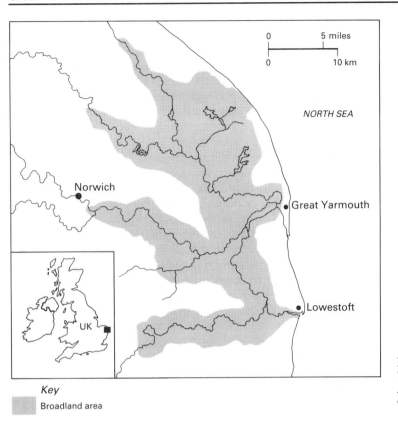

Figure 7.11
The Norfolk Broads, East Anglia, UK (after the Geographical Magazine *1979)*

rural medieval mining operations, the peat mine pits were sacrifice zones. Extraction of energy resources from the peat pits was essential to the well-being of the rural peasantry. A portion of the wildlife habitat of small game animals and fish was sacrificed in order to provide people with fuel. A significant proportion of the rural landscape was scarred by these sacrifice zones. This landscape disturbance was masked by the flooding of the peat pits, which initiated a new set of ecological processes.

Within the lakes and along the dikes (channels) that were cut to connect them and to provide easy cross-country transportation there evolved new plant and animal communities. Reeds grew along the banks and protected them from erosion. Bottom-growing water plants developed in the clear, unpolluted, sunlit, shallow river and lake waters. These plant communities provided the habitat for numerous fish and eels, the capture of which represented an important dietary addition to the local population. Wildfowl flourished in the marsh and swamp habitats adjacent to the Broads and frequently found their way into local cooking pots. Many fish species seasonally spawned and spent part of their life cycle in these wetlands before returning to the main water bodies. Over time, the aquatic and terrestrial habitats that evolved in and around the Broads came to be viewed as natural systems, valuable for the ecological diversity

that they brought to the region and the resources that they represented. The fact that they were largely the accidental consequence of
human activity, the wasted and cast-aside sacrifice zones of a bygone
era, was forgotten. After centuries, the memory of the Broads as
artificial landscapes has faded and they are regarded today as an
immutable part of the natural order. This evolutionary development,
despite its accidental nature, must be regarded as an outstanding
example of creative destruction both by virtue of longevity and
stability.

Today, this artificial natural habitat is once again faced with
destructive creation pressures, albeit from new sources. Digging
for peat is no longer a productive occupation, but other changes
threaten the viability of the Broads environment (O'Riordan 1979;
George 1992). Many of the wetlands located near the Broads and
dikes have been drained and converted into agricultural land.
Protection of these agricultural lands by flood barriers and levees
cuts off fish from access to the seasonal wetland habitat and confines them to the major waterways and Broads. As long as the
drainage schemes have been maintained as grazing dikes, with
limited use of fertilizers that are rich in phosphorus and nitrogen,
the impact on the aquatic environment of the Broads has been
minimal. However, the actual and impending conversion of much
of this land to winter wheat (Shaw 1986; Colman 1989) threatens
to increase nutrient loading in the Broads. Sewer discharges from
municipal sewage systems and rural septic systems are also major
contributors to eutrophication (Stansfield, Moss, and Irvine 1989),
which fosters algal blooms. These microphytic plants absorb sunlight, increase the turbidity of the water, shade bottom-growing
plants, and lead to the loss of habitat that is essential to fish
(Balls, Moss, and Irvine 1989). A contributing factor in this shading problem, which was not found in the region when sailboats
dominated river traffic, is the increased turbidity generated by
motorboats (Garrad and Hey 1988).

Because the Broads also function as a major recreation area, the
region attracts many tourists with an interest in boating.
Increasingly, this industry uses motorboats, as opposed to the
more traditional and environmentally benign sailboat. The wash
created by both commercial and private powerboats has a serious
impact on the banks of dikes and Broads as well as upon bottom
sediments. Boat wash has an enhanced impact for three reasons: (1)
decline in the water weeds and bottom plants due to pollution and
turbidity removes a buffer to undiminished wave energy buffeting
the bank and bottom; (2) indiscriminate mooring along the waterways breaks down the reed beds that protected the bank; and (3)
the invasion of the accidentally introduced coypu, a bank-burrow-
ing, vegetarian rodent, has lead to heavy grazing of much of the
bank and nearshore vegetation.

A lack of clear policy guidelines and political jurisdiction compounds the problems of management and control (O'Riordan 1990;
Shaw 1990). As a result, an act of accidental creative destruction,

which transformed medieval peat pits via flooding into productive lake, marsh, and swamp environments and has been sustainable for five centuries, is so threatened by contemporary destructive creation processes that it is in danger of being changed into a different, and less desirable, state. What this case suggests is that if one can take a time perspective of centuries, destructive creation can be converted into creative destruction – and vice versa!

Drainage, flood protection, and creative destruction

By necessity, the low-lying country of the Netherlands has created an environment secure from the ravages of North Sea flooding while at the same time areally expanding the nation through reclamation of wetlands and lands below sea level. While both reclamation and flood control projects have a long history in Holland, it was through the massive use of technology in the twentieth century that the Nederlanders finally created a safe haven for their industrial, recreational, and agricultural endeavors. The attainment of this stable condition has required major changes both in the terrestrial and coastal systems of the nation.

The Zuider Zee polders

There is a widespread saying, dating back at least to Voltaire, that God created all of the world, but that the Nederlanders built their own country. This expression refers first to the fact that approximately one-half of the country has been reclaimed from shallow sea floors just offshore from its natural coastline and wetlands; and second, that significant areas, especially in Holland, would be highly susceptible to coastal flooding if it were not for the multitude of engineering works, ranging in size from small to massive, that have been constructed since the sixteenth century (figure 7.12: see also Lambert 1971; van de Ven 1987). These works both have created and protected this lowland area from the damaging North Sea storms that have historically flooded large areas of this country. Despite centuries of hydraulic interventions, a massive 1953 winter gale resulted in 2,000 km^2 (772 sq. miles) of farmland and populated areas being flooded in northwestern Holland. Over 1,800 inhabitants were drowned as the North Sea's waters, pushed southward by the gale, inundated the low lying lands of Zeeland (Smiles and Huiskes 1981). This disaster acted as the catalyst for the Delta Plan, which, utilizing the technological advancements of the twentieth century, would culminate in the fortification of the country's coastline that had begun with small local projects centuries ago. Now not only would damage from moderate-intensity storms be curtailed, but with the massive construction of hydraulic storm surge dams across the estuaries of the delta, potential flooding from the rare but high-intensity storm would be avoided or greatly diminished.

Over hundreds of years, the natural landscape of the Netherlands has been transformed by human actions (figure 7.13). For defensive

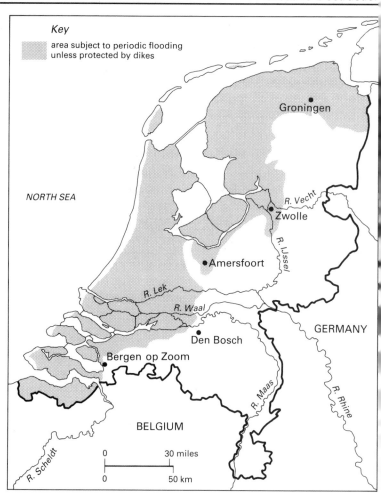

Figure 7.12
Low-lying areas vulnerable to flooding in the Netherlands (after Smits 1987)

purposes, by means of flood control, improved drainage, better transportation, and reclamation of the shallow sea floors, the Dutch have built thousands of kilometers of dikes and canals and have installed thousands of pumps throughout the nation to create the lands that comprise the modern Netherlands (Bardet 1987; van Meijgaard 1987; Schmal 1987). In this northwest corner of Europe, the end product of massive hydraulic engineering has been the creation of a complex set of artificial conditions necessary to sustain human life according to the desires of the inhabitants of this area (Volker 1982). Thus, much of the Netherlands is a classic example of creative destruction. This includes even some of the country's "higher" eastern regions (Veluwe and Twente) where former woodlands and marshes are today replaced by farmland, industrial parks, and urban and recreational areas that bear little resemblance to their natural state. The massive drainage of wetlands throughout this low lying nation, the removal of most of the original forests – and their replacement by farmland, pasture,

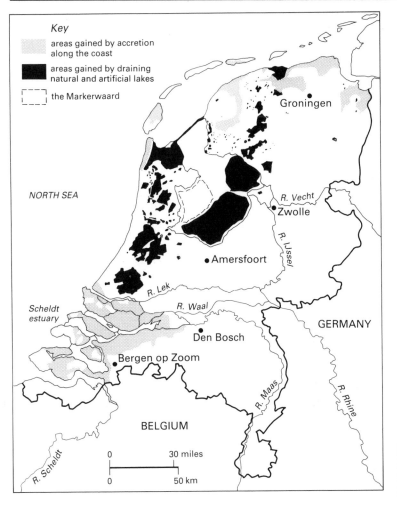

Figure 7.13
Lands gained by drainage in the Netherlands (after Smits 1987)

and urban land use – and the protection of the Meuse, Rhine, and Scheldt's deltas from the North Sea's storm floods have all contributed to the economic well-being of the Netherlands (Lambert 1971; Smiles and Huiskes 1981). Yet, the creation of the Dutch landscape that exists today, in response to the hundreds of years of land reclamation and coastal projects, is woven together by means of major environmental alteration and destruction of natural ecosystems (e.g., tidal flats, marshes, and heaths).

By the twentieth century the reclamation projects and coastal protection projects had evolved from the small incremental improvements of prior times. The pre-1900 projects primarily utilized animal and human labor for earthmoving operations and wind power for pumping. The massive landscape alterations and coastal controls initiated after the Zuider Zee Act of 1918 required modern technology in their reshaping of the landscape to better meet the land needs and flood protection requirements of the modern

Netherlands. Both the polder projects in the former Zuider Zee and the Eastern Scheldt Estuary Barrier Scheme altered the existing hydrologic conditions to a degree that was impossible prior to the technological advancements of the contemporary period. By creating 166,000 hectares (410,000 acres) of arable land (table 7.1) on below sea-level polders that bordered on or were located within the Zuider Zee, now Lake IJssel (see figure 7.14), as well as by protecting these new lands, along with all of the low-lying areas of central Netherlands, from North Sea flooding, the former salty and brackish Zuider Zee ecosystems were destroyed. In their place these areas

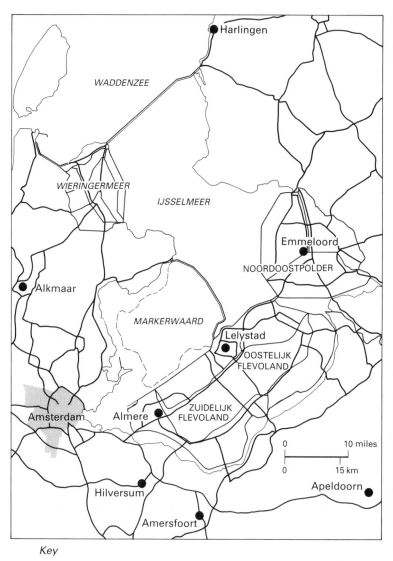

Key

——— motorway or main road

——— channel

Figure 7.14
*Lands recovered from the
Zuider Zee (after Smits 1987)*

Table 7.1 Land reclamation of the Zuider Zee/IJsselmeer

Polder	Year drained	Area (hectares)
Wieringermeer	1930	20,000
North Eastern Polder	1942	48,000
Eastern Flevoland	1957	54,000
South Flevoland	1968	44,000
Markerwaard	(Proposed but never drained)	60,000

Source: based on data derived from Graves (1990).

became the largely human-managed polder and freshwater systems of the Flevoland Region (van Duin and de Kaste 1984).

In the remainder of this section, we shall examine polder construction, the major method utilized in the Netherlands to create highly productive rural and urban areas from lands that formerly were below sea level. Impoldering, both in the Netherlands and worldwide, meets the criteria of creative destruction. In all cases, it alters the hydrologic systems associated with pre-polder conditions to such a degree as to result in the destruction of the original ecosystems and their replacement with systems that the inhabitants desire.

Polders are a very complicated form of land reclamation. Their construction is "carried out on lands with a high potential for agricultural development, and when impoldering has taken place they generally show a high productivity... Polders are found in deltas, low lying coastal areas, river valleys, marshes, swampy areas, and under former lakes, tidal embayments and gulf areas" (Volker, 1982: 2). Potential polderlands have high agricultural potential but, under natural conditions, due to waterlogging and flooding, normal agricultural practices are impossible and many other activities are greatly curtailed.

Impoldering includes a combination of strategies. In almost all cases, dikes need to be built to protect the lands from flooding. Second, diversion of local water systems from polder areas is required. Finally, the lowering of local water tables by pumping is needed. According to Volker (1982: 2) polders are defined as "...reclaimed level areas having a naturally high water table but where the surface and groundwater levels can be controlled." According to this definition, polders are found both near and below sea level worldwide. In various forms, they exist in Belgium, England, Denmark, Germany, Italy, Surinam, the United States, Egypt, Japan, India, Bangladesh, Southeast Asia and elsewhere (Darby 1940; Biswas 1970; Fukuda 1976). In all cases, impolderment results in the destruction of the natural conditions by altering the hydrologic, pedological, vegetation, and animal domains. In its place is substituted a set of conditions that meet the needs of humankind.

Because of advances in technology, by the turn of the twentieth century, it was possible for the Dutch to believe that the flooding

that had caused havoc to the nation from its earliest history could
be curtailed through massive public works. According to the 1918
Zuider Zee Act, it was decreed that the Zuider Zee should be cut off
from the Wadden Zee by the construction of an enclosing 30 km
(19 mile) long dam, stretching from Friesland to North Holland.
The aims of the decree primarily were: (1) to provide flood protection to the 300 km (190 mile) coastline, which was susceptible to
disastrous floodings when breaks in the dikes, primarily during
winter storms, occurred; (2) to create new farmlands for the country's agricultural overpopulation; (3) to increase food production;
and (4) to improve water management in the central Netherlands by
replacing the salt and brackish water of the Zuider Zee with the
fresh water of Lake IJssel (Smits 1970).

Through the impoldering process (Volker 1982), four large
polders have been created on former seafloor lands (table 7.1).
With the construction of the enclosing dike in 1932, the dredging
of canals, the emplacement of subterranean pipe drains (similar to
the tiles of the American Midwest), and continuous pumping of the
polder's waters, both new land and an artificial hydrologic system
have been created. Soil properties have been altered and new agricultural, recreational, and urban lands have been created. A complex cropping strategy increased the permeability of the soils and
reduced their pH to levels that satisfy the requirements of the crops
traditionally grown in the Netherlands (Smits 1970). New accessible recreational areas, both on Lake IJssel and in surrounding lands
through afforestation and landscaping projects, have been created
in the densely populated central Netherlands (Smits 1988).
However, this landscape creation has been accomplished at the
expense of ecological damage. The coastal ecosystems of the
former Zuider Zee are gone, replaced by the new terrestrial
polder ecosystems and the freshwater ecosystems of Lake IJssel.
Because of changing economic conditions, recreational needs, and
environmental concerns about the ecosystem changes inherent in
polder construction, the Markerwaard Polder has never been built
and thus another area of creative destruction has not occurred.

The Scheldt Estuary Barrier

The low-lying coastal and marshy lands of western Holland have
been particularly susceptible to storm surge disasters over its
modern history. Storm surges, in general, result from a combination of at least three major factors. First, a strong slow-moving
storm that is over a large body of water is a prerequisite. A minimum fetch of 645 km seems necessary for sufficient wind energy to
be transferred to the water to produce a possible surge (Thurman
1985). Under these conditions of high, steady winds and a large
body of water, strong wind-generated waves are produced. Second,
when the storm waves reach a coastal area during high tidal conditions, they interact with the tidal conditions to produce a higher
storm surge. During high spring tidal conditions (new and full
stages of the Moon) the above average high tides are ideal for

producing the highest storm surges. Third, the storm generating the waves must be moving toward the coastal area. This means that the wind-generated waves will pile up on the shoreline. If all of these conditions are met, the individual effects of each factor will reinforce each other and produce an exceptional storm surge. When these conditions occur south of the Faroe Islands, large quantities of North Atlantic Ocean water can be forced into the narrow confines of the North Sea and flood the low-lying coastal regions.

On January 31 – February 1, 1953, during particularly ideal storm surge conditions, over 4 billion m^3 (5.25 billion cu. yds) of North Atlantic Ocean water were pushed into the North Sea (Robinson 1953). A devastating storm surge, along with the high tide, produced waves over 5.5 m (18 ft) in height that blasted the Dutch coastline. More than 25,000 km^2 (10,000 sq. miles) of land was covered by the surging North Sea's waters in the Netherlands. This resulted in immense damage throughout western Holland. In response to this storm, the government of the Netherlands embarked on a massive public works program to protect this portion of the country in the same manner that the barrier dam now protected the coastal lands and polders of the former Zuider Zee.

Learning from their previous experiences, and because of the importance of ecological criteria in decision-making in the Netherlands during the 1970s, the construction of the Scheldt Estuary Barrier Scheme (figure 7.15), unlike the Zuider Zee strategy, did not lead to the building of permanent dams to close off completely the estuaries of western Holland. On this occasion, a marine ecosystem would not be converted to a lacustrine one. Through the construction of costly storm surge barriers, the estuarine conditions of the delta area and the wetlands of western Holland have been largely preserved, while protecting the low-lying southwestern areas.

The Oosterschelde barrier, the last component in a 30-year project to prevent flooding in the Oosterschelde estuary and western Holland, is comprised of a complex of dams, dikes, and channels. In contrast to the permanent barrier dam separating the Wadden Zee from Lake IJssel, the Oosterschelde barrier is made up of 62 hydraulic gates that can be lowered to the sea bed during storm conditions. Most of the time the gates are in their raised position, which permits the waters from the Oosterschelde estuary and other western coastal waters to continue to mix and flow into the North Sea. However, when storm conditions occur, the gates are lowered from the barrier. In their lowered format, the Oosterschelde barrier acts as a dam and prevents the North Sea's waters from flooding the low-lying areas inland from the barrier. When normal conditions return, the gates are raised, and once again the normal tidal rhythms of the estuarine area occur. However, to protect the rich estuarine environment, the project required an additional eight years to complete and the equivalent of over US $800 million was contributed to the project compared to the original proposed closed dam (Kohl 1986).

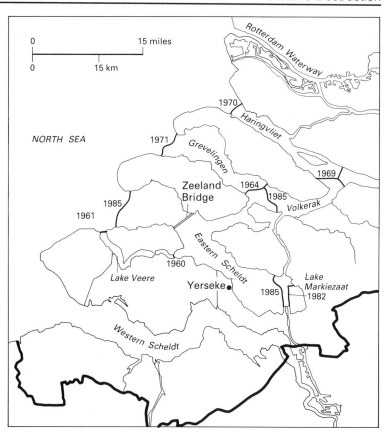

Figure 7.15
*The Scheldt Estuary barrier
scheme (after Smiles and
Huiskes 1981)*

At least in this case, the massive degradational costs of one type
of creative destruction strategy were considered to be too high, even
though it would have been economically less expensive than the
final adopted scheme of hydraulic storm surge barriers. The
adopted strategy is still one of creative destruction in that significant disruptions in the estuarine ecosystems took place during construction. Furthermore, even with the open barrier dam strategy,
there was a 60 percent reduction in the areas' salt marshes and a 45
percent reduction in tidal flats. Thus there were both large environmental as well as economic costs associated with the need to protect
the lowlands of western Holland from future North Sea storm
surges (Smiles and Huiskes 1981).

Irrigation: creative destruction in the Imperial Valley of California

The Imperial Valley, located in southeastern California, has been
an important irrigated agricultural area since the early 1900s
(figure 7.16). The transformation of this very arid and hot
corner of the United States from desert to irrigated agriculture
has been marked by major alterations in the surrounding lands.
Imperial Valley agriculture remains not only sustainable, but very

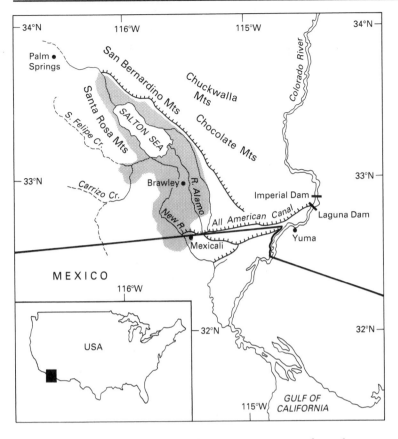

Figure 7.16
*The Imperial Valley and the
Salton Sea*

highly productive (Bureau of the Census 1987) after almost a
century of irrigation. This success has required the creation of
major sacrifice zones in the general vicinity of the irrigation perimeter as well as along the lower Colorado River from the Hoover
Dam to the Gulf of California. Along with other users of the
Colorado's waters, this sustainability has contributed to large
off-site alterations in the Gulf of California ecosystem, the lower
Colorado, and portions of the Salton Sink – the topographic area
in which the Imperial Valley is situated.

The development of Imperial Valley irrigated agriculture represents another example of creative destruction, albeit one that is in a
delicate adjustment. Imperial Valley agriculture – both its creation
and its maintenance – required high-intensity technological interventions. Especially critical are the series of dams built to control
both the Colorado's highly variable flows and high rates of sedimentation. An increasingly delicate balance endures from some of these
interventions, especially with regard to the interaction between the
escalating salinity of the lower Colorado and the pH value of the
valley's irrigated soils. Today, with expanding demands on the
Colorado's limited water supply, especially to meet urban
demands, major improvements are being implemented to increase
the efficiency of irrigation in this area (Committee on Western Water

Management 1992). The ramifications of these changes will continue the processes of creative destruction that were set into motion when individuals decided to convert this bleak and inhospitable area into highly productive agricultural lands. In the remainder of this section we explore some of the environmental alterations that have taken place to permit agriculture to prosper in the Imperial Valley.

Occupying a bowl-shaped depression, surrounded by a desolate desert area, and characterized by a furnace-like arid climate, the Salton Sink, in which the Imperial Valley is located, was referred to as the Valley of the Dead prior to the twentieth century (Worster 1985). This name harshly reflected the paucity of water and high summer temperatures experienced in this hot desert area. Average annual rainfall is less than 7 cm (3 in) per year and the whole depression is devoid of any significant nonsaline water (Seckler 1971). Temperatures over 40°C (100°F) are common during the region's summer months. The Imperial Valley is a rifting depression lying below sea level that, except in the south, is enclosed on all sides by low mountains. Its southern border is protected from the sea by the Colorado River's thick deltaic deposits. These deposits created a fan-like plug that cut off the Salton Sink depression from the Gulf. It is these deltaic deposits, comprised of fertile silts and clays (Robinson and Luthin 1976) and covering thousands of hectares, that formed the foundation for the area's agricultural development in 1901. However, before this could occur, because of the aridity of the area, the scarcity of useful water for irrigation had to be overcome. Fortunately, the Colorado River, flowing only 80 km (50 miles) to the east, had a large enough discharge to meet the water needs of the Imperial Valley. As the Colorado flows along a levee elevated above the Imperial Valley, its waters, when diverted, could flow by gravity into the area. The diversion of Colorado water into the Salton Sink, at the turn of the century, initiated creative destruction processes, some of which continue to this date.

In May 1901, under the aegis of the California Development Company, a diversion channel was cut into the riverbank of the Colorado across from Yuma (Seckler 1971: 194; Committee on Western Water Management 1992: 235). This diverted some of the river's waters through abandoned delta distributaries toward the Salton Sink (Reisner 1986). Using these waters, irrigated agriculture began to expand rapidly in the former Valley of the Dead. With its fertile soil, an apparent ample water supply, and the potential for a year- round growing season, 2,000 settlers were soon on the scene and 40,500 hectares (100,000 acres) of land were already under cultivation on the formerly barren lands by January 1902 (Reisner 1986). Settlement and irrigation continued to increase rapidly after the initial success of providing water to the area. By 1904, 7,000 settlers had moved into the area, and water demands for irrigation expansion grew continuously during this early period (Worster 1985).

Due to the heavy sediment load found in the river's waters, the diversion channel from the river immediately experienced heavy deposition throughout its length. Maintenance procedures were inadequate, as they could not keep pace with the deposition taking place within the channel. This threatened the water supply to the valley and hence the success of the newly established agriculture. In 1904, to offset the effects of the sedimentation, a new diversion channel was initiated further downstream from the original cut. However, before it was completed, a series of major floods occurred. In early 1905, the overbank flows reached the new diversion channel; and the crucial permanent controlling gate was not yet constructed.

Upon reaching this channel in February, the floodwaters quickly cut back toward the river's main channel and breached its levee. By the summer of 1905, the Colorado's major flow of 2,520 m^{3}s (90,000 ft^3/sec) had been diverted from the Gulf of California toward the Salton Sink. Now out of control, the Colorado flooded and destroyed large areas of the recently developed lands. The Salton Sea, occupying the lowest portion of this interior area, was created during the flood as an undesired by-product of the Imperial Valley irrigation development. From the land degradation perspective, all of the valley's lands permanently flooded by this 80 km long (50 mile) by 24 km wide (15 mile) interior sea were destroyed (Reisner 1986). From this perspective, the Salton Sea became an *unintended sacrifice zone* for Imperial Valley agriculture. Unlike the Aral Sea, which is being destroyed by irrigation follies, the Salton Sea was created by a different series of management problems.

In 1907, through a massive engineering and construction project, along with a rather tranquil period of river flows, the Colorado was returned to its former channel and agricultural development of the Valley once again was possible. However, by this time thousands of acres of farmland had been destroyed, more than four times the volume of earth excavated for the Panama Canal had been eroded and deposited in the sink, and the Salton Sea that permanently flooded some of the former agricultural lands had been created (Worster 1985). The sea's future was guaranteed by the continuous inflow of drainage waters from the surrounding irrigated lands, which was required to prevent the salinization of the valley's soils.

Water quality and salinization
Even before the Colorado's waters were used for irrigation, the lower Colorado had a naturally high concentration of total dissolved solids (TDS) compared to other major rivers. By 1968, because of the heavy utilization of its waters throughout the whole river basin, it had become even more saline. In addition to the natural inputs of soluble minerals from waters running off the region's alkaline soils and mineral springs, the Colorado receives additional mineral inputs from drainage of irrigated lands as well as

from mining, manufacturing, and municipal wastes. From its sources in the Rocky Mountains of Colorado and the Green River Basin in Utah and Wyoming until the Imperial Dam near the Mexican border, a 21-fold increase in its TDS, from 38 to 809 mg/l, occurs. Almost two-thirds of this increase is due to natural sources (Committee on Water 1968). By 1976, due to increases primarily linked to irrigation, the TDS level in the Colorado had increased to 877 mg/l. Some estimates indicate that with increasing water use and diversions the TDS level will reach a value of around 1,300 mg/l by the year 2000 (Robinson and Luthin 1976).

To prevent salinization of the Imperial Valley's soils through its Colorado River irrigation water, the accepted strategy has been to apply more water than was required to meet the evapotranspiration demands of the plants and climate. This excess of water prevents salts from accumulating in the top layers of soils by leaching the soil salts downward through the soil profile. A system of 2,240 km (1,400 miles) of pipelines throughout the Imperial Valley carries the salty drainage that results from this irrigation strategy to the saline rivers flowing into the Salton Sea. The increasing salinity of the Salton Sea resulting from these discharges, along with the inputs of untreated sewage, pesticide residues, and heavy metal contamination from Baja California (flowing in the New River), threatens the viability of the Salton Sea ecosystem (Committee on Western Water Management 1992). Already, high levels of toxins are found in the fish caught in this sea, making them unfit for consumption. The Salton Sea, while a sacrifice zone for the Imperial Valley's irrigation, has led to recreational benefits, and a new marine ecosystem developed after its creation in 1904. These positive facets of the region's creative destruction are now being threatened by the contemporary activities occurring within the Salton Sink and Mexico.

Because of increasing demands on the limited water supply of the Colorado, most notably urban growth and water transfers to Arizona, new strategies are required to utilize water more efficiently. Through water conservation, such as canal lining, building new regulating reservoirs, and canal spill interceptors, it will be possible to divert 13.1×10^6 m^3 (106,100 acre-ft) annually from irrigation to the Metropolitan Water District, which supplies water to southern California's urban areas, without reducing agricultural activities. However, this diversion will result in lowering the Salton Sea by 0.67 m (2 ft) and will likely exacerbate salinity problems in the Sea (Committee on Western Water Management 1992).

Today, about 203,000 hectares (500,000 acres) of irrigated land in the Imperial Valley generate $1 billion a year in agricultural products. Among these products are an array of vegetables and fruits, grains, and sugar beet (Bureau of the Census 1987). While some crops, such as lettuce, onions, carrots, and snap beans, might be reduced in future due to the increasing salinity of the irrigated waters, most crops will continue to prosper under the environmental conditions created by water transfers into the Imperial Valley

(Robinson and Luthin 1976). Technological improvements, such as desalination of drainage water, which are already under construction or in the planning stage, should insure the continued sustainability of this area.

However, to keep the Imperial Valley in operation over the past 90 years, a series of activities have been required, many of which have destroyed previously productive ecosystems. The lands flooded beneath the dams built on the Colorado to assure a reliable water supply and the pesticides that enter the Salton Sea, with their toxin residuals, are two examples of the destructive flip side of the creative destruction associated with the metamorphosis of this portion of the Sonoran desert into the highly productive lands of the Imperial Valley.

Summary

While more examples of creative destruction could be cited, the simple truth is that it is easier to find instances of destructive creation. Nonetheless, sufficient examples of creative destruction do exist in low-, medium-, and high-intensity resource-use systems, and in a sufficiently wide range of cultural settings, to engender confidence that the phenomenon is universal, albeit infrequently realized.

The low-intensity systems of hunters and gatherers, shifting agriculturalists, and traditional herders are time-honored adaptive cultural ecological systems that easily have met the test of sustainability for centuries, if not millennia. The ability of these low-intensity systems to endure is a result of two factors: (1) the slow rate of change that has been demanded of them, giving low-intensity systems opportunity to adjust to and to correct for adverse impacts generated by their livelihood activities; and (2) the long time intervals available in which to effect these adaptations. Limited contact with higher-intensity systems has made their adjustments easier, because the gradual pace of experiential knowledge acquisition and transmission that characterized these low-intensity systems was not pressured or overwhelmed by alternative forms of action and cognition. Minuscule rates of population growth also favored the adaptive application of creative destruction principles within low-intensity systems, because limited growth removed a major motivating factor encouraging rapid change in the system state and in the resource extraction level. Ironically, for many of the same reasons that were responsible for their success, low-intensity systems find it difficult to cope with the rapid change induced by moderate- and high-intensity systems. Lacking the ability to change rapidly, the efforts of low-intensity systems to adapt creatively to pressures that impact them often trigger the development of destructive creation. The insensitivity of practitioners of high-intensity systems to the needs and advantages of low-intensity systems often compounds these problems and creates intractable instances of destructive creation, some of which were noted in preceding chapters.

However, despite their decline – both in numbers of practitioners and in areal extent – low-intensity systems provide important insights into the principles of creative destruction. They do this by protecting their critical zones, rotating their use of resources in order to allow for resource base recovery, creating few sacrifice zones, capturing as many of the wastes that they generate as possible, and taking care to operate at all times within the constraints of their environment. It is this sensitivity to the *genius loci principle* that is perhaps the greatest achievement of low-intensity systems.

Moderate- and high-intensity systems are more likely than low-intensity systems to create sacrifice zones in order to increase the productivity of their critical zones. They accomplish this enhancement of productivity by increasingly powerful applications of labor and technology. Particularly as high-intensity systems substitute technology and machines for labor, the ability of the systems created to impact, often negatively, ecosystems far removed spatially from the source of the change is increased. Successful examples of creative destruction, such as the Dutch polders or Chinese integrated pond–dike systems, sacrifice some zones for the greater productivity of the new systems; but they are able to create stable new systems because they successfully enclose their new creation conceptually and pragmatically. By so doing, such systems are able to keep most wastes within the new system, where they are quickly noted and dealt with. Thus, the appearance of salinization in the Imperial Valley is minimized by the draining of excessive salty water into the Salton Sea, both figuratively and literally a sacrifice sink. In the most successful situations, waste products in one component of the system become productive inputs in another system segment. Mud from the Chinese carp pond becomes fertilizer for the nearby dike; and wastes from plants grown on the dike become food for the carp in the adjacent pond, the excrement from which helps to raise the nutrient level of the pond's mud. This recycling of energy and matter turns potential negatives into positives and prevents the system from tilting into a destructive creation mode.

Above all, moderate- and high-energy systems that are successful and sustainable – like the low-intensity systems that preceded them developmentally – are responsive to the *genius loci principle*, and build their intensification actions slowly upon the achievements of previous generations. Incremental improvement is always easier to absorb without serious disruption than is an abrupt shift to another exploitation system. Dutch impolderment in the former Zuider Zee is a good example of this process, for it is built upon centuries of drainage experience. Like the Chinese utilization of carp, which involved two millennia of experimentation before culminating in an interactive agriculture–aquaculture system, polders began as microscale drainage schemes, expanded to encompass larger areas when wind power was harnessed to drive pumps, and ultimately involved efforts to control and reclaim the floor of ocean embayments. The early twentieth century history of the development of the Imperial Valley, from a desert with minimal economic interest and potential into a new zone of highly productive irrigated agriculture, further illustrates the importance of time and incremental improvements in creative destruction fostered by high technology. During the initial stage of converting the area into a productive agricultural system, a series of poor management decisions resulted in the destruction of vast areas of productive land by flooding. The breaching of the diversion canal by the Colorado River resulted in much of the Imperial Valley being converted into an unforeseen and unintended sacrifice zone. The accidental creation of the Salton Sea, and the resulting loss of the lands submerged under its salty waters, is an example of what can occur when moderate- and high-intensity technologies are abruptly and carelessly utilized in new areas.

The destruction of a preexisting natural ecosystem and its replacement by a sustainable human-controlled environment is essential to system success, and is the hallmark of creative destruction. In all successful moderate- and high-intensity examples of creative destruction, a change in scale occurred at each major step in the transformation process. This change in scale was paralleled by a concomitant alteration and improvement in the technology that was required to maintain stability. The result achieved in any successful moderate- or high-intensity system is not guaranteed in perpetuity. However, if the system design is sound and it possesses reasonable flexibility, it should be able to adjust to changes within the system and to pressures brought to bear upon the system from without. This ability to continue to evolve adaptively is a key element in the sustainability of creative destruction.

EIGHT

Land Degradation and Creative Destruction: Retrospective

Introduction

The adverse changes in the environment that lower the overall productive potential of particular places are an inevitable consequence of human use of the Earth. As people work to derive a living from the resources of a region, the interaction that results between physical and human systems inevitably produces change. When these changes are not to the liking of humankind, when they reduce a population's long-term prospects of securing a stable existence, when they constrain present activities, and when they cannot be reversed within a 50-year period, we term the aggregate of these changes "land degradation." At issue is always the temporal and spatial extent as well as the magnitude of these negative developments, for some permanent degradation is acceptable if the ultimate result is a viable, long-lived human resource use system.

Nature, Society, and Variability

Nature itself is not a particularly benevolent entity. Anyone who has experienced the destructive power of a tornado, hurricane, or earthquake, or has viewed images of the devastation wrought by such events, is aware of how unkind nature can be to those unfortunate enough to have been caught by the dark side of natural forces. In nature a variety of other less high-intensity processes operate to alter the landscape. Drought, flood, soil erosion, fire, frost action, and disease operate to place constraints upon and cause changes in the quality of local environmental resources. These events can occur with or without any human interventions. Wild animals congregating around dry-season water resources, for instance, can place a great deal of stress on local vegetation and, at least temporarily, degrade it. A flood that changes the course of a river can destroy the basis for existence of plant and animal communities. A volcanic eruption can wipe out local ecosystems and, through the material spewed into the atmosphere, have a profound impact on the climate of distant places. Erosional processes eventually will transform hills into lowlands and sea-level rises will ultimately submerge coastlines. These types of events take place whether or not people want them; they are an integral part of the way in which nature operates. Even without human occupancy, the Earth would be a dynamic entity. In this sense, some cases of degradation can be viewed as natural processes that are unavoidable unless humans refrain from living in settings prone to experience these events. Given that many of these natural hazzards occur in localities that often have many favorable properties that facilitate the success of a number of activities, this is unlikely.

This background of degradation integral to nature is not something that humankind commonly contemplates. The low-frequency but high-intensity events, such as a hurricane or a volcanic eruption, impinge themselves upon human consciousness by virtue of their extreme character and awesome, awful impact. Despite the degree to which nature has been incorporated into human culture, and in a sense domesticated, extreme events remain outside human control and influence. Dams and flood control structures are created to cope with floods that can be expected to take place up to once every 100 years. Less frequent floods that are larger in magnitude are impossible to plan for effectively, because the structures that are required to cope with them are too large and too costly to justify constructing. The 1993 summer flood in the central portions of the Mississippi Valley (Iowa, Illinois, and Missouri) occurred notwithstanding a multitude of flood control measures. Furthermore, to capture a sufficient percentage of such floodwaters behind dams would create such large reservoirs that as much damage might be done to structures and ecosystems behind the dams as would occur if the floodwaters had poured unchecked downstream. It just has to be accepted that large floods will happen occasionally.

Earthquakes provide a similar example. San Francisco was largely destroyed by an earthquake in 1906. This did not stop the survivors or new inhabitants from rebuilding on the same hazardous site. This took place because people convinced themselves that technology could cope with any anticipated future earthquake of a similar magnitude, the likelihood of a similar event in the near future was remote, and/or because the immediate locational benefits of the central California harbor site outweighed potential future costs. Moreover, the longer the time span between extreme natural events, the more likely it is that people will manage to forget the intensity of the event and to plan their livelihood around the more benign conditions that characterize periods between catastrophes.

Even less prone to easy identification are the more modestly scaled, lower-intensity, higher-frequency occurrences such as drought or soil erosion. Much less spectacular, and slower to build to the point at which adverse impacts can be readily perceived, low-intensity events are easier to ignore in the short term, but likely have greater catastrophic implications for the long term if they are allowed to accelerate and proceed unchecked. According to the FAO, current rates of soil erosion will result in an area the size of Alaska (1,530,700 km^2, or 591,000 sq. miles) losing most of its agricultural value over the next 20 years (*Boston Globe* 1993). With the world's increasing population, if soil losses continue, at some point there might not be enough good land to meet the world's food needs. This has already occurred at the national level in some countries, such as Haiti. These low-intensity events can be influenced in their impact by: (1) altering the event, such as implementing successful conservation strategies; or (2) by altering the way in which society prepares for and responds to the occurrence rather than by changing the event itself. Society possesses a variety of ways to cope with and minimize the impact of low-intensity events that adversely affect humans.

The prime motivation in dealing with low-intensity events is humankind's inherent dislike of variability in the natural world. While variation may add spice to life, it can also prove disastrous to human survival if drought withers crops, floods wash away seeds, and rainfall runoff strips away soil from the landscape. When these events occur, human health and happiness diminish, and higher death rates are often the consequence. Variability, often in extreme forms, may be part of nature, but most individuals and societies prefer to deal with a limited range of expectable, dependable conditions that guarantee security of life, property, and food supply. In nature significant buffers usually exist, the purpose of which is to serve as shock absorbers for extreme events. These buffers provide important slack and lag time in the way nature operates. Traditional human systems always maintained substantial buffers in their use of the Earth. Long fallow cycles, low population densities, and the avoidance of fragile areas kept human exploitation of nature from accelerating low-intensity, high-frequency events and allowed room for natural processes to absorb the impact of high-intensity, low-frequency events. With these buffers in place, most traditional land-use systems avoided initiating land degradation. Sufficient flexibility existed in these systems such that when the resiliency of the system was threatened, feedback mechanisms curtailed the probability of irreversible negative change. Trends toward decreasing crop yields were countered by fallow. The slack in the system, in this case lands with high agricultural potential not actively being cultivated, was available for use when existing cropped lands no longer met current needs.

Contemporary resource-use systems increasingly remove these cushioning barriers between humankind and nature. Trucks now transport herds to seasonal grazing in a matter of hours where once it took days or weeks. Once in place, these herds consume the grass before it has a chance to set seed. The lag time permitted by animal-based and foot transport no longer protects the regenerative capabilities of many rangelands. We call this result overgrazing. Alternate-year fallow practices once insured time for groundwater levels to decline before the next year's application of irrigation water raised them again. Without construction of expensive drainage systems, irrigation every year causes a rapid rise in groundwater, which brings harmful salts into the root zone of cultivated plants. We call this outcome salinization. Economic and demographic pressures encourage farmers to cultivate lands the steepness or poor soil attributes of which make them high-risk environments. Unless soil conservation practices are employed at the same time, rapid downslope soil movement is almost inevitable. We call this process accelerated soil erosion. In some cases, even with soil conservation practices in place, the lands being cultivated are so perilous that excessive soil loss will occur.

Alterations of this type reduce the slack in the system and result in a decline in resilience. Not only do such changes increase vulnerability to the high-intensity, low-frequency event, but they also may adversely affect the ability of the resource-use systems established by humans to cope with the high-frequency, low-intensity event or process. This is ironic, because it is reasonable to expect that human systems adapted to high-frequency events would have to cope with frequent, if minor, variability in order to be successful. Problems occur in coping with high-frequency events because, despite the impressive gains that humans have made during the past two centuries in knowledge about how ecological systems function, there are many gaps in our understanding. One area in which our knowledge is limited is the identification of the precise point at which a system's resilience limit will be reached. Up to a certain limit, most systems possess considerable ability to absorb change without exhibiting a pronounced change in the state and composition of the system. Under economic and demographic pressure, people are often tempted to adjust upward the upper limit to which a system can be exploited. This boundary zone is a critical one, since it is linked to system stability. This upper limit is also not a sharp line in most cases, but rather is a relatively broad and ambiguous zone with imprecise boundaries, within which considerable variability may occur. Once exceeded by continuous and intense exploitation, a system may no longer have enough slack in order to be able to recover quickly from disturbance, thus losing an important dimension of its stability. Pushed to this limit, a system's capacity for resilience may be exceeded suddenly by a relatively minor fluctuation in its basic processes. When this happens, the system "flips" quickly into a different state and seeks stability around a very different set of characteristics. Thus, a lake that for many years shows no visible effect from herbicide and fertilizer pollutants washed in from adjacent farmlands, or from septic seepage from lakeside residents, may suddenly produce widespread algal blooms that threaten the survival of fish populations.

There are many reasons why humans push natural systems to the point at which land degradation occurs. The most common causes are twofold. First, there is a desire to remove variability from those portions of the natural world that are important to human livelihood, and to force those segments of the environment to produce at a level of intensity far beyond what they would produce under normal conditions. When efforts to do this are successful, habitat is altered in a creative and sustainable fashion and land degradation is minimized. Less successful efforts usually result in the acts of destructive creation that are a signature of much of the human use of the Earth. Second, population growth, with the concurrent demand for more goods and services, can also bring stress to bear upon natural systems. Meeting these demands exerts increasing pressures on natural resources. Eventually, a critical threshold is reached and degradational processes are initiated. With slack removed from the utilized systems, few options exist in many cases to meet the immediate needs of the inhabitants. Again, destructive creation is the result.

Nature, Society, and Technology

While our knowledge about variability and its impact on ecological systems is less than perfect, a great body of technical knowledge does exist about land degradation at a microscale. This knowledge is sufficiently extensive that we can say with confidence that most major forms of land degradation are preventable from a technical point of view. For example, soil erosion can be arrested by applying a variety of techniques, from contour plowing to bunds to terraces to establishment of a permanent vegetation cover. Similarly, desertification can be halted by fixing mobile soil surfaces, by limiting the concentration of people and animals, and by installing more efficient water-use technology. Salinization can be combated by less wasteful water application methods and by the installation of better drainage systems. If techniques are in existence with which all of these undesirable land degradation processes can be controlled, why does land degradation continue unabated? Indeed, why does land degradation appear to be accelerating in many places?

The reasons for the general failure to control land degradation are complex. However, four major factors are heavily implicated: (1) problems of social acceptability; (2) costs associated with land degradation control; (3) short-term planning horizons; and (4) subsidizing degradation by not incorporating the true environmental costs in the specific human activities that are utilizing the land resource.

Many efforts to halt land degradation and build sustainable agriculture fail because they inadequately fit into the local social context. The more sophisticated the technology employed, the more likely it is to require a level of skill that the local population does not possess. Combined with a failure to build upon the very real skills and institutions of local societies, there often is little correspondence between local capabilities and the requirements of introduced technology. The varying values held by indigenous communities and different development agencies can produce land degradation control projects that are not perceived by local societies as possessing particular merit. For example, foresters often believe in the effectiveness of tree plantations and shelterbelts as a way of increasing local fuel wood supplies as well as to oppose the spread of desertification. The techniques for establishing these plantations are well known. However, farmers and herders seldom can conceive of how these plantations will benefit them. They view the land devoted to trees as land lost to the farming and grazing uses that sustain them. As a consequence, even if such protective belts can withstand the rigors of a semiarid environment and can be successfully established, they require constant protection. As soon as the guard's back is turned, the woodlot is invaded. A difference in objectives and values fosters competition and conflict that often accelerates rather than arrests land degradation.

Similarly, Chinese pond–dike integrated systems may be a great success in modifying deltaic environments into extremely productive sustainable examples of creative destruction. However, the introduction of these systems into other regions with different cultural traditions is extraordinarily difficult. The lament of the technologist, that land degradation could readily be controlled if only people would adopt new practices and tools, ignores the powerful constraints placed on such innovations by local traditions and differently valued outcomes. Land degradation controls that are not rooted in local experience find limited application in many settings that superficially seem appropriate for their employment.

Furthermore, what may appear to be similar environments might well be significantly different in particular critical areas. Land clearing practices in a mid-latitude setting could be totally inappropriate in a tropical environment. When soils are compacted through the use of heavy equipment, frost action during the winter months can often restore a looser structure to mid-latitude soils. However, in a tropical habitat, where there is no frost to counter the compaction, the use of heavy equipment can have a more serious and long-lasting impact. Moreover, soil compaction could kill the worms, ants, and other subsurface-dwelling fauna that give tropical soils many of their favorable properties. Thus, the transfer of what at first glance appears to be appropriate technologies between seemingly "similar" environments in order to carry out analogous activities is a far more complex problem than might first appear to be the case.

Costs associated with arresting land degradation often inhibit successful application of existing technology. As a general rule, land degradation costs are seldom carefully calculated when a development project is proposed. If these costs were measured, many projects would never get off the drawing board. Given the analytical boundaries that are used, it is often easy to ignore the cost of land degradation. When boundaries are set narrowly, costs that fall off the project site can be easily overlooked. These distant sacrifice zones have a high potential for catastrophe because no one pays attention to them until land degradation is well under way. Unlike the sacrifice zones that are incorporated into the structure of creative destruction, such as the hillsides degraded by the Nabateans in order to increase the productivity of the valley bottoms, distant sacrifice zones are disaggregated from corresponding benefit zones. In these instances, the losers in the process of change are so distant (including economic and social as well as geometric distance) from the populations and habitats that gain benefits that the degraded areas do not enter into the calculation of benefits and costs.

Frequently, there is no existing agency that looks out for the welfare of the environment. Where environmental protection agencies do exist, their work is easily compromised by the need to consider potential job losses if major polluting industries are forced to change their practices. The fear that serious efforts to control land degradation will cause companies to close or move is an important inhibiting factor in continued inattention to environmental decay.

Equally significant are the major financial commitments that are required if stringent environmental protection standards are enforced. New technology is often costly and its widespread adoption is a slow process under current financial practices, since land degradation costs are rarely included in the price of environmentally deleterious activities. For example, with the lack of environmental accounting, a farmer irrigates his land without having to include the effect that this activity has on the lowering of water quality. The farmers down stream will receive waters with greater salinity due to upstream irrigation activities. Under contemporary conditions, the costs of this lower water quality, such as lower crop yields or salinization, are borne by the downstream users and not by the culprits who cause the resource to degrade. Under current accounting practices, many environmentally unsound practices are not incorporated into the cost of the activity and few incentives exist for remedial actions short of legislation. In the meantime, land degradation continues unchecked in many places.

Land degradation is quite independent of the particular political or economic philosophy that guides national planning and development. Socialist and capitalist economies alike are culprits in degrading their land, air, and water resources. Australian or American herders are as prone to overgrazing problems as are their Tanzanian or Chinese counterparts. Land degradation associated with industrial activities is as widespread in Eastern Europe, where it occurred under state socialism, as it is in Latin America, where it occurs largely due to private-sector business activity. Even the major driving variables are often the same regardless of culture, economy, and political ideology. Among the most important factors that promote land degradation processes are the need for short-term survival and actions that favor one segment of the population and economy over others. Most people who deforest an area for fuel wood or excessively coppice trees to provide emergency fodder for their animals are aware of the implications of their actions. They know that if they persist they will seriously degrade their habitat. However, in most instances they see no viable alternative to the practices that they are following.

These degradational activities often are linked to fluctuations in the environment, such as drought. The excessive pressure placed on the habitat by traditional herders is envisaged as temporary. Cutting tree branches for fodder is intended to be a transitory practice that will be halted once pasture conditions improve. However, changes independent of the practice of temporary fodder coppicing often increase stress on the system. Population growth may place more herders and animals on the range. Agricultural expansion may confine herders and animals to a smaller area. In many cases, this contraction in rangeland robs herders of the critical land reserve areas needed in their land-use system during drought stress periods. Under these new circumstances, the previously successful, time-honored practices that resulted in sustainable land use are carried out under new constraints, which undermine the viability of previous adjustments that were environmentally sound. Under contemporary conditions, the previous adjustments, which worked well when more extensive land resources were available, often fail to maintain resource base stability and may result in severe land degradation. If people see no alternative in the present because their survival is threatened, they will do what is necessary to get through the bad period and hope that both environmental and social conditions will improve in the future. Unfortunately, as lag times and unused "slack" capacity decrease within land-use systems and degradation occurs, the likelihood of improving conditions decreases in the absence of significant, sustained, and substantial outside financial and technical assistance.

Commercial ranchers adjust similarly when drought or a fall in prices impacts their operations. They are reluctant to destock, hoping that in the near future conditions will improve. Animals are concentrated on the best portions of the range, where grass and water are most abundant. If drought does not end or meat prices do not rise, the result is serious and prolonged land pressure and degradation of the most productive portions of the habitat. When people and place collide in a struggle for survival, it is the norm that the quality of the resource base is the first victim.

Favoritism of one economic sector or segment of the population over others often compounds degradation. This segmental approach to land and resource management leads to a cascade effect, whereby the changes introduced into one area spill over into other zones. These spillover effects have a particularly serious impact on the zones affected because they are seldom adequately considered when evaluations of the primary initiating area are undertaken. Thus agricultural expansion into drier zones during periods of greater than anticipated moisture can compromise sustainable use of these dry zones over the long term. Caught without adequate vegetation cover during a more typical period of drier conditions, increasingly rapid rates of soil loss may occur. Because these zones often are important fallback reserves during drought periods, many rural populations find that traditional famine foods are no longer available. In consequence, the ability of a region and its people to cope with hard times using their own resources is sacrificed to progress in other economic sectors and populations. Creation of unintentional sacrifice zones has serious long-term implications for the stability and resilience of the environments that are impacted. Generally, they are the losers in the process of change, and degradation of the land that sustains them in the short run is the consequence.

To most observers, the rate of land degradation changes appears to be increasing in all habitats and in all socioeconomic systems. While bright spots of creative destruction that result in sustainable development can be identified, in today's environment these positive developments are more the exception than the rule. The reason for this acceleration in the rate and scale of adverse environmental change can be found in the way in which nature, society, and technology interact. Over time, humankind has improved its technological capabilities and knowledge base to an enormous degree. This has made it possible to increase the intensity with which people use particular habitats. As the intensity of environmental exploitation has shifted from the low-intensity systems that characterized most of human history, there has been a corresponding increase in the rate and scale of impacts that cause land degradation. Under low- and most medium-intensity systems, considerable slack existed in human use of resources. Grazing reserves that today are converted to agricultural fields once were available lands for sustaining herds during drought. Lower human population densities obviated the need to seek new ways to exploit old areas and to extend use to new regions. Less investment in rigid engineering works and infrastructure and fewer absolutist tenure practices promoted greater flexibility within and between communities and livelihood systems in their use of resources. A slower pace of change meant that there was more time in which to respond to feedback that warned resource users that land degradation problems were emerging. Also in past times, when systems were severely stressed, often food shortages or famines resulted in a declining human population. While not a particularly benevolent adjustment, from the land degradation perspective it did contribute to lowering human-induced stresses on the natural environment and hence helped to minimize land degradation. Today, food aid often sustains inhabitants in areas where their existence is threatened due to food shortages that result from meteorological variability or other negative factors. One result is that traditional feedback mechanisms that worked to curtail stresses are short-circuited and critical thresholds are reached that initiate land degradation.

Despite considerable contemporary sophistication in understanding the interconnection of cultural ecological systems in the abstract, we are frequently surprised by where, when, and how rapidly unexpected environmental changes will appear. Our ancestors faced fewer problems because both the pace and the intensity of change were less rapid and less powerful. Modern high-intensity technologies are the primary cause of rapid and widespread land degradation. The power and scale of these technologies makes them simultaneously appealing and difficult to cope with. By their power to modify the environment for human benefit, modern technologies promise great improvements in the human condition. This lure encourages people, planners, and politicians to move rapidly to adopt changes the long-term implications of which are only partially understood. As humankind ratchets up the intensity with which land resources are exploited, the risks are increased. A non-benevolent nature, variable and pulsating in its basic structure, can spring surprises that compromise human use systems. These unexpected events in both time and space have the potential to reveal hidden flaws in human use of the Earth and to stimulate abrupt land degradation and habitat transformation.

Modern technology is not the villain in a bad morality play; rather, it is the often careless way in which the human use of modern technology takes place that causes land degradational problems. These difficulties in land degradation could be avoided at all scales of resource-use intensity and the rate of unwanted negative change reduced to manageable proportions if a set of basic generic principles were followed.

Commandments for the Minimization of Land Degradation and for Sustainable Development

Eight principles extracted from our review of the linked processes of creative destruction and destructive creation govern the wise use of resources and diminish the prospects for land degradation.

The first principle is the commitment to *plan holistically*. Many land degradation problems arise because planning for environmental management and economic development is vested in an array of different agencies and institutions. Broad overview authority to reduce competition between different user groups seldom exists, even in centrally planned economies. As a consequence, unintentional sacrifice zones are created, often in distant places far removed from the source of the original environmental impact. Thus, the response to local air pollution was the erection of taller smokestacks in order to insert the pollutants higher into the atmosphere where they stood a better chance of being transported away from the production facility. The consequential acid rain, falling upon distant areas, was not part of the calculation of the plant managers who built the smokestack. These managers did not intend to create acid rain and cause land degradation; in many instances the original decision-makers had no concept nor concern about how acid rain was produced or its effect on distant lands and waters. The fact that the impacted area was far removed from the original source of the contaminants simply made it easier to ignore potential bad consequences. The decision to build the taller smokestacks solved the immediate local problem of negative pollution impacts on the local environment. Furthermore, there was also the belief that atmospheric and aquatic systems had an unlimited ability to absorb refuse without negative consequences. In combination, these factors made it easy to ignore possible consequences until a serious problem existed.

The land degradation problems made possible by the absence of holistic planning are worsened by the competition that often takes place between different sectors of the economy. The approach to economic development that characterizes much of the globe is a segmental one. Each component of the economy has its set of goals and plans, and undertakes activities that may be in each segment's perceived short-term best interest. This pursuit of separate sectorial gain occurs with only the most limited reference to the needs and activities of other segments of the economy or other areas. For example, industrial enterprises in most countries have a higher priority in national development efforts than do other economic activities. As a result, industrial firms are able to follow waste disposal practices that do serious harm to the environment. The leaching of hazardous chemicals into surface and groundwater can not only reduce drinking water supplies but also – before being detected – can result in serious illness to the consumers. The export of wastes into the general environment in pursuit of the chimera of economic development is responsible for much of the environmental damage sustained in industrial and industrializing countries, regardless of political and economic systems. When authoritarian political systems limit dissent, whatever planning agencies may exist find themselves powerless to control and limit land degradation. The lack of environmental accounting in most industrial projects only exacerbates this trend. These problems of segmental development also are found within individual sectors. For instance, animal husbandry, dry farming of cereals, and irrigation farming are all agricultural activities. In most countries, each activity is administered through a separate component of the Ministry of Agriculture. However, in countries with limited capital resources, decisions that favor irrigation – often made because this form of agriculture produces the high yields that seem to merit intensive capital investment, or because external funding agencies generally favor these projects for a number of reasons – starve other agricultural sectors of the resources that they need in order to grow. In some cases, the irrigation projects actually have direct negative effects on the other agricultural sectors, such as by the flooding of their lands by needed reservoirs.

Primarily newcomers, rather than the local inhabitants, become involved in irrigation farming schemes. When excessive concentration of activity in irrigation squeezes out the agriculturalists who practice dry farming, for their survival they expand into the zones formerly occupied by nomads. This forces pastoralists to place more pressure on the rangeland that remains within their control. The absence of development inputs in these less favored sectors is paralleled by the emergence of land degradation as each community places increased pressure on its resource base without the benefit of the capital resources needed to intensify resource use productively. Failure to meet development goals and to avoid land degradation is attributed to lapses internal to the less favored sectors rather than to the inadequate, segmental planning process that is not structured in ways that view economy and environment in a holistic fashion.

The second sustainable development principle is the need to *avoid arbitrary boundaries*. Arbitrary boundaries are the product of a flawed, segmental planning process. Erection of rigid analytic and pragmatic limits to development projects and planning activities establishes absolute domains that are unrealistic. In the traditional world, different livelihood activities often overlapped in space, although they seldom overlapped in time. Thus, farmers cultivated fields watered by a river's annual flood while herders were engaged in exploiting distant seasonal pastures. When the crops were harvested, and the farmer had no major productive use for the land, herders returned to graze on the post-harvest stubble. This nonexclusive, overlapping use of resources worked to everyone's benefit. Fixing a rigid boundary between spaces devoted to farming and those used for animal husbandry reduced the resilience of both livelihood systems. In the long run, the collapse of overlapping pastoral and agricultural ecologies may have worked to undermine the stability of sedentary farming systems, because it promoted processes that encourage overgrazing and soil erosion in pastoral areas. Increased frequency of dust storms, silted reservoirs, more extreme river regimes, and clogged canals were often the consequence – a set of feedback loops that have serious consequences for sedentary agriculture. Usually, it is essential to maintain greater flexibility in conceiving system structure and to preserve overlapping ecologies that encourage multipurpose use of the same space in different seasons. Yet it is often the essential ingredient in avoiding the elimination of one group's critical zone in an effort to enhance growth and development in another sector of the society and economy.

The presence of rigid boundaries makes it difficult to *calculate all costs* because off-site costs are generally overlooked. Yet possessing the ability to assess, at least qualitatively, all of the costs associated with environmental change is essential to avoiding land degradation. Confining planning and environmental assessment within strictly fixed boundaries makes it relatively easy to enumerate the benefits and costs that fall within the boundary zone. However, there are always important cascade effects that occur outside the spatial zone under consideration. In theory, establishing a boundary that defines the primary object of analysis should serve not only as a device for calculating gains and losses within the bounded region but also should make it easier to note the exchanges that occur across the boundary. In practice, the erection of a boundary quickly evolves into a conceptual barrier that relegates to inconsequential status events that are taking place beyond the boundary. It then becomes easy to ignore land degradational impacts that affect distant areas. This is a serious violation of the third principle, which insists on the importance of including all costs generated by a given set of actions as well as the benefits. Benefits close to home are easy to note; costs that are distant are easy to overlook.

It is the failure to include these costs in environmental and economic assessments that creates distant, unintended sacrifice zones. Because these sacrifice zones are not recognized, they can rapidly spiral out of control and attain a severe degree of land degradation before steps can be taken to arrest or reverse the process. Uncalculated land degradation costs also have a serious impact on societies that occupy degrading, sacrificed areas, because the exported costs imposed upon them create a category of human losers who are impoverished by changes that benefit other areas and peoples.

It is important to note that these sacrifice zones are fundamentally different from the primary sacrifice zones that are integral to creative destruction. Zones within an area of transition to more sustainable status, which are sacrificed for the benefit of part of the habitat, are deliberately degraded and the losses incurred are consciously balanced off against the gains that result. Thus Nabatean hillsides, degraded of their soil and water resources for the benefit of valley-bottom fields, or swamp habitats destroyed to create the Chinese pond–dike system, were transformations that were deliberately undertaken by the people involved. Because the transformation proceeded incrementally, sufficient time was allowed to make corrections if defects emerged in the evolving system. As a consequence, the losses that were sustained in the zones that were sacrificed were less than the gains concentrated in the primary production zones. The lesson to be learned from this experience is that it is vitally important not to export costs blindly to other parts of the ecumene. Only the careful calculation of all costs sustained within and without the boundaries of favored zones through a holistic planning process can begin to deal satisfactorily with land degradation.

Given a holistic planning process that avoids establishing arbitrary boundaries and calculates land degradation costs for all areas, minimization of land degradation is made more likely by practices that *retain buffers*. Buffers are important recovery mechanisms in most low-intensity traditional land-use systems and nature. These devices permit lag time to exist as an integral part of these systems and of nature's structure. Lag time is vital because it insures that a stressed land area has sufficient time to recover before it is

subjected to renewed stress. Fallow cycles are widespread important buffers utilized in many low-intensity land-use systems. They
allow sufficient time between episodes of human utilization for the
recovery of soil fertility, structure, and vegetation cover. In ancient
Mesopotamia, irrigation was alternated with fallow on an annual
basis. This rhythm of use and buffer combated salinization by
preventing a permanent rise in the groundwater level. The result
was a sustainable system. For almost every low-intensity agricultural system, there is a minimum fallow buffer below which a
system cannot fall without land degradation occurring or a
major shift to a new mode of production taking place. These
fallow periods were very long in the case of slash and burn
agricultural systems in tropical forested areas, but were greatly
reduced in more intensive cultivation systems. Nonetheless, in
successful agricultural systems this reduction in slack time was
always accompanied by sufficient inputs to enable the farming
system to maintain basic fertility. Inclusion of fallow periods
and fallow land within the agricultural system also left open
niches that other groups, mainly mobile pastoralists, could
exploit. This, in turn, reinforced the principle of nonrigid, flexible
boundaries that permit overlapping ecologies to survive and flour-
ish side by side.

Increasingly in the modern world, intensification of production
systems removes buffers in the interest of increased production.
This removal of nature's shock absorbers makes it difficult for
livelihood systems to cope with unexpected random events that
are of low frequency but high intensity. Deprived of its capacity
to deal with the unexpected by the removal of buffers, sudden shifts
in the system state of human habitats can occur. While recognizing
the difficulty of retaining buffers in their original form, the need to
structure buffer substitutes into the human-modified landscape is
considerable. Without the provision of adequate buffers, it is difficult to employ successfully the remaining four principles for the
minimization of land degradation.

The most important of these principles is the need to *protect
critical zones*. In every livelihood system these is a critical zone
without which that system cannot easily survive. Defense of these
crucial resources is essential if the livelihood system is to remain
viable. In some instances critical resources for one component of the
ecumene are sacrificed in order to concentrate benefits in another
sector. An example of this process is the decision to build the High
Dam at Aswan, which trapped behind the dam fresh water and
sediments formerly carried in the Nile's floodwaters. One ramification of this decision was that it altered both the salinity and the
nutrient balance of the eastern Mediterranean. This set into motion
a set of serious consequences for the aquatic food chains upon
which the Mediterranean Sea's fishermen depended. In this case,
construction of the dam for the benefit of most of the Egyptian
people was incompatible with the protection of a critical zone for
the offshore fishery.

When changes that are deemed essential do take place, it is imperative that adequate substitutes for the lost critical resources are provided. For pastoralists, for instance, retention of adequate natural grazing around dry season water sources is a particularly critical resource zone. Without access to these grazing resources, pastoralists are unable to maintain their herds. When changes occur that deny access to such resources, or when herders are confined to a portion of their normal range for the entire year, both land degradation through overgrazing and, episodically during drought, catastrophic herd losses and famine are outcomes that one can anticipate. These problems can be avoided in large measure by integrating agricultural and pastoral activities more closely. Of course, to do this could place some constraints on agricultural expansion, a policy that most governments and development agencies have not been willing to consider. When a more holistic perspective pervades planning, it is possible to make available post-harvest residues and supplemental fodder to compensate for lost critical dry season pasture zones. This can only occur through an application of the principle of critical zone protection or of its corollary, critical zone resource substitution.

Equally important is the need to *capture wastes* from one region, livelihood, or process and use these wastes as important inputs into other activities. The Chinese integrated pond–dike system is based on this principle of recycling organic wastes from production systems located on the land and using them to promote fertilization in the ponds. Carefully maintained but artificially high fertility levels in the ponds sustain the aquatic food chains upon which fish culture depends. Aquatic plants and pond mud, in turn, are used as a fertilizer source for the terrestrial agricultural system. This very tight linking of system components through the use of wastes from one part of the system as a basic input to other parts is rare. However, it establishes a model by which to avoid one of the worst effects of land degradation: the export of undesirable materials into other sectors. Nabatean run-on farming employed this principle with great success by trapping behind valley check dams soil and water transported from adjacent hillsides. The waste output from one part of the environment became the foundation input for productivity in another part. As was the case with the Nabateans, deliberate encouragement of this waste input led to the creation of sacrifice zones that were essential to long-term system success. When wastes cannot be captured in this way, there is a very high potential for land degradation. Thus, soil eroded from upslope agricultural zones can produce excessive deposition along valley bottoms that kills crops planted in these low-lying areas. Excessive soil loss transported in stream systems can result in sediments that silt harbors, increase turbidity in estuarine systems, and suffocate coral reefs. Harbor facilities are degraded, fishing productivity is undermined, and wave erosion can accelerate as a consequence of this form of excessive waste. Inattention to holistic planning and a failure to capture wastes productively produces serious environmental and economic damage. In essence, the principle of waste capture is an attempt to transform a negative – the production of wastes that pollute and degrade – into a positive – the generation of inputs for agrarian and other systems – that has a high probability of achieving long-term sustainability.

In managing resources with the minimization of land degradation as a primary objective, it is extremely important to balance long-term needs against short-term benefits. This is reflected in the popular expression, *look before you leap*. The struggle for survival leads many people in high-risk habitats to engage in practices that damage and degrade the environment. Expansion of agriculture into marginal habitats, for instance, may promise short-term gains in crop production, but these actions carry with them serious implications for longer-term soil erosion, fertility decline, and, all too often, a collapse of an area's agricultural livelihood. The abrupt introduction of new technology into places for which that technology seems appropriate, but where experience in using the technology is limited, poses great risks for degradation. For this reason, the extension of mechanized dry farm cereal cultivation into pastoral zones carries with it a high risk of wind-generated soil erosion as well as a loss of primary rangeland productivity. Similarly, the exuberant adoption of large-scale groundwater extraction technology, such as central-pivot irrigation, carries with it grave risks of mining groundwater resources when extraction rates exceed recharge. The flexibility and potential land resources use of future generations are greatly constrained when short-term benefits result in a deteriorating resource base.

Application of the "look before you leap" principle makes it possible to assess the potential impacts of proposed changes more accurately. It gives a higher degree of assurance that long-term costs will be given equal consideration with the more easily calculated short-term benefits of a given development. Proceeding cautiously also increases the likelihood that feedback messages can be observed in time to take corrective action. This implies that it is essential to build into every habitat appropriate monitoring systems that report qualitatively the rate and trend of habitat modification. In this way there is reason to believe that the feedbacks that stabilize and sustain a cultural ecological system can be supported, and that those that drive a system into rapid, destructive decline can be avoided.

The eighth and final principle is the commandment to *build on the genius loci and traditional institutions* that affect environment positively. The genius loci principle assumes that each place is special and has its own unique array of opportunities and constraints. Being attentive to the spirit of place is not a mystical appeal to a form of modern geomancy; rather, it is a recognition that the diversity of cultures and environments that characterize humankind's global expression require particularistic and local attention. Important lessons can be learned outside local areas and applied, either in principle or through direct transfer, to the

solution of local problems and to the enhancement of local productivity. However, for success to occur, for land degradation to be
avoided, the frame of reference always must consider the local
habitat and culture. Without consulting intimately and in detail
with the peculiarities and potentials of particular places, serious
insults to local environmental stability are the norm rather than
the exception.

It follows that the starting point for efforts to exploit the local
resource base more intensively must be the traditional institutions
that have developed in that space for extended periods. If they offer
nothing else, these local cultural ecological systems have a depth of
experiential knowledge about place that transcends the limited
instrumental record of modern technoscience. The principles that
these resource-use systems developed form a sound basis upon
which to erect new ways of using the environment. Granted that
changing conditions in environment and society make it impossible
to operate old systems in the traditional way, at issue is whether
development must begin with a clean slate that carries no messages
from the past, or whether it can build on older ways of knowing
and doing to blend the traditional with the modern. We believe that
the record of both past experience and modern development clearly
shows that land degradation can best be controlled by building on
the traditional, being sensitive to the spirit of place, and garnishing
the best features of the indigenous with appropriate inputs of the
modern. In this way one may not attain the quantum leaps in
production so fervently desired by all, but the results, while
slower, likely will build incrementally toward much more sustainable ways of using our environment.

Population growth, technology transfers, and implementation of
new technologies will continue to exert pressure on the Earth's
resource base if solutions to immediate problems are not concep-
tualized within a long-term framework. Furthermore, if environmental accounting and off-site costs do not become integral
components of the evaluation process prior to instituting change,
unintentional land degradation will likely continue. The eight principles just summarized need to be considered prior to making decisions affecting land and water resources. Corrective strategies are
always more costly both in environmental, human, and economic
terms than prevention of degradation. Paying lip service to environmental concerns is becoming common. The challenge is now to
convert this increasing awareness into an integral component of the
planning processes.

There are signs that the principles presented in this book are
beginning to filter down into development strategies. There is an
increasing responsiveness to incorporating environmental considerations in development schemes. Given the awareness of the negative impacts of land degradation in terms of production, economic
costs, and overall quality of life, this sensitivity to land resource use
will have payoffs if it is actively pursued in the decision-making
processes. For example, since 1987 the World Bank has begun to

articulate the need to incorporate the links between economics and environment into development strategies if sustainability, the antithesis of land degradation, is to be attained. However, without further development of environmental accounting in the evaluation of land-use changes, economic considerations will likely remain dominant in most plans. Finally, since a significant proportion of land degradation results from a multitude of individual decisions, and significant gaps remain in our ability to integrate the demands of society, technology, and environment, land degradation will likely continue to haunt humankind in the immediate future.

References

Acheson, J. M., 1988, *The Lobster Gangs of Maine,* Hanover, New Hampshire: University Press of New England.

Ackefors, H. and Enell, M., 1990, Discharge of nutrients from Swedish fish farming to adjacent sea areas. *Ambio,* 19(1), 28–35.

Adams, M. E., 1983, Nile Water: a crisis postponed? *Economic Development and Cultural Change,* 31(3), 639–43.

Adams, R. McC., 1965, *Land behind Baghdad: a history of settlement on the Diyala Plains,* Chicago: University of Chicago Press.

——, 1966, *The Evolution of Urban Society: Early Mesopotamia and Prehistoric Mexico,* Chicago: Aldine.

——, 1978, Strategies of maximization, stability, and resilience in Mesopotamian society, settlement and agriculture. *Proceedings of the American Philosophical Society,* 122(5), 329–35.

——, 1981, *Heartland of Cities: Surveys of Ancient Settlement and Land Use on the Central Floodplain of the Euphrates,* Chicago: University of Chicago Press.

Adams, R. McC. and Nissen, H. J., 1972, *The Uruk Countryside: the Natural Setting of Urban Societies,* Chicago: University of Chicago Press.

Adams, W. M., 1989, Dam construction and the degradation of floodplain forest on the Turkwel River, Kenya. *Land Degradation & Rehabilitation,* 1(3), 189–98.

Alford, T. L., 1969, The West as a desert in American thought prior to Long's expedition. *Journal of the West,* 3(4), 515–25.

Anderson, J. K., 1985, *Hunting in the Ancient World,* Berkeley: University of California Press.

Antonsson-Ogle, B., 1990, Who uses forest foods? *Forests, Trees and People Newsletter,* No. 8, Swedish University of Agricultural Sciences, Uppsala, 21.

Arkell, T., 1991, The decline of pastoral nomadism in the western Sahara. *Geography,* 76(2), 162–6.

Arnold, J. E. M., 1992, *Community Forestry – Ten Years in Review,* Community Forestry Note 7, FAO, Rome.

Artz, N. E., Norton, B. E., and O'Rourke, T. J., 1986, Management of common grazing lands: Tamahdite, Morocco. In NRC, *Proceedings of the Conference on Common Property Management,* Washington, DC: National Academy Press, 259–80.

Babbitt, B., 1991, Age-old challenge: water and the West. *National Geographic,* 179(6), 2–34.

Baines, J. and Malek, J., 1980, *Atlas of Ancient Egypt*, New York: Facts on File Publications.

Balls, H., Moss, B., and Irvine, K., 1989, The loss of submerged plants with eutrophication I. Experimental design, water chemistry, aquatic plant and phytoplankton biomass in experiments carried out in ponds in the Norfolk Broadland. *Freshwater Biology*, 22(1), 71–87.

Bardet, J., 1987, Civiele techniek. In *De Physique Esistentie dezes Lands: Jan Blanken* (Rijksmuseum, Amsterdam Essaybundel), Danhoff, Zuidhorn: AMA boeken, 113–40.

Barnes, C. P., 1938, Land unfit for farming in the humid areas. In *Soils and Man: Yearbook of Agriculture, 1938*, Washington, DC: US Department of Agriculture.

Barrow, C. J., 1991, *Land Degradation*, Cambridge: Cambridge University Press.

Barrow, E. G. C., 1990, Usufruct rights to trees: the role of *ekwar* in dryland central Turkana, Kenya. *Human Ecology*, 18(2), 163–76.

Barth, F., 1961, *Nomads of South Persia: the Basseri Tribe of the Khamseh Confederacy*, Prospect Heights, Illinois: Waveland Press, (reissued 1986).

Bascom, J. B., 1990, Border pastoralism in eastern Sudan. *Geographical Review*, 80(4), 416–30.

Beadle, L. C., 1974, *The Inland Waters of Tropical Africa*, London: Longman.

Beck, L., 1981, Government policy and pastoral land use in southwest Iran. *Journal of Arid Environments*, 4, 253–67.

Beckerman, S., 1987, Swidden in Amazonia and the Amazon Rim. In B. L. Turner II and S. B. Brush (eds), *Comparative Farming Systems*, New York: Guilford, 55–94.

Bedrani, S., 1983, Going slow with pastoral cooperatives. *Ceres*, 16(4), 16–21.

——, 1991, Legislation for livestock on public lands in Algeria. *Nature & Resources*, 27(4), 24–30.

Bell, B., 1975, Climate and the history of Egypt: the Middle Kingdom. *American Journal of Archaeology*, 79, 223–69.

Bencherifa, A. and Johnson, D. L., 1991, Changing resource management systems and their environmental impacts in the Middle Atlas Mountains of Morocco. *Mountain Research and Development*, 11(3), 183–94.

Bennett, H. H., 1929, Facing the erosion problem. *Science*, 69, 48.

——, 1939, *Soil Conservation*, New York: McGraw-Hill.

Bennett, J. W., 1976, *The Ecological Transition: Cultural Anthropology and Human Adaptation*, New York: Pergamon Press.

Berkes, F., 1986, Marine inshore fishery management in Turkey. In NRC, *Proceedings of the Conference on Common Property Management*, Washington, DC: National Academy Press, 63–83.

——(ed.), 1989, *Common Property Resources: Ecology and Community-based Sustainable Development*, London: Belhaven Press.

Berkes, F., Feeny, D., McCay, B. J., and Acheson, J. M., 1989, The benefits of the commons. *Nature*, 340, 91–3.

Bernus, E. 1980, Desertification in the Aghazer and Azawak region: case study presented by the government of Niger. In J. Mabbutt and C. Floret (eds), *Case Studies on Desertification*, Paris, Unesco: 115–46.

——, 1990, Dates, dromedaries, and drought: diversification in Tuareg pastoral systems. In J. G. Galaty and D. L. Johnson (eds), *The World of Pastoralism*, New York: Guilford, 149–76.

Berry, L., 1983, *East African Country Profile – Sudan*, Program for International Development, Clark University, Worcester, Massachusetts.

Biswas, A. K., 1970, *History of Hydrology*, Amsterdam: North Holland.

Blaikie, P. and Brookfield, H., 1987, *Land Degradation and Society*, London: Methuen.

Borchert, J. R., 1989, *Americas's Northern Heartland*, Minneapolis: University of Minnesota Press.

Borgstrom, G., 1973, *World Food Resources*, New York: Intext Educational Publishers.

Boston Globe, 1990, California's well runs dry. *The Boston Globe*, April 25, 1990, 3.

——, 1993, Failing soil threatens food supply, UN warns. *The Boston Globe*, July 14, 1993, 9.

Bourn, D., 1978, Cattle, rainfall and tsetse in Africa. *Journal of Arid Environments*, 1, 49–61.

Bowden, M. J., 1969, The perception of the Western interior of the United States, 1800–1870: a problem in historical geosophy. *Proceedings of the Association of American Geographers*, 1, 16–21.

——, 1975. Desert wheat belt, Plains corn belt: environmental cognition and behavior of settlers in the Plains margin, 1850–99. In B. W. Blouet and M. P. Lawson (eds), *Images of the Plains: the Role of Human Nature in Settlement*, Lincoln: University of Nebraska Press, 189–201.

Bowerstock, G. W., 1983, *Roman Arabia*, Cambridge, Massachusetts: Harvard University Press.

Bowonder, B. and Ramana, K. V., 1986, *Environmental Degradation and Economic Development: a Case Study of a Marginally Productive Area*, Centre for Energy, Environment & Technology, Administrative Staff College of India, Bella Vista, Hyderabad.

Braeman, J., 1986, The Dust Bowl: an introduction. *Great Plains Quarterly*, 6(2), 67–8.

Breman, H. and de Wit, C. T., 1983, Rangeland productivity and exploitation in the Sahel. *Science*, 221, No. 4618 (September 30), 1341–7.

Brown, L. D. and Oliver, J. E., 1976, Vertical crustal movements from leveling data and their relation to geologic structure in the eastern United States. *Review of Geophysics and Space Physics*, 14, 13–35.

Brown, L. R., 1992, Launching the environmental revolution. In L. R. Brown (ed.), *State of the World–1992*, New York: W. W. Norton, 174–90.

Brown, L. R. and Wolf, E. C., 1984, *Soil Erosion: Quiet Crisis in the World Economy*, Worldwatch Paper 60, Washington, DC.

Bunyard, P., 1986, The death of the trees. *The Ecologist*, 16(1), 4–13.

Bureau of the Census, 1987, *1987 Census of Agriculture*, Vol. 1, Geographic Area Series, Part 5, California, Washington, DC: US Government Printing Office.

Burns, G., Billard, T. C., and Matsui, K. M., 1990, Salinity threat to Upper Egypt. *Nature*, 344, 25.

Butzer, K. W., 1971, *Environment and Archaeology: an Ecological Approach to Prehistory*, Chicago: Aldine–Atherton, second edition.

——, 1976, *Early Hydraulic Civilization in Egypt: a Study in Cultural Ecology*, Chicago: University of Chicago Press.

Cairns, J. Jr. and Pratt, J. R., 1990, Biotic impoverishment: effects of anthropogenic stress. In G. M. Woodwell (ed.), *The Earth in Transition*, Cambridge: Cambridge University Press, 495–505.

Campbell, D. J., 1981, Land-use competition at the margins of the rangelands: an issue in development strategies for semi-arid areas. In G. Norcliffe and T. Pinfold (eds), *Planning African Development*, Boulder, Colorado: Westview Press, London: Croom Helm, 39–61.

——, 1986, The prospect for desertification in Kajiado District, Kenya. *The Geographical Journal*, 152(1), 44–55.

Carlstein, T., 1982, *Time Resources, Society and Ecology: on the Capacity for Human Interaction in Space and Time, Vol. I: Preindustrial Societies*, London: George Allen & Unwin.

Carr, A. F., 1967, *So Excellent a Fishe: a Natural History of Sea Turtles*, Garden City, New York: Natural History Press.

Carr, C. J., 1977, *Pastoralism in Crisis: the Dasanetch and their Ethiopian Lands*, Research Paper No. 180, Department of Geography, University of Chicago.

Carter, W. E., 1969, *New Lands and Old Traditions: Kekchi Cultivators in the Guatemalan Lowlands*, Gainesville: University of Florida Press.

Center for Rural Affairs, 1988, *Beneath the Wheel of Fortune: the Economic and Environmental Impacts of Center Pivot Irrigation Development on Antelope County, Nebraska*, Water Policy and Practices Project, Whitehill, Nebraska.

Central Arid Zone Research Institute (CAZRI), 1976, *International Cooperation to Combat Desertification: Luni Development Block – a Case Study on Desertification*, Jodhpur, India: Central Arid Zone Research Institute.

Chandler, T., 1987, *Four Thousand Years of Urban Growth: an Historical Census*, Lewiston/Queenston: St. David's University Press.

Changnon, S. A., 1983, Record dust storms in Illinois: causes and implications. *Journal of Soil and Water Conservation*, 38, 58–63.

Chasin, B. H. and Franke, R. W., 1980, *Seeds of Famine: Ecological Destruction and the Development Dilemma in the West African Sahel*, Montclair, New Jersey: Allanheld, Osmun.

Chisholm, A. and Dumsday, R., 1987, *Land Degradation: Problems and Policies*, New York: Cambridge University Press.

Chorley, R. J., Schumm, S. A., and Sugden, D. E., 1984, *Geomorphology*, New York: Methuen.

Coates, D. R. and Vitek, J. D., 1980, *Thresholds in Geomorphology*, Boston: Allen & Unwin.

Colacicco, D., Osborn, T., and Alt, K., 1989, Economic damage from soil erosion. *Journal of Soil and Water Conservation*, 44(1), 35–9.

Collier, J. G. and Davies, L. M., 1986, *Chernobyl*, Central Electricity Generating Board (UK) Report, London.

Colman, D., 1989, Economic issues from the Broads grazing marshes conservation scheme. *Journal of Agricultural Economics*, 40(3), 336–44.

Committee on Selected Biological Problems in the Humid Tropics, 1982, *Ecological Aspects of Development in the Humid Tropics*, Washington, DC: National Academy Press.

Committee on Water, 1968, *Water and Choice in the Colorado River Basin*, Washington, DC: National Academy of Sciences.

Committee on Western Water Management, Water Science and Technology Board, National Research Council, 1992, *Water Transfers in the West: Efficiency, Equity, and the Environment*, Washington, DC: National Academy Press.

Conacher, A. J., 1990, Salt of the earth. *Environment*, 32, 4–9, 40–2.

Conklin, H. C., 1954, An ethnological approach to shifting cultivation. *Transactions of the New York Academy of Science*, Series 2, 17, 133–42.

——, 1980, *Ethnographic Atlas of Ifugao: a Study of Environment, Culture, and Society in Northern Luzon*, New Haven: Yale University Press.

Cook, K., 1984, Conservation and the 1985 Farm Bill: round 1. *Journal of Soil and Water Conservation*, 39(3), 179–81.

Cook, K. A., 1989, The environmental era of U.S. agricultural policy. *Journal of Soil and Water Conservation*, 44(5), 362–6.

Cox, A. and Hart, R. B. H., 1986, *Plate Tectonics: How it Works*, Oxford: Blackwell.

Cox, S. J. B., 1985, No tragedy on the commons. *Environmental Ethics*, 7(1), 49–61.

Craven, A. O., 1926, *Soil Exhaustion as a Factor in the Agricultural History of Virginia and Maryland, 1606–1860*, Urbana, Illinois: University of Illinois Press.

Cronon, W., 1983, *Changes in the Land: Indians, Colonists, and the Ecology of New England*, New York: Hill and Wang.

Dahl, G., 1991, The Beja of Sudan and the famine of 1984–1986. *Ambio*, 20(5), 189–91.

Darby, H. C., 1940, *The Draining of the Fens*, Cambridge: Cambridge University Press.

Davis, S. H., 1977, *Victims of the Miracle*, Cambridge: Cambridge University Press.

Day, G. M., 1935, The Indian as an ecological factor in the northeastern forest. *Ecology*, 32, 329–46.

Deevey, E. S. Jr., 1960, The human population. *Scientific American*, 203(3), 194–204.

Denevan, W. M., 1981, Swiddens and cattle versus forest: the imminent demise of the Amazon rain forest reexamined. In V. H. Sutlive, N. Altshuler, and M. D. Zamora (eds), *Where Have All the Flowers Gone? Deforestation in the Third World*, Department of Anthropology, College of William and Mary, Williamsburg, Virginia, 25–44.

Deqi, J. and Li, X., 1990, Soil and water conservation in Shaanxi Province. In E. Baum, P. Wolff, and M. A. Zöbisch (eds), *Experience with Available Conservation Technologies*, Deutsches Institut für tropische und subtropische Landwirtschaft, Witzenhausen, Germany, 83–110.

DESFIL/PID, 1985, *Development Strategies for Fragile Lands*, Washington DC: US Agency for International Development.

Detwyler, R. T., 1971, *Man's Impact on the Environment*, New York: McGraw-Hill.

Douglas, I. and Spencer, T., 1985, *Environmental Change and Tropical Geomorphology*, London: George Allen & Unwin.

Draz, O. 1974, *Report to the Government of the Syrian Arab Republic on Range Management and Fodder Development*, UNDP No. Ta 3292, Rome: FAO.

——, 1977, *Role of Range Management in the Campaign against Desertification: the Syrian Experience as an Applicable Example for the Arabian Peninsula*, Regional Preparatory Meeting for the Mediterranean Area, Algarve, 28 March – 1 April 1977, International Cooperation to Combat Desertification (UNCOD), UNCOD/MISC/13, Mimeo.

——, 1990, The *hema* system in the Arabian Peninsula. In National Research Council, Office of International Affairs, Board on Science and Technology for International Development, *The Improvement of Tropical and Subtropical Rangelands*, Washington, DC: National Academy Press, 321–31.

Dregne, H. E., 1976, Desertification: symptom of a crisis. In P. Paylore and R. A. Haney, Jr. (eds), *Desertification: Process Problems, Perspectives*, Arid/Semi-arid Natural Resources Program, University of Arizona, Tucson, 11–24.

——, 1977a, Generalized Map of the Status of Desertification of Arid Lands. In *Status of Desertification in the Hot Arid Regions*, UN Conference on Desertification, Nairobi.

——, 1977b, Development strategies for arid land use. In Y. Mundlak and S. F. Singer (eds), *Arid Zone Development: Potentialities and Problems*, Cambridge, Massachusetts: Ballinger, 255–61.

van Duin, R. H. A. and de Kaste, G., 1984, *The Pocket Guide to the Zuyder Zee Project*, Ministry of Transport and Public Works/ Rijksdienst voor de IJsselmeerpolders, Lelystad, The Netherlands.

Dunwell, S., 1978, *The Run of the Mill*, Boston: David R. Godine.

Eckholm, E., 1975, *The Other Energy Crisis: Fuelwood*, Worldwatch Paper 1, Washington, DC.

Eckholm, E. and Brown, L. R., 1977, *Spreading Deserts – the Hand of Man*, Worldwatch Paper 13, Washington, DC.

Economist, 1991, The Green counter-revolution. *The Economist*, 319, No. 7703 (April 20), 85–6.

Edwards, M., 1993, A broken empire. *National Geographic*, 183(3), 2–53.

El-Swaify, S. A., Moldenhauer, W. C., and Lo, A. (eds), 1985, *Soil Erosion and Conservation*, Ankeny, Iowa: Soil Conservation Society of America.

Ellis, W. S., 1990, A Soviet sea lies dying. *National Geographic*, 177(2), 73–93.

Elmgren, R., 1989, Man's impact on the ecosystem of the Baltic Sea: energy flows today and at the turn of the century. *Ambio*, 18(6), 326–32.

Embleton, C. and King, C. A. M., 1975, *Periglacial Geomorphology*, New York: Halsted Press.

Emel, J. and Brooks, E., 1988, Changes in form and function of property rights institutions under threatened resource scarcity. *Annals of the Association of American Geographers*, 78(2), 241–52.

Emel, J., Roberts, R., and Sauri, D., 1992, Ideology, property, and ground-water resources: an exploration of relations. *Political Geography*, 11(1), 37–54.

Englebert, E. A. with Scheuring, A. F., 1982, *Competition for California Water: Alternative Solutions*, Berkeley: University of California Press.

——, 1984, *Water Scarcity: Impacts on Western Agriculture*, Berkeley: University of California Press.

Estioko-Griffin, A. and Griffin, P. B., 1981, Woman the hunter: the Agta. In F. Dahlberg (ed.), *Woman the Gatherer*, New Haven: Yale University Press, 121–51.

Evans, N. A., 1972, Transmountain water diversion for the High Plains. In D. MacPhail (ed.), *The High Plains: Problems of Semiarid Environments*, Colorado State University, Fort Collins, 42–59.

Evans, R., 1993, Sensitivity of the British landscape to erosion. In D. S. G. Thomas and R. J. Allison (eds), *Landscape Sensitivity*, Chichester: John Wiley, 189–210.

Evenari, M., 1977, Ancient desert agriculture and civilizations: Do they point the way to the future? In Y. Mundlak and S. F. Singer (eds), *Arid Zone Development: Potentialities and Problems*, Cambridge, Massachusetts: Ballinger, 83–97.

——, 1981, Twenty-five years of research on runoff desert agriculture in the Middle East. In L. Berkofsky, D. Faiman, and J. Gale (eds), *Settling the Desert*, Jacob Blaustein Institute for Desert Research, Ben Gurion University of the Negev, Sede Boqer, Israel, 5–27.

Evenari, M., Shanan, L., and Tadmor, N., 1971, *The Negev: the Challenge of a Desert*, Cambridge, Massachusetts: Harvard University Press.

——, 1982, *The Negev: the Challenge of a Desert*, Cambridge, Massachusetts: Harvard University Press, second edition.

Evenari, M., Shanan, L., Tadmor, N. and Aharoni, Y., 1961, Ancient agriculture in the Negev. *Science*, 133, No. 3457 (March 31), 979–96.

Ezcurra, E., 1990, *De las chinampas a la megalópolis: el medio ambiente en la Cuenca de México*, Fondo de Cultura Económica, serie, La ciencia desde México, México, D. F.

Falloux, F. and Mukendi, A. (eds), 1987, *Desertification Control and Renewable Resource Management in the Sahelian and Sudanian Zones of West Africa*, World Bank Technical Paper No. 70, World Bank, Washington, DC.

FAO (Food and Agriculture Organization of the United Nations), 1990, *Forest Products Yearbook – 1988*, Rome: FAO.

Farnworth, E. G. and Golley, F. B. (eds), 1974, *Fragile Ecosystems: Evaluation of Research and Application in the Tropics*, New York: Springer-Verlag.

Fernside, P. M., 1990, Deforestation in Brazilian Amazonia. In G. M. Woodwell (ed.), *The Earth in Transition*, Cambridge: Cambridge University Press.

Fenneman, N. M., 1938, *Physiography of the Eastern United States*, New York: McGraw-Hill.

Fernea, R. A., 1973, *Nubians in Egypt: Peaceful People,* Austin: University of Texas Press.

Ferrians, O. J., Jr., Kachadoorian, R., and Greene, G. W., 1969, Permafrost and Related Engineering Problems in Alaska. *Geological Survey Professional Paper 678*, Washington, DC: US Government Printing Office.

Fisher, I., 1992, Digging out and digging in. *The New York Times*, 141, No. 49,001 (June 18), pp. B1, B5.

Fisher, W. B., 1993, Egypt: physical and social geography. In *The Middle East and North Africa 1982–1983*, London: Europa, 360–408.

Flawn, P. T., 1970, *Environmental Geology: Conservation, Land-use Planning, and Resource Management*, New York: Harper & Row.

Fleischer, S., Hamran, S., Kindt, T., Rydberg, L., and Stibe, L., 1987, Coastal eutrophication in Sweden: reducing nitrogen in land runoff. *Ambio*, 16(5), 246–51.

Forests Industries, 1984, Miracle vine threatens southern U.S. forests. *Forests Industries*, 111 (July), 58.

Forrester, F., 1978, Land subsidence. In R. O. Utgard, G. D. McKenzie, and D. Foley (eds), *Geology in the Urban Environment*, Minneapolis: Burgess, 82–4.

Fortes, M. D., 1988, Mangrove and seagrass beds of East Asia: habitats under stress. *Ambio*, 17(3), 207–13.

Frankfort, H., 1956, *The Birth of Civilization in the Near East*, Garden City, New York: Doubleday Anchor.

Fuchs, V., 1986, *The Health Economy*, Cambridge, Massachusetts: Harvard University Press.

Fukuda, H., 1976, *Irrigation in the World: Comparative Developments*, Tokyo: University of Tokyo Press.

Gadgil, M. and Iyer, P., 1989, On the diversification of common-property resource use by Indian society. In F. Berkes (ed.), *Common Property Resources: Ecology and Community-based Sustainable Development*, London: Belhaven, 240–55.

Galaty, J. G. and Johnson, D. L. (eds), 1990, *The World of Pastoralism*, New York: Guilford.

Gallais, J., 1972, Essai sur la situation actuelle des relations entre pasteurs et paysans dans le Sahel ouest-africain. In *Etudes de géographie tropicale offertes à Pierre Gourou*, Paris: Mouton, 301–13.

Garrad, P. N. and Hey, R. D., 1988, River management to reduce turbidity in navigable Broadland rivers. *Journal of Environmental Management*, 27(3), 273–88.

Geertz, C., 1966, *Agricultural Involution: the Process of Ecological Change in Indonesia*, Berkeley and Los Angeles: University of California Press.

George, C. J., 1972, The role of the Aswan High Dam in changing the fisheries of the southeastern Mediterranean. In M. T. Farvar and J. P. Milton (eds), *The Careless Technology: Ecology and International Development*, Garden City, New York: Natural History Press, 159–78.

George, M., 1992, *The Land Use, Ecology and Conservation of Broadland*, Chichester: Packard.

Gibson, M., 1974, Violation of fallow and engineered disaster in Mesopotamian civilization. In T. E. Downing and M. Gibson (eds), *Irrigation's Impact on Society*, Anthropological Papers of the University of Arizona, No. 25, University of Arizona Press, Tucson, 7–19.

Gilles, J. L., Hammoudi, A., and Mahdi, M., 1986, Oukaimedene, Morocco: a high mountain *agdal*. In NRC, *Proceedings of the Conference on Common Property Resource Management*, Washington, DC: National Academy Press, 281–304.

Gittus, J. H., et al., 1988, *The Chernobyl Accident and Its Consequences*, London: United Kingdom Atomic Energy Authority, second edition.

Glantz, M. H., 1990, Running on empty. *The Sciences*, 30(6), 16–20.

Goldammer, J. G., 1993, Historical biogeography of fire: tropical and subtropical. In P. J. Crutzen and J. G. Goldammer (eds), *Fire in the*

Environment: the Ecological, Atmospheric, and Climatic Importance of Vegetation Fires, Chichester: John Wiley, 297–314.

Goudie, A., 1981, *The Human Impact: Man's Role in Environmental Change*, Oxford: Blackwell.

Gradwohl, J. and Greenberg, R., 1988, *Saving the Tropical Forests*, Washington, DC: Earthscan Publications.

Grainger, A., 1900, *The Threatening Desert: Controlling Desertification*, London: Earthscan Publications.

Graves, W. (ed.), 1990, *Atlas of the World*, Washington, DC: National Geographic Society, sixth edition.

Grove, A. T., 1982, Egypt has too much water. *Geographical Magazine*, 54(8), 437–41.

Gunnarsson, Á., 1973, *Volcano Ordeal by Fire in Iceland's Westmann Islands*, Reykjavík: Iceland Review.

Hack, T. T., 1960, Interpretation of erosional topography in humid temperate regions. *American Journal of Science*, 256A, 80–97.

Hallberg, R. O., 1991, Environmental implications of metal distribution in Baltic Sea sediments. *Ambio*, 20(7), 309–16.

Hambidge, G., 1938, Soils and men – a summary. In *Soils and Men, Yearbook of Agriculture*, US Department of Agriculture, Washington, DC, 1–44.

Hansson, S., 1987. Effects of pulp and paper mill effluents on coastal fish communities in the Gulf of Bothnia, Baltic Sea. *Ambio*, 16(6), 344–8.

Hansson, S. and Rudstam, L. G., 1990, Eutrophication and Baltic fish communities. *Ambio*, 19(3), 123–5.

Hardin, G., 1968, The tragedy of the commons, *Science*, 162, No. 3859 (December 13), 1243–8.

Hare, F. K., 1977, Climate and desertification. In *Desertification: Its Causes and Consequences*, Oxford: Pergamon Press, 63–113.

Harris, B. L., Habiger, J. N., and Carpenter, Z. L., 1989, The conservation title: concerns and recommendations from the Great Plains. *Journal of Soil and Water Conservation*, 44(5), 371–5.

Hart, J. F., 1972, The Middle West. *Annals of the Association of American Geographers*, 62(2), 258–82.

Haynes, K. E. and Whittington, D., 1981, International management of the Nile – Stage three? *Geographical Review*, 71(1), 18–31.

Healy, R. G. and Sojka, R. E., 1985, Agriculture in the South: conservation's challenge. *Journal of Soil and Water Conservation*, 40(2), 189–94.

Heath, R. C., Fosworthy, B. L., and Cohen, P., 1966, The changing pattern of ground-water development on Long Island, New York. USGS Circular 524.

Hecht, S. B., 1981, Deforestation in the Amazon Basin: magnitude, dynamics and soil resource effects. In V. H. Sutlive, N. Altschuler, and M. D. Zamora (eds), *Where Have All the Flowers Gone? Deforestation in the Third World*, Department of Anthropology, College of William and Mary, Williamsburg, Virginia, 61–108.

——, 1989, The sacred cow in the green hell: livestock and forest conversion in the Brazilian Amazon. *The Ecologist*, 19(6), 229–34.

Hecht, S. and Cockburn, A., 1989, *The Fate of the Forest*, London: Verso.

Hecht, S., Anderson, A. B., and May, P., 1988, The subsidy from nature: shifting cultivation, successional palm forests, and rural development. *Human Organization*, 47(1), 25–35.

Hefny, K., 1982, Land-use and management problems in the Nile Delta. *Nature and Resources*, 18(2), 22–7.

Helbaek, H., 1960, Ecological effects of irrigation in ancient Mesopotamia. *Iraq*, 22, 186–96.

Hellier, C., 1988, The mangrove wastelands. *The Ecologist*, 18(2), 77–9.

Henning, D. and Flohn, H., 1977, Climate aridity index map. *UN Conference on Desertification, Publication A/CONF. 74/31*, Nairobi: UNEP.

Hewes, L., 1951, The Northern wet prairie of the United States: nature, sources of information, and extent. *Annals of the Association of American Geographers*, 41(4), 307–23.

Hewes, L. and Frandson, P. E., 1952, Occupying the wet prairie: the role of artificial drainage in Story County, Iowa. *Annals of the Association of American Geographers*, 42(1), 24–50.

Hingston, F. J. and Galaitis, V., 1976, The geographical variation of salt precipitated over Western Australia. *Australian Journal of Soil Research*, 14, 313–35.

Hjort af Ornas, A., 1990, Pastoral and environmental security in East Africa. *Disasters*, 14(2), 115–22.

Hobbs, J. J., 1989, *Bedouin Life in the Egyptian Wilderness*, Austin: University of Texas Press.

Holling, C. S., 1973, Resilience and stability of ecological systems. *Annual Review of Ecology and Systematics*, 4, 1–23.

Horowitz, M. M., 1981, Social analysis of desertification and its control in Kordofan and Darfur provinces, Republic of the Sudan, Khartoum; cited in Olsson, L. (1985), op. cit.

Horsley, A. D., 1986, *Illinois: a Geography*, Boulder, Colorado: Westview Press.

Hudson, N. W., 1976, *Soil Conservation*, London: Batsford.

Huggett, R., 1990, *Catastrophism*, London: Edward Arnold.

Hundley, N., Jr., 1992, *The Great Thirst*, Berkeley: University of California Press.

Huszar, P. C. and Young, J. E., 1984, Why the great Colorado plowout? *Journal of Soil and Water Conservation*, 39(4), 232–5.

al-Ibrahim, A., 1991, Excessive use of groundwater resources in Saudi Arabia: impacts and policy options. *Ambio*, 20(1), 34–7.

Ibrahim, F. N., 1984, *Ecological Imbalance in the Republic of the Sudan – with Reference to Desertification in Darfur*, Bayreuther Geowissenschaftliche Arbeiten Vol. 6, Bayreuth: Druckhaus Bayreuth.

Illinois Department of Agriculture, 1988, *Illinois Agricultural Statistics – Annual Summary*, Springfield, Illinois.

Ingold, T., 1987, *The Appropriation of Nature: Essays on Human Ecology and Social Relations*, Iowa City: University of Iowa Press.

Jacobsen, T., 1960, The waters of Ur. *Iraq*, 22, 174–85.

Jacobsen, T. and Adams, R. M., 1958, Salt and silt in ancient Mesopotamian agriculture. *Science*, 128, No. 3334 (November 21), 1251–8.

Jäger, J. and Barry, R. G., 1990, Climate. In B. L. Turner II et al. (eds), *The Earth as Transformed by Human Action*, Cambridge: Cambridge University Press with Clark University, 335–51.

James, B., 1993, Redefining rural France. *France Magazine*, No. 27 (Summer), 14–17.

James, D., 1992, Some principles and practices of desert revegetation seeding. *Arid Lands Newsletter*, 32 (Spring/Summer), 22–7.

James, P. E., 1929, The Blackstone Valley: a study in chorography in southern New England. *Annals of the Association of American Geographers*, 19(2), 67–109.

Jansson, A. M. and Jansson, B. O., 1988, Energy analysis approach to ecosystem redevelopment in the Baltic Sea and Great Lakes. *Ambio*, 17(2), 131–6.

Janzen, J., 1983, The modern development of nomadic living space in southeast Arabia – the case of Dhofar (Sultanate of Oman). *Geoforum*, 14(3), 289–309.

Jennings, J. N., 1952, *The Origin of the Broads*, Royal Geographical Research Series, No. 2, London: Royal Geographical Society.

Jenny, H., 1941, *Factors of Soil Formation: a System of Quantitative Pedology*, New York: McGraw-Hill.

——, 1984, My friend the soil. *Journal of Soil and Water Conservation*, 39(3), 158–61.

Jodha, N. S., 1985, Population growth and the decline of common property in Rajasthan, India. *Population and Development Review*, 11(2), 247–64.

——, 1988, The effects of climatic variation on agriculture in dry tropical area regions of India. In M. L., Parry, T. R., Carter, and N. T. Konijn (eds), *The Impact of Climatic Variations on Agriculture*, Vol. 2, Dordrecht, The Netherlands: Kluwer Academic, 503–16.

Johansson, P. O., 1990, Valuing environmental damage. *Oxford Review of Economic Policy*, 6(1), 34–50.

Johnson, D. L., 1969, *The Nature of Nomadism: a Comparative Study of Pastoral Migrations in Southwestern Asia and Northern Africa*, Research Paper No. 118, Department of Geography, University of Chicago.

——, 1973, The response of pastoral nomads to drought in the absence of outside intervention. UN Special Sahelian Office (ST/SSO/18), New York.

——, 1993a, Pastoral nomadism and the sustainable use of arid lands. *Arid Lands Newsletter*, 33 (Spring/Summer), 26–34.

——, 1993b, Nomadism and desertification in Africa and the Middle East. *GeoJournal*, 31(1), 51–66.

Johnson, D. L. and Whitmore, T., 1987, Old World population reconstructions: the Nile and Mesopotamia. 83rd Annual Meeting of the Association of American Geographers, Portland, Oregon, Mimeo.

Johnson, H. B., 1975, Rational and ecological aspects of the quarter section: an example from Minnesota. *Geographical Review*, 47, 66–87.

——, 1976, *Order Upon the Land*, Oxford: Oxford University Press.

Johnson, M. B., 1992, Tree legumes for reforestation and afforestation of arid and savannah lands. *Arid Lands Newsletter*, 32 (Spring/Summer), 28–31.

Jordan, C. F. (ed.), 1989, *An Amazonian Rain Forest*, MAB Vol. 2, Paris: Unesco/Carnforth, UK: Parthenon Publishing.

Jordan, W. R. III, Gilpin, M. E., and Aber, J. D. (eds), 1988, *Restoration Ecology: a Synthetic Approach*, Cambridge: Cambridge University Press.

Joshua, W. D., 1977, Soil erosive power of rainfall in the different climatic zones of Sri Lanka. *Symposium Proceedings of Erosion and Solid Matter*

Transport in Inland Waters, IAHS–AISH (International Association of Hydrological Sciences) Publication 122, 51–61.

Judson, S., 1981, Erosion of the land. In B. J. Skinner (ed.), *Use and Misuse of the Earth's Surface*, Los Altos, California: William Kaufman, 130–9.

Kaatz, M. R., 1955, The Black Swamp: a study in historical geography. *Annals of the Association of American Geographers*, 45(1), 1–35.

Kadri, J., 1991, *The Good, the Bad, and the Ugly: 1990 Wastewater Treatment Plant Performance Survey*, Providence, Rhode Island: Save The Bay.

Kassas, M., 1972, Impact of river control schemes on the shoreline of the Nile Delta. In M. T. Farvar and J. P. Milton (eds), *The Careless Technology: Ecology and International Development*, Garden City, New York: Natural History Press, 179–88.

Kates, R. W. and Haarmann, V., 1992, Where the poor live: Are the assumptions correct? *Environment*, 34(4), 4–11, 25–8.

Kemp, W. B., 1971, The flow of energy in a hunting society. *Scientific American*, 225, 104–15.

Kennedy, D. and Riley, D., 1990, *Rome's Desert Frontier from the Air*, Austin: University of Texas Press.

al-Khashab, W. H., 1958, *The Water Budget of the Tigris and Euphrates Basin*, Research Paper No. 54, Department of Geography, University of Chicago.

Khogali, M. M., 1991, Desertification, famine, and the 1988 rainfall – the case of Umm Ruwaba District in the northern Kordofan region. *GeoJournal*, 25(1), 81–9.

Kiersch, G. A., 1965, The Vaiont Reservoir disaster. *Mineral Information Service*, 18(7), 129–38.

Kisangani, E., 1986, A social dilemma in a less developed country: the massacre of the African elephant in Zaire. In NRC, *Proceedings of the Conference on Common Property Resource Management*, Washington, DC: National Academy Press, 137–60.

Kishk, M. A., 1986, Land degradation in the Nile Valley. *Ambio*, 15(4), 226–30.

Kivell, P. T., Parsons, A. J., and Dawson, B. R. P., 1989, Monitoring derelict urban land: a review of problems and potentials of remote sensing techniques. *Land Degradation & Rehabilitation*, 1(1), 5–21.

Kohl, L., 1986, The Oosterschelde Barrier – man against the sea. *National Geographic*, 170(4), 526–37.

Kondratyev, K. Y., 1988, *Climate Shocks: Natural and Anthropogenic*, New York: John Wiley.

Kotlyakov, V. M., 1991, The Aral Sea Basin: a critical environmental zone. *Environment*, 33(1), 4–6.

Kovda, V. A., 1980, *Land Aridization and Drought Control*, Boulder, Colorado: Westview Press.

Kramer, S. N., 1959, *History Begins at Sumer*, Garden City, New York: Doubleday Anchor.

——, 1961, Mythology of Sumer and Akkad. In S. N. Kramer (ed.), *Mythologies of the Ancient World*, Garden City, New York: Doubleday Anchor, 93–137.

——, 1963, *The Sumerians: Their History, Culture, and Character*, Chicago: University of Chicago Press.

Kromm, D. E. and White, S. E. (eds), 1992, *Groundwater Exploitation in the High Plains*, Lawrence: University of Kansas Press.

Kuhlmann, D. H. H., 1988, The sensitivity of coral reefs to environmental pollution. *Ambio*, 17(1), 13–21.

Kumar, M. and Bhandari, M. M., 1992, Impact of protection and free grazing on sand dune vegetation in the Rajasthan Desert, India. *Land Degradation & Rehabilitation*, 3(4), 215–27.

——, 1993a, Human use of the sand dune ecosystem in the semiarid zone of the Rajasthan Desert, India. *Land Degradation & Rehabilitation*, 4(1), 21–36.

——, 1993b, Impact of human activities on the pattern and process of sand dune vegetation in the Rajasthan Desert. *Desertification Bulletin*, No. 22, 45–54.

Kunstadter, P., 1987, Swiddeners in transition: Lua' farmers in northern Thailand. In B. L. Turner II and S. Brush (eds), *Comparative Farming Systems*, New York: Guilford, 130–55.

Lal, R., 1976, Soil erosion on alfisols in western Nigeria – effects of rainfall characteristics. *Geoderma*, 16, 389–401.

—— (ed.), 1988, *Soil Erosion Research Methods*, Ankeny, Iowa: Soil and Water Conservation Society, 150–60.

Lambert, A. M., 1971, *The Making of the Dutch Landscape*, London: Seminar Press.

Lambert, J. M., 1960, *The Making of the Broads: a Reconsideration of their Origin in the Light of New Evidence*, London: John Murray.

Lamprey, H. F., 1975, Report on the desert encroachment reconnaissance in northern Sudan. UNESCO/UNEP, Nairobi.

Lapidus, I. M., 1981, Arab settlement and economic development of Iraq and Iran in the age of the Umayyad and Early Abbasid caliphs. In A. L. Udovitch (ed.), *The Islamic Middle East, 700–1900: Studies in Economic and Social History*, Princeton, New Jersey: Darwin Press, 177–208.

——, 1988, *A History of Islamic Societies*, Cambridge: Cambridge University Press.

Larsen, C. E. and Evans, G., 1978, The Holocene geological history of the Tigris–Euphrates–Karun delta. In W. C. Brice (ed.), *The Environmental History of the Near and Middle East Since the Last Ice Age*, London: Academic Press, 227–44.

Lawson, W. E., Walsh, L. M., Stewart, B. A., and Boelter, D. H. (eds), 1981, *Soil and Water Resources: Research Priorities for the Nation*, Madison, Wisconsin: Soil Science Society of America.

Leach, G. and Mearns, R., 1990, *Beyond the Fuelwood Crisis*, London: IIED.

Lee, R. B., 1968, What hunters do for a living, or, how to make out on scarce resources. In R. B. Lee and I. DeVore (eds), *Man the Hunter*, Chicago: Aldine–Atherton, 30–43.

——, 1969, !Kung bushman subsistence: an input–output analysis. In A. P. Vayda (ed.), *Environment and Cultural Behavior: Ecological Studies in Cultural Anthropology*, Garden City, New York: Natural History Press, 47–79.

Lee, R. B. and DeVore, I. (eds), 1968, *Man the Hunter*, Chicago: Aldine–Atherton.

Lepsch, I. F., Buol, S. W., and Daniels, R. B., 1977, Soil–landscape relationships in the occidental plateau of São Paulo state, Brazil: II. Soil morphology, genesis, and classification. *Soil Science Society of America Journal*, 41(1), 109–15.

Le Strange, G., 1905, *The Lands of the Eastern Calilphate: Mesopotamia, Persia, and Central Asia from the Moslem Conquest to the Time of Timur*, Cambridge: Cambridge University Press.

Lewis, J., 1990, The Ogallala Aquifer: an underground sea. *EPA Journal*, 16(6), 42–44.

Lewis, L. A., 1981, The movement of soil materials during a rainy season in western Nigeria. *Geoderma*, 25, 13–25.

——, 1985, Assessing soil loss in Kiambu and Murangi Districts, Kenya. *Geografiska Annaler*, 67A(3–4), 273–84.

——, 1992 Terracing and accelerated soil loss on Rwandan steeplands: a preliminary investigation on the implications of human activities affecting soil loss. *Land Degradation & Rehabilitation*, 3(4), 241–6.

Lewis, L. A. and Berry, L., 1988, *African Environments and Resources*, Boston: Unwin Hyman.

Lewis, L. A. and Coffey, W. J., 1985, The continuing deforestation of Haiti. *Ambio*, 14(3), 158–60.

Lewis, L. A. and Nyamulinda, V., 1989, Les relations entre les cultures et les unités topographiques dans les régions agricoles de la Bordure du Lac Kivu et de l'Impara au Rwanda: quelques stratégies pour une agriculture soutenue. *Bulletin Agricole Rwanda*, 3 (Juillet), 143–9.

Lewis, N. and Brubaker, K. L., 1989, *Bring Back the Blackstone*, Providence, Rhode Island: Save the Bay.

Linear, M., 1985, The tsetse war. *The Ecologist*, 15(1/2), 27–35.

Little, P. D., 1987, Land use conflicts in the agricultural/pastoral borderlands: the case of Kenya. In P. D. Little and M. M. Horowitz (eds), with A. E. Nyerges, *Lands at Risk in the Third World: Local-Level Perspectives*, Boulder, Colorado: Westview Press, 195–212.

Lo, C. P., 1990, People and environment in the Zhu Jiang Delta of South China. *National Geographic Research*, 6(4), 400–17.

Malayang III, B. S., 1991, Tenure rights and exclusion in the Philippines. *Nature & Resources*, 27(4), 18–23.

Malingreau, J. P. and Tucker, C. J., 1987, The Contribution of AVHRR data for measuring and understanding global processes: large-scale deforestation in the Amazon Basin. *Proceedings of the International Geoscience and Remote Sensing Society*, 1, 443–50.

Manshard, W., 1974, *Tropical Agriculture: a Geographical Introduction and Appraisal*, London: Longman.

Marsh, G. P., 1864 (edited edition, 1965), *Man and Nature: Or, Physical Geography as Modified by Human Action*, Cambridge, Massachusetts: Belknap Press of Harvard University Press.

Martin, P., 1973, The discovery of America. *Science*, 179, No. 4077 (March 9), pp. 969–74.

Martin, R., 1989, *A Story that Stands Like a Dam*, New York: Henry Holt.

Mather, E. C., 1972, The American Great Plains. *Annals of the Association of American Geographers*, 62(2), 237–57.

McCabe, J. T., 1990, Turkana pastoralism: a case against the tragedy of the commons. *Human Ecology*, 18(1), 81–103.

McManus, J. W., 1988, Coral reefs of the ASEAN region: status and management. *Ambio*, 17(3), 189–93.

McNeill, W. H., 1976, *Plagues and Peoples*, Garden City, New York: Anchor Doubleday.

McPhee, J. A., 1989, *The Control of Nature*, New York: Farrar Straus Giroux.

Meggers, B. J., 1971, *Amazonia: Man and Culture in a Counterfeit Paradise*, Arlington Heights, Illinois: AHM Publishing.

van Meijgaard, C. H., 1987, Jan Blanken en de landsverdediging. In *De Physique Existentie dezes Lands: Jan Blanken* (Rijksmuseum, Amsterdam Essaybundel), Danhoff, Zuidhorn: AMA boeken, 41–58.

Meyer, A. H., 1936, The Kankakee "Marsh" of northern Indiana and Illinois. *Michigan Papers in Geography*, 6, 359–96.

Micklin, P. P. 1991, Touring the Aral: visit to an ecologic disease zone. *Soviet Geography*, 32(2), 90–105.

Miller, G. T., 1993, *Environmental Science*, Belmont, California: Wadsworth, fourth edition.

Milliman, J. D., Broadus, J. M., and Gable, F., 1989, Environmental and economic implications of rising sea level and subsiding deltas: the Nile and Bengal examples. *Ambio*, 18(6), 340–5.

Ministry for Economic Policy and Development of the Czech Republic, 1991, *Northern Bohemia Revitalization Project*, Prague.

Mohr, E. C. J., van Baren, F. A., and van Schuylenborgh, J., 1972, *Tropical Soils*, The Hague: Mouton/Ichtiar Baru: Van Hoeve.

Molinelli, J. A., 1984, Geomorphic processes along the Autopista Las Americas in North Central Puerto Rico, Ph.D. dissertation, Clark University, Worcester, Massachusetts.

Moorehead, R., 1989, Changes taking place in common-property resource management in the inland Niger Delta of Mali. In F. Berkes (ed.), *Common Property Resources: Ecology and Community-based Sustainable Development*, London: Belhaven Press, 256–72.

Moroney, M. G., 1984, *Iraq After the Muslim Conquest*, Princeton, New Jersey: Princeton University Press.

Moroz, A., 1988, Morye prosit vody. *Komosomolskaya Pravda*, January 28, cited in Precoda (1991), op. cit.

Moss, B., 1979, An ecosystem out of phase. *Geographical Magazine*, 52(1), 47–9.

Mould, R. F., 1988, *Chernobyl: the Real Story*, Oxford: Pergamon Press.

Mumford, L., 1961, *The City in History: its History, Its Origins, Its Transformations, and Its Prospects*, New York: Harcourt, Brace and World.

Munslow, B., Katarere, Y., Ferf, A., and O'Keefe, P., 1987, *The Fuelwood Trap*, London: IIED.

National Geographic Society, 1992, Geographica. *National Geographic*, 181(2), xiv.

Nebel, B. J., 1990, *Environmental Science*, Englewood Cliffs, New Jersey: Prentice-Hall.

Neel, J. V., 1970, Lessons from a "primitive" people. *Science*, 170, No. 3960 (November 30), 815–22.

Negev, A., 1986, *Nabatean Archaeology Today*, New York: New York University Press.

Nicholson, S. E., Kim, J., and Hoopingarner, J., 1988, *Atlas of African Rainfall and its Interannual Variability*, Department of Meteorology, Florida State University, Tallahasse.

Nietschmann, B., 1972, Hunting and fishing focus among the Miskito Indians, eastern Nicaragua. *Human Ecology*, 1(1), 41–67.

——, 1973, *Between Land and Water: The Subsistence Ecology of the Miskito Indians, Eastern Nicaragua*, New York: Seminar Press.

Nissen, H. J., 1988, *The Early History of the Ancient Near East 9000–2000 B.C.*, translated by E. Lutzeier and K. J. Northcott, Chicago: University of Chicago Press.

NRC (National Research Council), Office of International Affairs, Board on Science and Technology for International Development, Panel on Common Property Resource Management, 1986, *Proceedings of the Conference on Common Property Resource Management*, April 21–26, 1985, Washington, DC: National Academy Press.

OECD, 1987, *The Radiological Impact of the Chernobyl Accident in OECD Countries*, Paris: Nuclear Energy Agency, OECD.

OIA, 1978, *Dryland Agriculture: What is It?*, Office of International Agriculture, Oregon State University, Corvallis, Oregon.

O'Keefe, P., Raskin, P. and Bernow, S. 1984, *Energy Development in Kenya: Opportunities and Constraints*, Energy, Environment, and Development in Africa Series, Vol. 1, Scandinavian Institute of African Studies and Beijer Institute, Stockholm.

O'Leary, M., 1984, Ecological villains or economic victims: the case of the Rendille of northern Kenya. *Desertification Control Bulletin*, No. 11 (December), 17–21.

Olsson, K. and Rapp, A., 1991, Dryland degradation and conservation for survival. *Ambio*, 20(5), 192–5.

Olsson, L., 1985, *An Integrated Study of Desertification*, Department of Geography, The University of Lund, Sweden.

O'Riordan, T., 1979, Signs of disaster and a policy for survival. *Geographical Magazine*, 52(1), 50–9.

——, 1990, Environmental assessment and a future strategy for the Broads. *Planner*, 76(7), 7–10.

Oswalt, W. H., 1973, *Habitat and Technology: the Evolution of Hunting*, New York: Holt, Rinehart and Winston.

OTA (Office of Technology Assessment), 1988, *Enhancing Agriculture in Africa: a Role for US Development Assistance*, Washington, DC: US Government Printing Office.

Oterbridge, T., 1987, The disappearing Chinampas of Xochimilco. *The Ecologist*, 17(2), 76–83.

Parizek, R. R., 1971, Impact of highways on the hydrogeologic environment. In D. R. Coates (ed.), *Environmental Geomorphology*, Binghamton, New York: State University of New York at Binghamton, 155–99.

Parsons, J. J., 1962, *The Green Turtle and Man*, Gainesville: University of Florida Press.

Pattison, W. D., 1957, *Beginnings of the American Rectangular Land Survey System, 1784–1800*, Research Paper No. 50, Department of Geography, University of Chicago.

Pawluk, R. R., Sandor, J. A., and Tabor, J. A., 1992, The role of indigenous soil knowledge in agricultural development. *Journal of Soil and Water Conservation*, 47(4), 298–302.

Pereira, H. C., 1973, *Land Use and Water Resources*, Cambridge: Cambridge University Press.

Perevolotsky, A., 1987, Territoriality and resource sharing among the bedouin of southern Sinai: a socio-ecological interpretation. *Journal of Arid Environments*, 13, 153–61.

Peters, E. L., 1968, The tied and the free: an account of a type of patron-client relationship among the bedouin pastoralists of Cyrenaica. In J.-G. Peristiany (ed.), *Contributions to Mediterranean Sociology: Mediterranean Rural Communities and Social Change*, Paris: Mouton, 167–88.

Pilkey, O. H. and Neal, W. J., 1992, Save beaches, not buildings. *Issues in Science and Technology*, 8(3), 36–41.

Pinchot, G., 1947, *Breaking New Ground*, New York: Harcourt Brace.

Piper, S., 1989, Measuring particulate pollution damage from wind erosion in the western United States. *Journal of Soil and Water Conservation*, 44(1), 70–5.

de Ploey, J., 1990, Modelling the erosional susceptibility of catchments in terms of energy. *Catena*, 17, 175–83.

Pollack, A., 1992. Drug industry going back to nature. *The New York Times*, March 5, pp. D1, D9.

Postel, S., 1988, Global view of a tropical disaster. *American Forests*, 94(11–12), 25–9, 69–71.

Postel, S. and Ryan, J. C., 1991, Reforming forestry. In L. R. Brown et al. (eds), *State of the World – 1991*, New York: W. W. Norton, 74–92.

Power, J. F. and Follett, R. F., 1987, Monoculture. *Scientific American*, 256(3), 79–86.

Powers, W. E., 1966, *Physical Geography*, New York: Appleton–Century–Crofts.

Powers, W. L., 1987, The Ogallala's bounty evaporates. *Science of Food and Nutrition*, 5(3), 2–5.

Precoda, N., 1991, Requiem for the Aral Sea. *Ambio*, 20(3–4), 109–14.

Press, F. and Siever, R., 1974, *Earth*, San Francisco: W. H. Freeman.

Proctor, J., 1983, Mineral nutrients in tropical forests. *Progress in Physical Geography*, 7, 422–31.

Pyne, S. J., 1993, Keeper of the flame: a survey of anthropogenic fire. In P. J. Crutzen and J. G. Goldammer (eds), *Fire in the Environment: the Ecological, Atmospheric, and Climatic Importance of Vegetation Fires*, Chichester: John Wiley, 245–66.

Rauschkolb, R. S., 1971, Land degradation, *FAO Soil Bulletin*, No. 13, Rome: FAO.

Redman, C. L., 1978, *The Rise of Civilization: From Early Farmers to Urban Society in the Ancient Near East*, San Francisco: W. H. Freeman.

Reed, E. A., 1920, *Tales of a Vanishing River*, New York: John Lane.

Reed, L. A., 1980, Suspended sediment discharge in five streams near Harrisburg, Pennsylvania, before, during, and after highway construction. *Geological Survey Water-supply Paper 2072*, US Geological Survey, Washington, DC.

Reining, D., 1967, Rock, sand and gravel resources – strong but challenged. Preprint of paper presented at annual meeting of the American Institute of Mining and Metallurgical Engineers, Los Angeles, February 19–23.

Reisner, M., 1986, *Cadillac Desert*, New York: Viking Penguin.

Renard, K. G., Foster, G. R., Weesies, G. A., and Porter, J. P., 1991, RUSLE – Revised Universal Soil Loss Equation. *Journal of Soil and Water Conservation*, 46(1), 30–33.

Reynolds, D. M. and Myers, M. (eds), 1991, *Working in the Blackstone River Valley: Exploring the Heritage of Industrialization,* Woonsocket, Rhode Island: Rhode Island Labor History Society.

Richards, J. F., 1990, Agricultural impacts in tropical wetlands: rice paddies for mangroves in South and Southeast Asia. In M. Williams (ed.), *Wetlands: a Threatened Landscape*, Oxford: Blackwell, 217–33.

Richards, P. W., 1952, *The Tropical Rainforest*, Cambridge: Cambridge University Press.

Richards, P., 1985, *Indigenous Agricultural Revolution: Ecology and Food Production in West Africa*, London: Hutchinson/Boulder, Colorado: Westview Press.

Richards, W. Q., 1984, Reconstructing the national conservation program. *Journal of Soil and Water Conservation*, 39(3), 156–7.

Riebsame, W. E., 1986, The Dust Bowl: historical image, psychological anchor, and ecological taboo. *Great Plains Quarterly*, 6(2), 127–36.

Riehl, H. and Meitin, J., 1979, Discharge of the Nile River: a barometer of short period climatic variation. *Science*, 206, No. 4423 (December 7), 1178–9.

Robinson, A. H. W., 1953, The storm surge of 31st January – 1st February, 1953. *Geography*, 38, 134–41.

Robinson, F. E. and Luthin, J. N., 1976, Adaption to increasing salinity of the Colorado River. *California Water Resources Center University of California, Davis, Contribution No. 160.*

Robinson, G. D. and Speiker, A. M., 1978, Nature to be Commanded . . . . *US Geological Survey Professional Paper 950*, Washington DC: US Government Printing Office.

Rostlund, E., 1957, The myth of a natural prairie belt in Alabama: an interpretation of historical records. *Annals of the Association of American Geographers*, 47(4), 392–411.

Rubin, R., 1991, Settlement and agriculture on an ancient desert frontier. *Geographical Review*, 81(2), 197–205.

Ruddle, K., 1974, *The Yukpa Cultivation System: a Study of Shifting Cultivation in Columbia and Venezuala*, Ibero-Americana Vol. 52, Berkeley: University of California Press.

—— ,1989, Solving the common-property dilemma: village fisheries rights in Japanese coastal waters. In F. Berkes (ed.), *Common Property Resources*, London: Belhaven Press, 168–84.

Ruddle, K. and Manshard, W., 1981, *Renewable Natural Resources and the Environment: Pressing Problems in the Developing World*, Dublin: Tycooly International for the United Nations University.

Ruddle, K. and Zhong, G., 1988, *Integrated Agriculture–Aquaculture in South China: The Dike–Pond System of the Zhujiang Delta*, Cambridge: Cambridge University Press.

Ruddle, K., Furtado, J. I., Zhong, G. F., and Deng, H. Z., 1983, The mulberry dike – carp pond resource system of the Zhujiang (Pearl River) Delta, People's Republic of China: I. Environmental context and system overview. *Applied Geography*, 3, 45–62.

Rutkowski, A., 1991, Unnatural disasters. *World Press Review*, 38 (June), 45.

Ruttan, V. W., 1989, Biological and technical constraints on crop and animal productivity: report on a dialogue, *Staff Paper Series P89–45*, Department of Agriculture and Applied Economics, University of Minnesota, St. Paul.

Salati, E., Dallolio, S., Matsui, E., and Gat, J., 1979, Recycling of water in the Amazon Basin: an isotope study. *Water Resources Research*, 15, 1250–8.

Salati, E., Dourojeanni, M. J., Novaes, F. C., De Oliveira, A. E., Perritt, R. W., Schubart, H. O. R., and Umana, J. C., 1990, Amazonia. In B. L. Turner II et al. (eds), *The Earth as Transformed by Human Action*, Cambridge: Cambridge University Press with Clark University, 479–93.

Sauer, C. O., 1956, The agency of man on the Earth. In W. L. Thomas, Jr. (ed.), *Man's Role in Changing the Face of the Earth*, Chicago: University of Chicago Press, 49–69.

Schick, A. P., Lekach, J., and Hassan, M. A., 1987, Vertical exchange of coarse bedload in desert streams. In L. E. Frostick and I. Reid (eds), *Desert Sediments: Ancient and Modern*, Geological Society of London, Special Publication 35, Oxford: Blackwell, 7–16.

Schmal, H., 1987, De ontwikkeling van de infrastructuur van het einde van de achttiende tot het midden van de 19de eeuw. In *de Physique Existentie dezes Lands: Jan Blanken* (Rijksmuseum, Amsterdam Essaybundel), Danhoff, Zuidhorn: AMA boeken, 95–112.

Schneider, W. J., Rickert, D. A., and Speiker, A. M., 1973, Role of water in urban planning and management. *USGS Circular 601-H*, Washington, DC.

Schumm, S. A., 1956, The evolution of drainage systems and slopes in badlands at Perth Amboy, New Jersey. *Bulletin of the Geological Society of America*, 67, 597–646.

Seckler, D., 1971, *California Water*, Berkeley: University of California Press.

Sharon, D., 1972, The spottiness of rainfall in a desert area. *Journal of Hydrology*, 17, 161–76.

Sharp, R. P. and Carey, D. L., 1976, Sliding stones, Racetrack Playa, California. *Bulletin of Geological Society of America*, 87, 1704–17.

Shaw, J. M., 1986, Managing change – the Broadlands experience. *Planner*, 72(10), 15–17.

—— ,1990, Environmental government: the example of the Broads. *Planner*, 76(7), 11–14.

Shaxson, T. F., Hudson, N. W., Sanders, D. W., Roose, E., and Moldenhauer, W. C., 1989, *Land Husbandry, a Framework for Soil and Water Conservation*, Ankeny, Iowa: Soil and Water Conservation Society.

Sheppard, T., 1912, *The Lost Towns of the Yorkshire Coast*, London: A. Brown.

Sheridan, D., 1981, *Desertification of the United States*, Washington, DC: Council on Environmental Quality.

Shinn, E., 1989, What is really killing the corals? *Sea Frontiers*, 35(2), 72–81.

Shoup, J., 1990, Middle Eastern sheep pastoralism and the hima system. In J. G. Galaty and D. L. Johnson (eds), *The World of Pastoralism*, New York: Guilford, 195–215.

Sigalov, V., 1987, Aral: a view from space (translation of: "Aral: vzglyad iz kosmosa"). *Izvestiya*, June 23.

Simmons, I. G., 1989, *Changing the Face of the Earth: Culture, Environment, History*, Oxford: Blackwell.

Simon, A. and Hupp, C. R., 1986, Channel Evolution in Modified Tennessee Channels. In *Nevada Proceedings of the 4th Federal Interagency Sedimentation Conference, Las Vegas*, Interagency Advisory Committee on Water Data (US Geological Survey), Reston, Virginia, 5–71 – 5–82.

Simons, M., 1992, Pollution blights investment, too, in East Europe. *The New York Times*, 141 (May 13), pp. A1, A12.

Skempton, A. W. and Hutchinson, J. N., 1969, Stability of Natural Slopes and Embankment Foundations. In *Proceedings, 7th International Conference on Soil Mechanics and Foundation Engineering*, State of the Art Volume, Mexico City.

Skidmore, E. L. and Woodruff, N. P., 1968, Wind erosion forces in the United States and their use in predicting soil loss. Handbook No. 346, US Department of Agriculture, Washington, DC.

Skinner, B. J. and Porter, S. C., 1987, *Physical Geology*, New York: John Wiley.

Smiles, M. and Huiskes, A. H. L., 1981, Holland's Eastern Scheldt Estuary Barrier Scheme: some ecological considerations. *Ambio*, 10(4), 158–65.

Smith, N. J. J., 1982, *Rainforest Corridors: the Transamazonian Colonization Scheme*, Berkeley: University of California Press.

Smits, H., 1970, Land reclamation in the former Zuyder Zee in the Netherlands. *Geoforum*, 4, pp. 37–44.

—— ,1988, *Land Reclamation in the Former Zuyder Zee in the Netherlands*, revised by de J. Jong, Lelystad, Netherlands: Ministerie van Verkeer en Waterstaat.

Sollod, A. E., 1990, Rainfall, biomass and the pastoral economy of Niger: assessing the impact of drought. *Journal of Arid Environments*, 18, 97–107.

Sonner, S., 1992, US tally of forest growth is faulted. *The Boston Globe*, 241, No. 168 (June 16), 3.

Spencer, J. E. and Hale, G. A., 1961, The origins, nature, and distribution of agricultural terracing. *Pacific Viewpoint*, 2(1), 1–10.

Splinter, W. E., 1976, Center-pivot irrigation. *Scientific American*, 234(6), 90–9.

Stanley, D. J. and Warne, A. G., 1993, Nile Delta: recent geological evolution and human impact. *Science*, 260, No. 5108 (April 30), 628–34.

Stansfield, J., Moss, B., and Irvine, K., 1989, The loss of submerged plants with eutrophication III. Potential role of organochlorine pesticides: a palaeoecological study. *Freshwater Biology*, 22(1), 109–32.

Stern, M., 1985, Census from Heaven? Ph.D. dissertation, Department of Physical Geography, University of Lund, Sweden.

Stewart, O. C., 1956, Fire as the first great force employed by man. In W. L. Thomas, Jr. (ed.), *Man's Role in Changing the Face of the Earth*, Chicago: University of Chicago Press, 115–33.

Stocks, B. J. and Trollope, W. S. W., 1993, Fire management: principles and options in the forested and savanna regions of the world. In P. J. Crutzen and J. G. Goldammer (eds), *Fire in the Environment: The*

Ecological, Atmospheric, and Climatic Importance of Vegetation Fires, Chichester: John Wiley, 315–26.

Stockton, C. W. and Meko, D. M., 1983, Drought recurrence in the Great Plains as reconstructed from long-term tree ring records. *Journal of Climate and Applied Meteorology*, 22(1), 17–29.

Stolgitis, J., 1991, *Briefings: Rhode Island Division of Fish and Wildlife*, Providence, Rhode Island: Rhode Island Division of Fish and Wildlife.

Sutton, R. K., 1977, Circles on the Plains: center pivot irrigation. *Landscape*, 22(1), 3–10.

SWCS (Soil and Water Conservation Society), 1984, Out of the Dust Bowl. *Journal of Soil and Water Conservation*, 39(1), 6–17.

Tank, R. W. (ed.), 1983, *Environmental Geology*, New York: Oxford University Press.

Thesiger, W., 1964, *The Marsh Arabs*, London: Longman.

Thomas, D. S. G. (ed), 1989, *Arid Zone Geomorphology*, London: Belhaven Press.

Thornbury, W. D., 1965, *Regional Geomorphology of the United States*, New York: John Wiley.

Thurman, H. V., 1985, *Introductory Oceanography*, Columbus, Ohio: Charles E. Merrill.

Torrens, I. M., 1984, What goes up must come down: the acid rain problem, *The OECD Observer*, No. 127 (July), 9–19.

Tricart, J., 1965, *Le modèle des régions chaudes, forêts et savanes*, Paris: Société d'Édition d'Enseignement Supérieur.

Trimble, S. W., 1974, *Man-induced Soil Erosion on the Southern Piedmont*, Ankeny, Iowa: Soil Conservation Society of America.

Trussell, D., 1989, The arts and planetary survival, *The Ecologist*, 19(5), 170–6.

Tuan, Y. F., 1970, Our treatment of the environment in ideal and actuality, *American Scientist*, 58, 244–9.

UNDP/FAO (United Nations Development Programme/Food and Agriculture Organization of the United Nations), 1978, *Control of Waterlogging and Salinity in the Areas West of the Noubaria Canal, Egypt: Project Findings and Recommendations*, Rome: FAO/UNDP.

UNEP, 1980, *United Nations Environmental Program Annual Review*, Nairobi: UNEP.

—— ,1991, *Environmental Data Report*, Cambridge, Massachusetts: Blackwell, third edition.

UNEP/GEMS (United Nations Environment Program/Global Environment Monitoring System), 1991, Earthwatch Global Environment Monitoring System WHO/UNEP. In *Report on Water Quality: Progress in the Implementation of the Mar del Plata Action Plan and a Strategy for the 1990s*, Geneva/Nairobi: UNEP/GEMS.

United States Department of the Interior, 1967, *Impact of Surface Mining on the Environment*, Washington, DC: US Government Printing Office.

USDA (United States Department of Agriculture), 1983, *Conversion of Southern Cropland to Southern Pine Plantings: Conversion for Conservation Feasibility Study*, Washington, DC: USDA.

—— ,1988, *National Resources Inventory*, Washington, DC: Soil Conservation Service.

USGAO (United States General Accounting Office), 1984, *Department of Energy Acting to Control Hazardous Wastes at its Savannah River*

Nuclear Facilities (GAO/RCED–85–23), Washington, DC: The United States General Accounting Office.

van de Ven, G. P., 1987, Blanken en de Waterstaat. In *de Physique Existentie dezes Lands: Jan Blanken* (Rijksmuseum, Amsterdam Essaybundel), Danhoff, Zuidhorn: AMA boeken, 59–94.

Vogel, H., 1987, Terrace farming in Yemen. *Journal of Soil and Water Conservation*, 42(1), 18–21.

Volker, A., 1982, Polders: an ancient approach to land reclamation. *Nature and Resources*, 18(4), 2–13.

Wagstaff, J. M., 1985, *The Evolution of Middle Eastern Landscapes: an Outline to A.D. 1840*, Totowa, New Jersey: Barnes & Noble.

Walls, J., 1980, *Land, Man, and Sand*, New York: Macmillan.

Walters, S.E., 1970, *Water for Larsa: an Old Babylonian Archive Dealing with Irrigation*, New Haven: Yale University Press.

Ward, F., 1990, Florida's coral reefs are imperiled. *National Geographic*, 178(1), 115–32.

Warren, A. and Maizels, J. K., 1977, Ecological change and desertification. In UNCOD Secretariate (compilers and eds), *Desertification: Its Causes and Consequences*, New York: Pergamon Press, 169–260.

Waterbury, J., 1979, *Hydropolitics of the Nile Valley*, New York: Syracuse University Press.

Watson, A. M., 1983, *Agricultural Innovation in the Early Islamic World: the Diffusion of Crops and Farming Techniques 700–1100*, Cambridge: Cambridge University Press.

Watson, R. M., 1989, The green menace creeps north. *Garden*, 13, 8–9.

Watters, R. F., 1971, *Shifting Cultivation in Latin America*, FAO Forestry Development Paper No. 17, Rome: FAO.

Westphal, E., 1975, *Agricultural Systems in Ethiopia*, Wageningen, The Netherlands: Center for Agricultural Publishing and Documentation.

White, G. F., 1988, The environmental effects of the High Dam at Aswan. *Environment*, 30(7), 5–11, 34–40.

White, J. M., 1991, *The Journeying Boy*, New York: Atlantic Monthly Press.

White, L. Jr., 1967, The historical roots of our ecological crisis. *Science*, 155, No. 3767 (March 10), 1203–7.

Whitmore, T. M., Turner II, B. L., Johnson, D. L., Kates, R. W., and Gottschang, T. R., 1990, Long-term population change. In B. L. Turner II et al. (eds), *The Earth as Transformed by Human Action*, Cambridge: Cambridge University Press, 25–39.

Whittington, D. and Guariso, G., 1983, *Water Management Models in Practice: a Case Study of the Aswan High Dam*, Amsterdam: Elsevier.

Whittington, D. and Haynes, K. E., 1985, Nile water for whom? Emerging conflicts in water allocation for agricultural expansion in Egypt and Sudan. In P. Beaumont and K. McLachlan (eds), *Agricultural Development in the Middle East*, Chichester: John Wiley, 125–49.

WHO, 1992, *Our Planet, Our Health*, Geneva: World Health Organization.

Wiessner, P., 1982, Risk, reciprocity and social influences on !Kung San economics. In E. Leacock and R. Lee (eds), *Politics and History in Band Societies*, Cambridge: Cambridge University Press/Paris: Editions de la Maison des Sciences de l'Homme, 61–84.

Wikjman, A. and Timberlake, L., 1985, Is the African drought an act of God or of man? *The Ecologist*, 15(1/2), 9–18.

Wilkie, R. W. and Tager, J., 1990, *Historical Atlas of Massachusetts*, Amherst: The University of Massachusetts Press.

Willcocks, W., 1989, *Egyptian Irrigation*, London: Spon.

Williams, G. P. and Guy, H. P., 1973, Erosional and depositional aspects of Hurricane Camille in Virginia, 1969. *US Geological Survey, Professional Paper 804*, Washington, DC: US Government Printing Office.

Williams, J. and Williams, E. H., 1988, Coral reef bleaching. *Sea Frontiers*, 34(2), 80–7.

Williams, O. B., 1978, Desertification in the pastoral rangelands of the Gascoyne Basin, Western Australia. *Search*, 9(7), 257–61.

Williams, T., 1992, The elk-ranch boom. *Audubon*, 94(3), 14–20.

Wischmeier, W. H. and Smith, D. D., 1978, *Predicting Rainfall Erosion Losses – a Guide to Conservation Planning*, Agricultural Handbook 537, Washington, DC: US Department of Agriculture.

Withers, B. and Vipond, S., 1980, *Irrigation: Design and Practice*, Ithaca, New York: Cornell University Press, second edition.

Woodwell, G. M. (ed.), 1990, *The Earth in Transition*, Cambridge: Cambridge University Press.

Wooley, C. L., 1965a, *The Sumerians*, New York: W. W. Norton.

——, 1965b, *Ur of the Chaldees*, New York: W. W. Norton.

World Resources Institute, 1992, *World Resources 1992–93*, New York: Oxford University Press.

Worster, D., 1979, *Dust Bowl*, New York: Oxford University Press.

——, 1985, *Rivers of Empire: Water, Aridity, and the Growth of the American West*, New York: Pantheon.

Wyatt, A. W., 1988, Estimated net depletion shown. *The Cross Section*, 34(1), 2.

Young, A., 1972, *Slopes*, Edinburgh: Oliver and Boyd.

Young, J. E., 1992, Mining the Earth. In L. R. Brown et al. (eds), *State of the World – 1992*, New York: W. W. Norton.

Younkin, L. M., 1974, *Prediction of the Increase in Suspended Sediment Transport Due to Highway Construction*, Lewisburg, Pennsylvania: Bucknell University.

Index